MACHINISTS LIBRARY Volume I

Basic Machine Shop

by Rex Miller

Macmillan Publishing Company
New York
Collier Macmillan Publishers
London

FOURTH EDITION

Macmillan Publishing Company
866 Third Avenue, New York, N.Y. 10022
Collier Macmillan Canada, Inc.

Library of Congress Cataloging-in-Publication Data

Miller, Rex, 1929-
Machinists library.

Reprint. Originally published: 4th ed. Indianapolis:
T. Audel, c1983.
Includes indexes.
Contents: v. 1. Basic machine shop—v. 2. Basic machine shop.
1. Machine-shop practice. I. Title.
TJ1160.M566 1986 621.9'02 86-8723
ISBN 0-672-23381-9 (v. 1)
ISBN 0-672-23382-7 (v. 2)

Macmillan books are available at special discounts for bulk purchases for sales promotions, premiums, fund-raising, or educational use. For details, contact:

Special Sales Director
Macmillan Publishing Company
866 Third Avenue
New York, N.Y. 10022

10 9 8 7 6 5 4 3

Printed in the United States of America

Foreword

The purpose of this book is to aid in providing a better understanding of the fundamentals and principles of machine shop practice for those persons desiring to become machinists. The beginning student or machine operator cannot make adequate progress in his trade until he possesses a knowledge of the basic principles involved in the proper use of the various tools and machines commonly found in the machine shop.

One of the chief objectives has been to make the book clear and understandable to both students and workers; therefore, illustrations have been used generously to present the how-to-do-it phase of many of the machine shop operations and to enable the reader to become familiar with the tools and machines usually found in the machine shop. We believe that the material presented will be helpful to both the machine shop instructor and individual students or workers who desire to improve themselves in their trade.

The properties and uses of metals and materials are presented in a manner that is easily understood. The proper use of hand tools and machines, as well as care and safety in using them, has been stressed throughout the book. The principles of sharpening and grinding are dealt with thoroughly, and cutting tools and cutters for the various machines are profusely illustrated.

This book is presented at a time when there is a definite trend toward expanded opportunities for vocational training for youth. An individual who is ambitious to perfect himself in the machinist trade will find the material presented in an easy-to-understand manner—whether it is a young man studying alone in his spare time or an apprentice working under close personal supervision on the job.

REX MILLER

Contents

CHAPTER 6

CHAPTER 7

CHAPTER 8

CHAPTER 9

CHAPTER 10

CHAPTER 11

CHAPTER 12

CHAPTER 13

CHAPTER 14

CHAPTER 15

CHAPTER 16

CHAPTER 17

CHAPTER 18

CHAPTER 19

CHAPTER 1

Blueprint Reading

The blueprint is a print of a drawing in which the lines are white on a blue background. Formerly, almost all working drawings used in industrial shops were blueprints. A blueprint is made from a tracing. The tracing and a sensitized sheet of paper are placed over a framed glass plate. The assembly is then tightly pressed against the glass by pressure of a spring clamp exerted on the back. When exposed to light, the paper turns a pale green color except in places covered by the lines of the drawing, which exclude the light. When sufficiently exposed, the paper is immersed in water, and that portion of the paper acted upon by the light turns blue.

Many other process prints of working drawings have been developed, such as brown prints, blue line prints, ozalid prints, photostat prints, lithoprints, and many other copying methods, all of which are used for the same purpose as the blueprint. Many of them are referred to as "blueprints."

Regardless of the method of reproduction of the working drawing or print, blueprint reading skill is a basic essential for any machinist. It is not necessary to become a mechanical draftsman in order to learn blueprint reading. However, in studying mechanical drawing, the student can learn many things that are necessary in reading blueprints.

DESCRIPTIVE DRAWINGS

A study of descriptive geometry can be very helpful in understanding orthographic drawings, a knowledge of which is essential for blueprint reading. Descriptive geometry deals with the projection drawing method of representing points, lines, objects, and the solution of related problems. The descriptive method of drawing is based on parallel projections to a plane by rays perpendicular to the plane.

Orthographic Projection

All working drawings on blueprints are made by orthographic projection. Orthographic, by definition, means of, or pertaining to, perpendicular lines. Therefore, *orthographic projection* means a projection in which the projecting lines are perpendicular to the plane of projection.

Through long usage and common consent, however, orthographic projection has been accepted to mean the combination of two or more such views; hence the definition follows: Orthographic projection is a means of representing the exact shape of an object in two or more views on planes, which are generally at right angles to each other, by extending perpendiculars from the object to the planes.

Orthographic Projection Views

In the orthographic method, several views are required to show an object completely (Fig. 1-1). The number of views required depends on the shape of the object, and the position of the views relative to each other depends on the quadrant selected for the projection.

Front View—This view is developed by projection from the horizontal plane into a vertical plane (Fig. 1-2); or it represents the

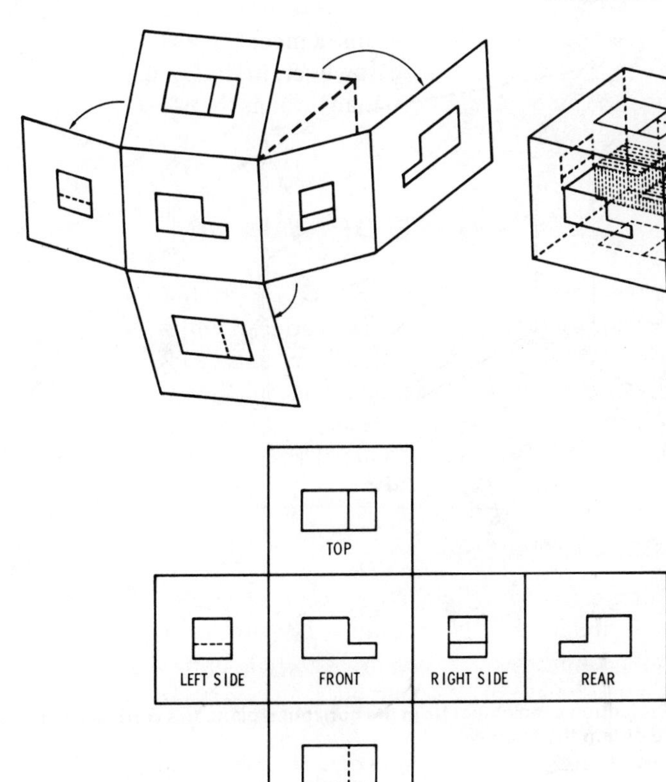

Fig. 1-1. Six-view orthographic projection.

object as seen from the front. The front view shows the true height and the true width of an object.

Top View—This view is developed by projection from the horizontal plane to another horizontal plane above the object (Fig. 1-3) to represent the object as it would appear when looking at the object directly from above. The top view shows the true depth of the object from front to back, in addition to the true width already shown in the front view. The top view is always drawn directly over the front view.

Side View—The development of this view is illustrated in Fig. 1-4, in which the pane of glass is placed to the right of the object in

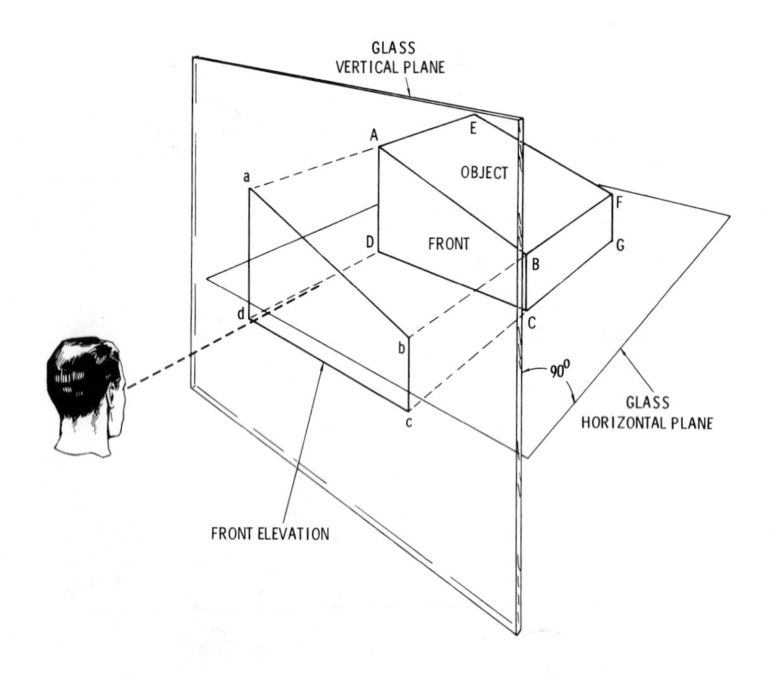

Fig. 1-2. Projection of an object from the horizontal plane to a vertical plane to obtain the front view.

a vertical position and parallel to the right side of the object. This view represents the object as seen directly from the right side of the object. The side view shows the true depth of the object, as shown in the top view, and the true height of the object, as shown in the front view. The side view is always shown directly to the right side of the front view.

Arrangement of Views—Since several orthographic views are necessary to show an object completely, the order in which the different views are laid out on a flat sheet of paper is worthless unless they are grouped properly. Usually, it is unnecessary to show all six views of an object for a working drawing. It may be desirable to show a left-side view, or a rear view in certain instances.

As illustrated in Fig. 1-5, a rectangular object may be "unfolded" in such a manner that views come into the same plane as the paper.

When the various sides have been opened fully, they are in the same plane as the front view, and the assembly is in proper order (Fig. 1-5).

The front view of an object is selected, usually, as the view most truly representing the object, because this view gives both the true height and the true width of the object. Therefore, the front view is usually drawn first in constructing a working drawing.

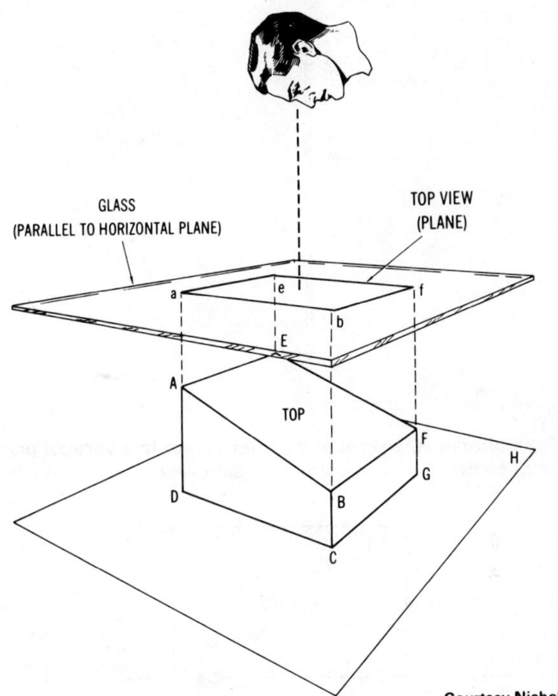

Fig. 1-3. Projection of the top of the horizontal plane above to obtain the top view.

Then the top view is located immediately above the front view. Lines or points of importance in illustrating true width can be projected by perpendiculars from the front view into the top view without actual measurement. The completed top view will show the actual depth of the object from front to back, in addition to the actual width and height shown in the front view.

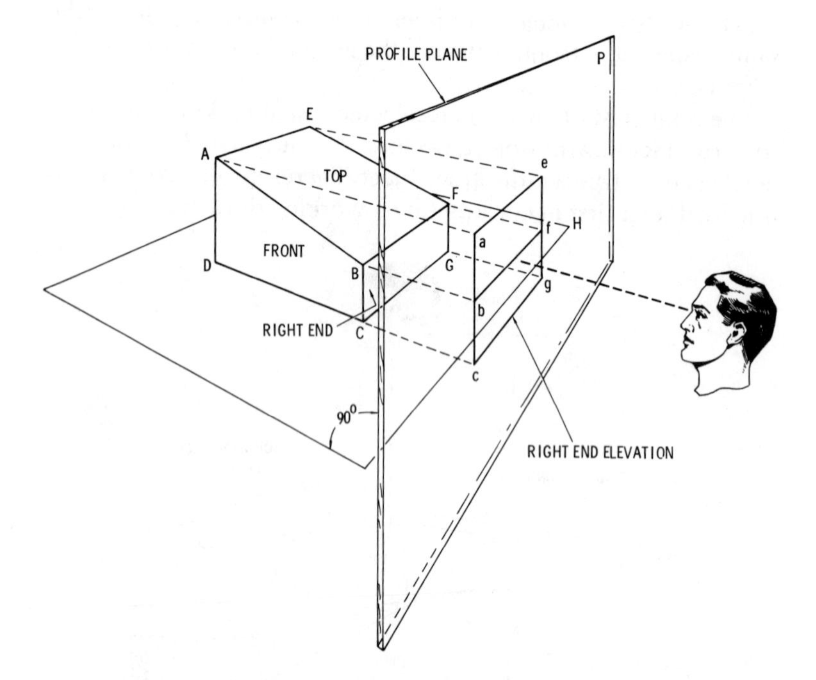

Fig. 1-4. Projection of the right end of the object to a vertical plane at the right of the object to obtain the side view.

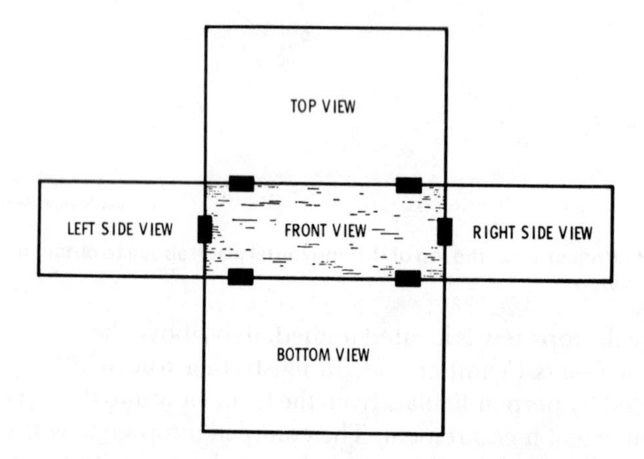

Fig. 1-5. Illustration of five of the views of a rectangular object as drawn in orthographic projection.

The side view is located immediately to the right of the front view. This view illustrates the true height of the object projected from the front view, and the true depth as taken from the top view. If a left-side view is also necessary, it is placed to the immediate left of the front view, as shown in Fig. 1-5.

The front, top, and side views are commonly drawn as shown in Fig. 1-6. Only two views may be necessary for objects that are symmetrical, but irregularly shaped objects usually require the three views.

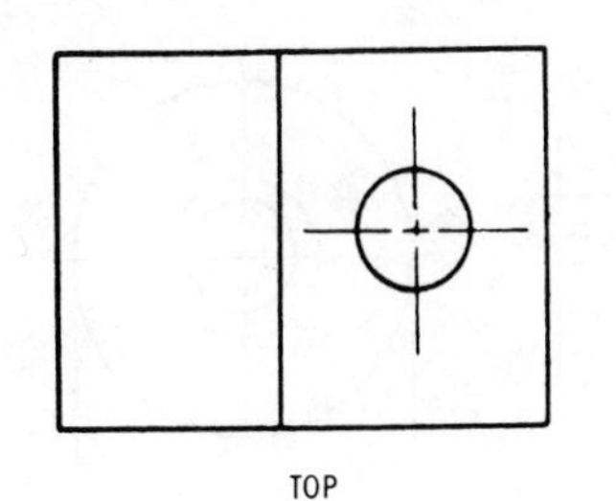

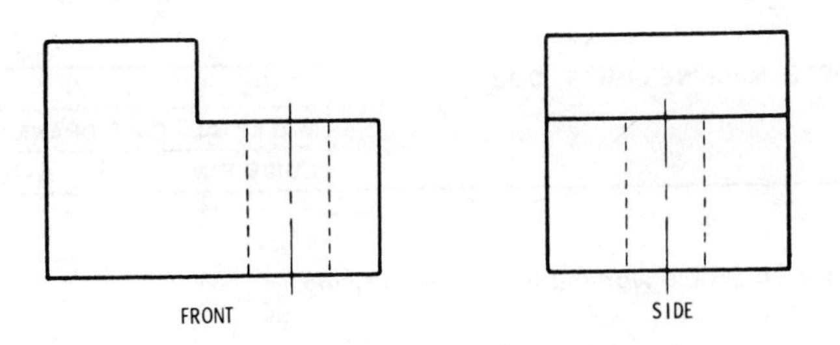

Fig. 1-6. Typical arrangement of the top, front, and side views in an orthographic projection.

WORKING DRAWINGS

A working drawing is usually in the front of the blueprint or other print (Fig. 1-7). Working drawings are orthographic projections that are completely dimensioned and contain any notes or other data necessary for the mechanic to do the indicated job without having to ask questions (blueprint reading).

Assembly Drawings

A drawing of an object having many parts, that is, showing it as a whole with its parts assembled in their proper positions, is called an assembly drawing, which is one kind of working drawing (Fig. 1-8). This kind of drawing shows the mechanic how to put together the various parts that make up the object.

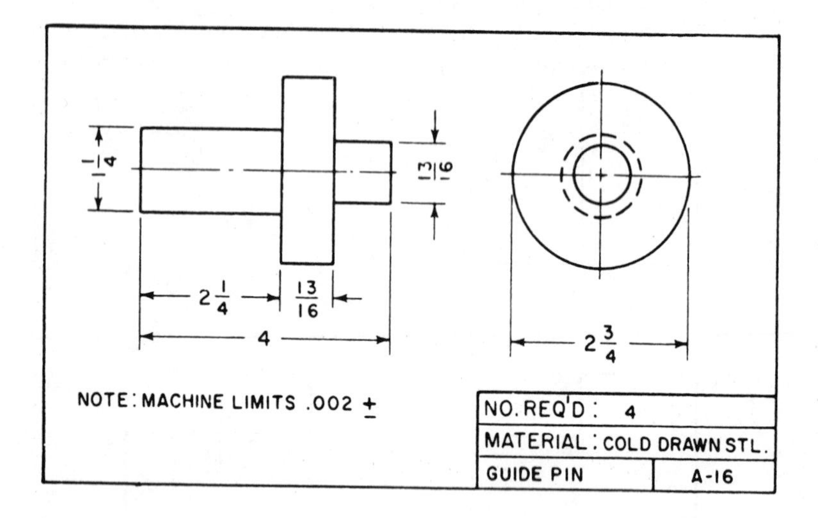

Fig. 1-7. A simple working (machine) drawing.

Detail Drawings

The simplest form of working drawing is the detail drawing (Fig. 1-9). This must be a drawing of a single object, or piece, along with all the information necessary for making it. The detail drawing must be a complete and accurate description of the piece, with carefully selected views and well-located dimensions of the piece. Detail drawings having only the dimensions and information needed by a particular workman are sometimes made for the different workmen, such as the patternmaker, machinist, or welder. A detail drawing on a separate sheet may be made for each part of the machine to be manufactured.

Sections and Sectional Views

Sectional views may be necessary in many drawings to bring out and fully dimension the piece (Fig. 1-10). Usually, one sectional view is sufficient, but several sections may be required for some irregular objects. An object may be completely represented with one cross-sectional view or with only one-half of a cross section.

Frequently, one sectional view may be used in place of a full view and a section. Some objects may require more than one section for showing it completely. A sectional view with all information and dimensions necessary to machine the object becomes a working drawing for the machinist.

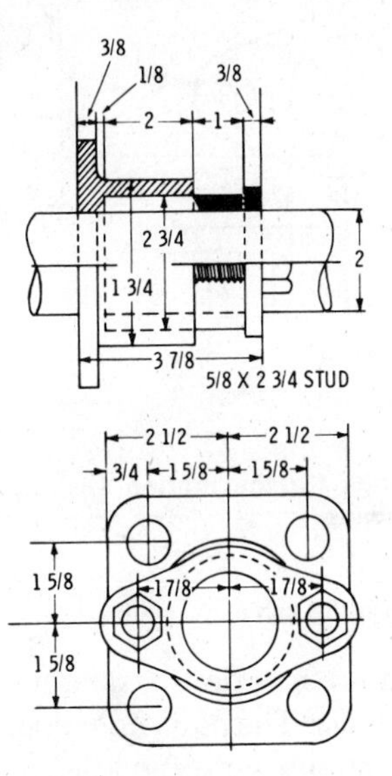

Fig. 1-8. Assembly drawing of a marine stuffing box for a yacht.

GRAPHICAL SYMBOLS

In addition to the conventional lines used by draftsmen to convey ideas on blueprints, various combinations of light and heavy lines, spacing, etc., are used to indicate not only a section but also the kind of metal used in construction of the piece.

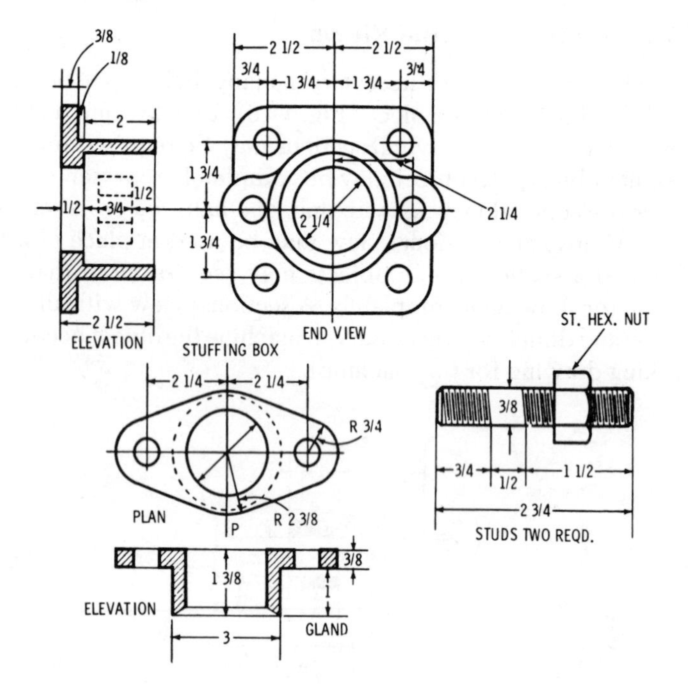

Fig. 1-9. Detail drawings for the marine stuffing box illustrated in the assembly drawing.

Representation of Metals

Section lines are used in various arrangements to represent the various metals (Fig. 1-11). The plain light line sectioning, similar to that used for cast iron and with the kind of metal specified by abbreviations, may be used to designate the kind of metal in a piece.

Abbreviation on Drawings

Abbreviations used on drawings to indicate the various metals are as follows:

C.I. .... Cast iron
M.C. or Mal. Cast .... Malleable casting

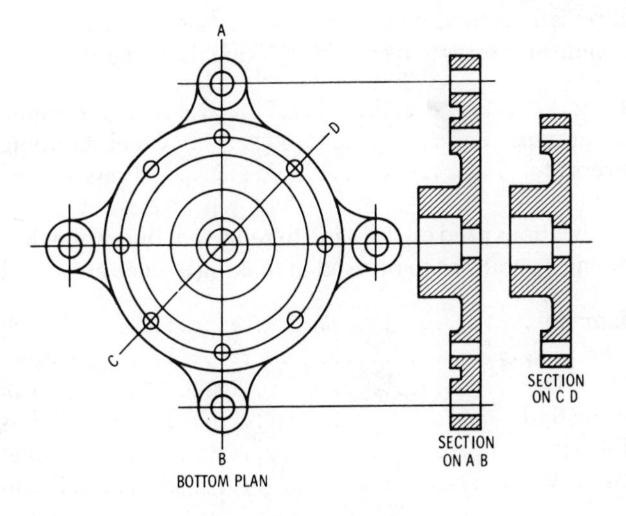

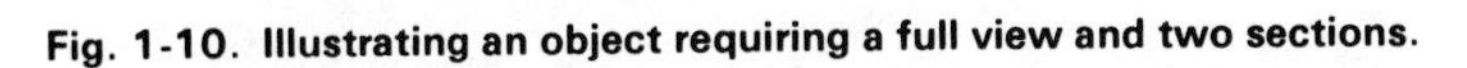

Fig. 1-10. Illustrating an object requiring a full view and two sections.

S.C. or St. Cast	Steel casting
Brs. Cast	Brass casting
Brz. Cast	Bronze casting
T.S.	Tool steel
M.S.	Machine steel

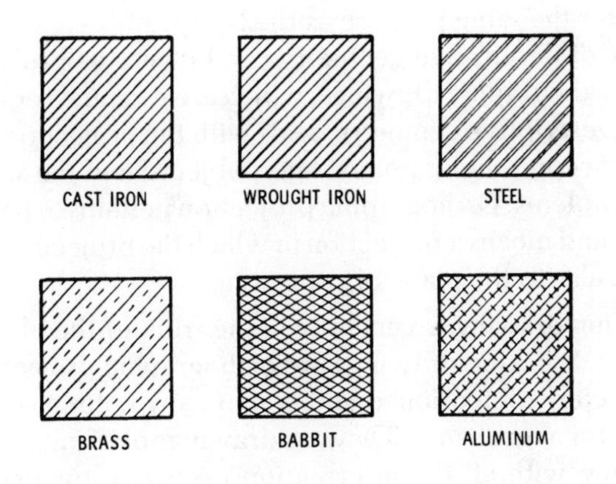

Fig. 1-11. Arrangements of section lines to indicate various metals.

Abbreviations used on drawings to represent the various kinds of machine processes to be used in making the piece are as follows:

Forg .... Forging
St. Stamp............................ Steel stamping
Press St.............................. Pressed steel

Abbreviations used on drawing to represent the various kinds of machining operations to be used in making a piece are as follows:

f. or fin .... Finish
Thr. .... Thread
D. or Dr. .... Drill
R. or Rad.................................. Radius
Fil.. Fillet
Rm .... Ream

SUMMARY

Blueprint drawings are white lines on a blue background. Many other process prints or working drawings have been developed, such as brown prints, blue line prints, ozalid prints, photostat prints, lithoprints, and many other copying methods, all of which are used for the same purpose as the blueprint.

A study of descriptive geometry can be very helpful in understanding orthographic drawings, which are essential for blueprint reading. Descriptive geometry deals with the projection drawing method of representing points, lines, objects, and the solution of related problems. Orthographic projection pertains to perpendicular lines, and means a projection in which the projecting lines are perpendicular to the plane of projection.

A working drawing is generally in the front of the blueprint or other print. Working drawings are orthographic projections that are completely dimensioned and contain any notes or other data necessary for a mechanic. They are drawings of a single object, or piece, along with all the information necessary for making the object.

REVIEW QUESTIONS

1. What is a detail drawing?
2. What is a descriptive drawing and an orthographic projection?
3. Explain the working drawing.
4. How are blueprints made?
5. What is an assembly drawing?
6. What are the three views of an orthographic drawing?
7. Why is dimensioning important on a drawing?
8. What is a graphic symbol?
9. What symbol does the drafter use for cast iron? bronze? forging? radius?
10. Why are blueprints called "blueprints"?
11. Are all "blueprints" blue in today's workshop?
12. How does a study of geometry help in the reading and drawing of orthographic projections?

CHAPTER 2

Benchwork

The term benchwork relates to work performed by the mechanic at the machinists' bench with hand tools rather than machine tools. It should be understood that the terms benchwork and vise work mean the same thing; the latter, strictly speaking, is the correct term, as in most cases the work is held by the vise, while the bench simply provides an anchorage for the vise and a place for the tools. However, both terms are used almost equally. Today, work at the bench is not performed as much as formerly; the tendency, with the exception of scraping, is to do more and more benchwork with machines.

EQUIPMENT

Operations that can be performed at the bench may be classed as:

1. Chipping.
2. Sawing.
3. Filing.
4. Scraping.

The Bench

The prime requirements for a machinists' bench are that it should be strong, rigid, and of the proper width and height that the work can be performed conveniently. Correct height is important, and this will depend on the vise type used, that is, how far its jaws project above the bench. The location of the bench is important. It should be placed where there is plenty of light.

Bench Tools

A great variety of tools is not necessary for benchwork; they may be divided into a few general classes:

1. Vises.
2. Hammers.
3. Chisels.
4. Hacksaws.
5. Files.
6. Scrapers.

Vises

By definition, a vise is a clamping device, usually consisting of two jaws that close with a screw or a lever, and commonly attachable to a workbench; it is used for holding a piece of work firmly. There is a great variety of vises on the market, and they may be classed as:

1. Blacksmith.
2. Machinist.
    a. Plain.
    b. Self-adjusting.
    c. Quick acting.
    d. Swivel.

3. Combination.
4. Pipe.

The machinists' vise (Fig. 2-1) is usually provided on machine shop workbenches. Several types are provided; some of their features are parallelism, swivel action, and quick-acting jaws. These vises will withstand terrific abuse and are well adapted for a heavy and rough class of work.

For shops having frequent use for round stock and pipe, the combination vise is well adapted. It is shown in Fig. 2-2. A regular pipe fitters' vise is shown in Fig. 2-3. Vise jaws have faces covered with cross cuts in order to grip the work more firmly. It is evident that a piece of finished work held in such a manner would be seriously marred. This trouble may be avoided by using false jaws of brass or Babbitt metal, or by fastening leather or paper directly to the steel jaws.

Hammers

These tools find frequent use in benchwork. Machinists' hammers may be classed with respect to the peen as:

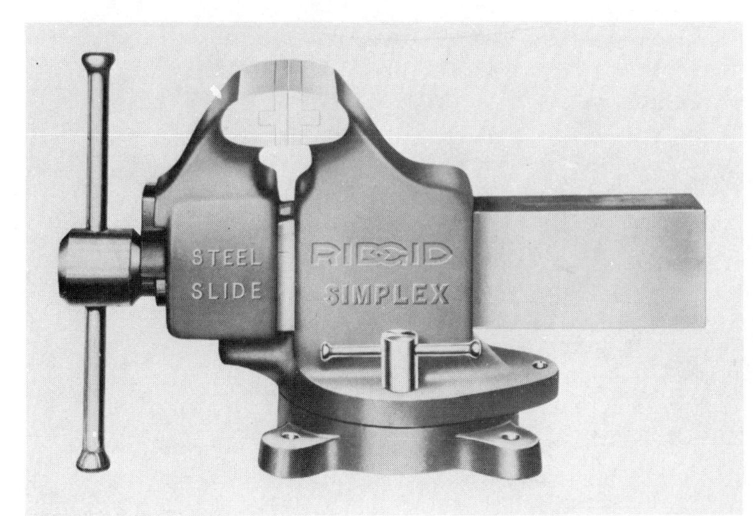

Courtesy Ridge Tool Company

Fig. 2-1. Machinists' vise.

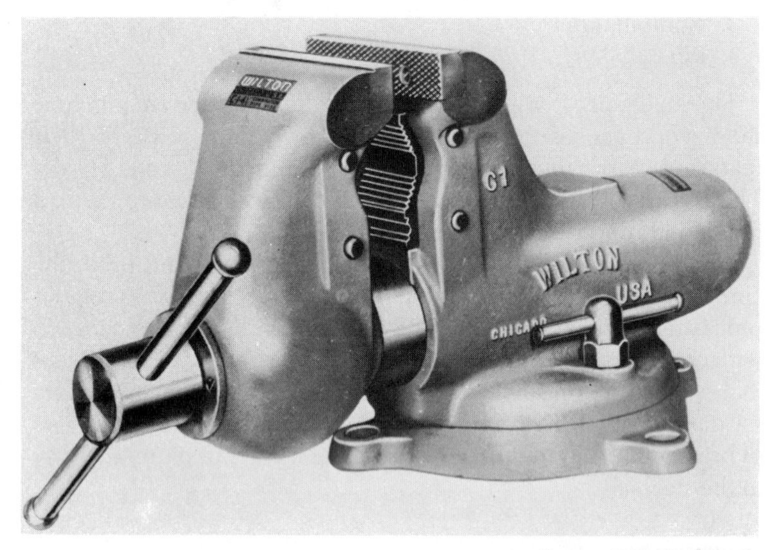

Courtesy Wilton Tool Mfg. Co.

Fig. 2-2. Combination vise. The inner teeth are for holding either pipe or round stock.

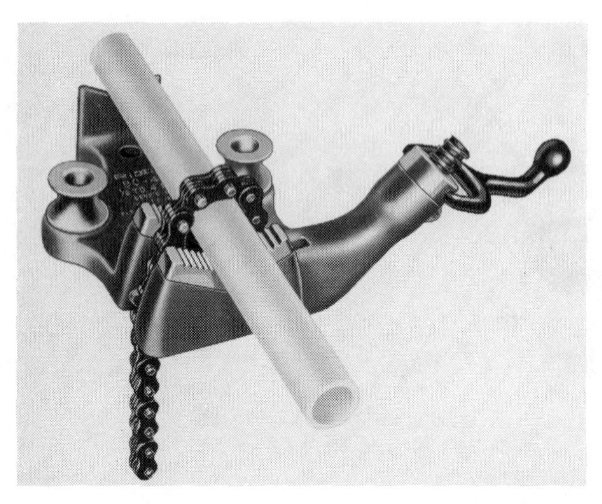

Courtesy Ridge Tool Company

Fig. 2-3. Pipe vise.

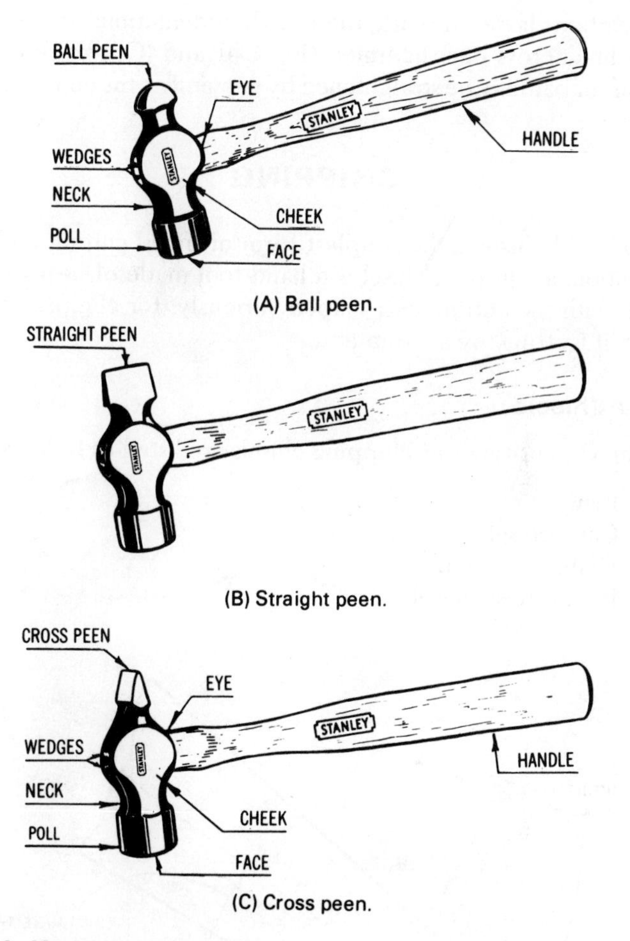

(A) Ball peen.

(B) Straight peen.

(C) Cross peen.

Fig. 2-4. Machinists' hammers.

1. Ball peen.
2. Straight peen.
3. Cross peen.

By definition, peening is the operation of hammering metal so as to indent or compress it, in order to expand or stretch that portion of the metal adjacent to the indentation. These hammers are shown in Fig. 2-4. The ball peen hammer (Fig. 2-4A), with its spherical end, is generally used for peening or riveting operations.

For certain classes of work, the straight indentations of either the straight- or cross-peen hammers (Fig. 2-4B and 4C) are preferable. A shaft or bar may be straightened by peening on the concave side.

CHIPPING

The cold chisel is the simplest form of metal cutting tool. By definition, a chipping chisel is a hand tool made of heat-treated steel, with the cutting end shaped variously, for chipping metal when it is struck by a hammer.

Cold Chisels

The various types of chipping chisels are listed as follows:

1. Flat.
2. Cape chisel.
3. Diamond-point.
4. Round-nose chisel.

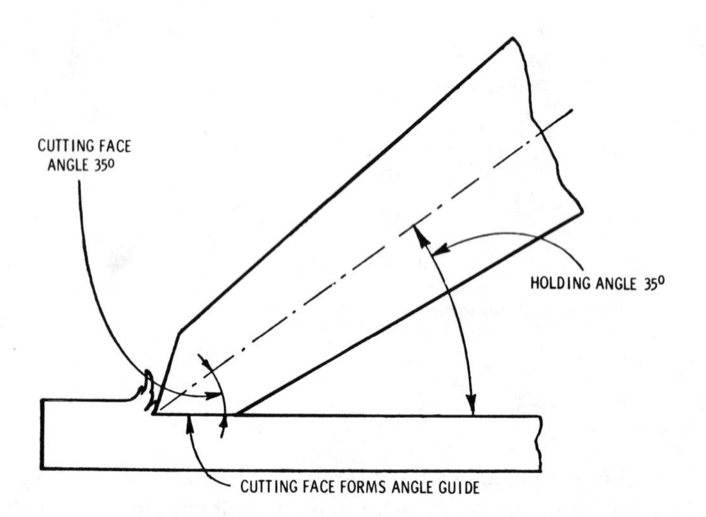

Fig. 2-5. Detail of cutting end of cold chisel. Note the angle of application and the angle guide.

One of the first operations that a student or apprentice must learn in becoming a machinist is how to chip metal. This involves learning how to hold the chisel and how to use the hammer.

Flat chisel—The work is placed firmly in the vise with the chisel held in the left hand. The chisel must be held firmly at the proper angle (Fig. 2-5) to the work. The lower face of the cutting edge of the chisel acts as a guide, while the wedging action of the metal being chipped tends to guide the chisel on a straight line. The cutting edge of the chisel is ground at an included angle of 60° to 70° (Fig. 2-7A). The cutting face is the guide to hold the chisel at the correct angle (Fig. 2-6).

The flat chisel is used for surfaces having less width than the castings and for all general chipping operations (Fig. 2-7). The cutting edge generally is about one-eighth of an inch wider than the stock from which it is forged.

The beginning machinist learns to vary the chipping angle more or less as demanded by the nature of the work (Fig. 2-5). The first exercise in chipping is usually a broad surface on which both the

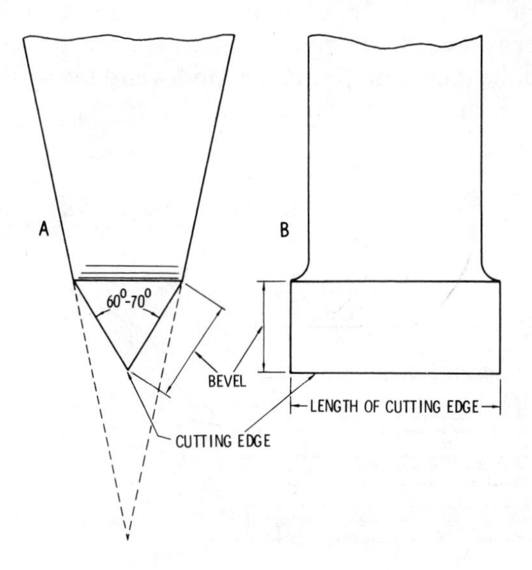

Fig. 2-6. Cutting end of cold chisel showing bevel angle (A) and length of cutting edge (B).

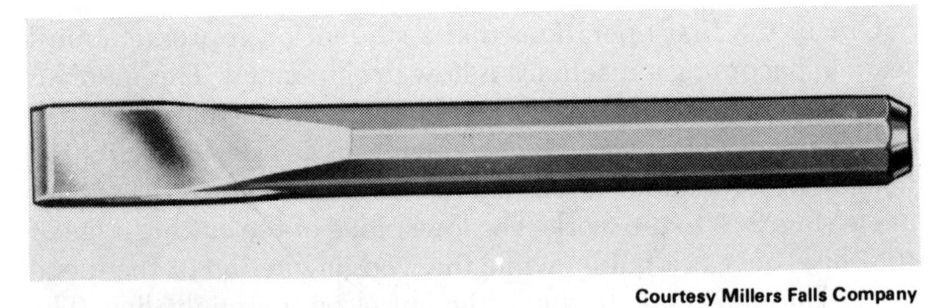

Courtesy Millers Falls Company

Fig. 2-7. Cold chisel used in chipping operations.

cold chisel and the cape chisel are used. First, grooves are cut in the piece to be chipped with the cape shield (Fig. 2-8), and the raised portions are removed with a flat chisel (Fig. 2-9).

In chipping, the worker should always chip toward the stationary jaw of the vise because its resistance to the blows of the hammer is greater. Start with a light chip, and watch only the cutting edge of the chisel. Chamfer the front and back edges of the work to avoid risk of breaking off the stock below the chipping line and to facilitate starting the chisel.

Use a 1- to 1¾-lb. hammer for ordinary chipping work. Grasp the hammer near the end of the handle, with the fingers around the handle and the thumb projecting on top toward the striking end.

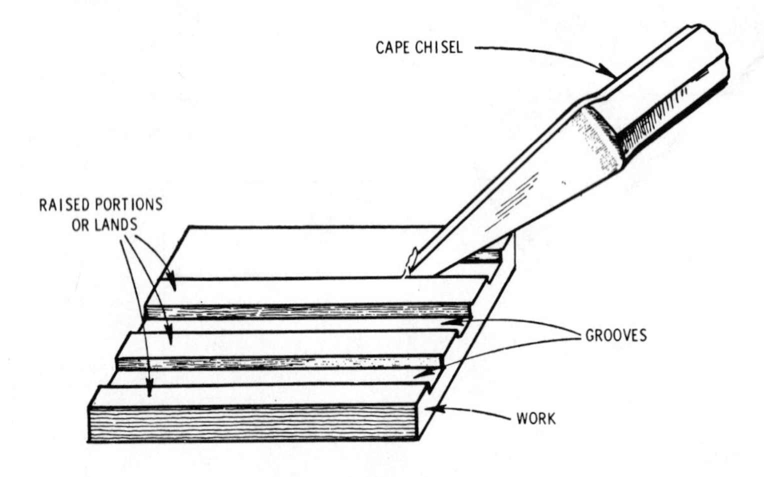

Fig. 2-8. Cape chisel used to cut grooves.

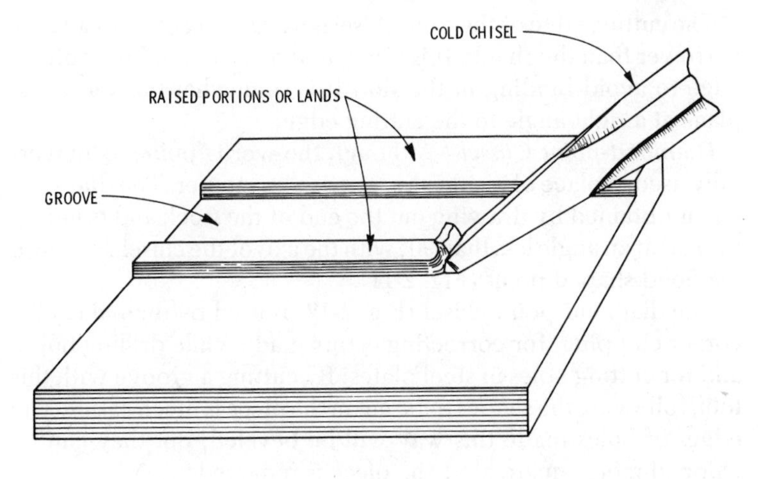

Fig. 2-9. Using the cold chisel to remove "lands" in chipping a broad flat surface.

The chisel should be held firmly with the second and third fingers, and the little finger should be used to guide the chisel as required. The first finger and the thumb should be left slack; they are then in a state of rest, with the muscles relaxed. The fingers are less liable to injury if struck with the hammer than if they were closed rigidly about the chisel. Reset the chisel to its proper position after each blow.

Cape Chisel—A cape chisel (Fig. 2-10) is used to facilitate work in removing considerable metal from a flat surface, or to break up surfaces too wide to chip with a cold chisel alone. It is also used, along with other chisels to cut keyways and channels.

Courtesy Millers Falls Company

Fig. 2-10. Cape chisel.

The cutting edge of the cape chisel is usually an eighth of an inch narrower than the shank. It is thin enough just behind the cutting edge to avoid binding in the slot. It is somewhat thicker in the plane at a right angle to the cutting edge.

Diamond-point Chisel—Although the word "point" is universally used in place of "end," the term is a misnomer. The diamond end is obtained by drawing out the end of the stock and grinding the end at an angle less than 90° with the axis of the chisel, leaving a diamond-shaped point (Fig. 2-11).

The diamond-point chisel (Fig. 2-12) is used by diemakers for corner chipping, for correcting errors made while drilling holes, and for cutting holes in steel plates. By cutting a groove with this tool, following the shapes to be cut in the piece is much easier. The edges of holes made this way will be beveled, but they can be chipped more square after the piece is removed.

Round-nose Chisel—This chisel is sometimes called a round-nose cape chisel (Fig. 2-13). The nose itself is cylindrical in section at the cutting end with tangential sides intersecting at the extremity. The tool is ground at an angle of 60° with its axis.

These chisels are called center chisels when they are used to "draw" the starting of drilling holes, in order to bring them into

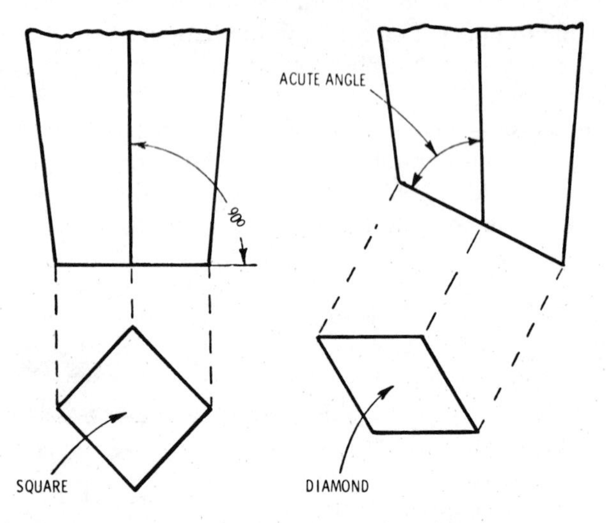

Fig. 2-11. Detail of cutting end of square and diamond-point chisels.

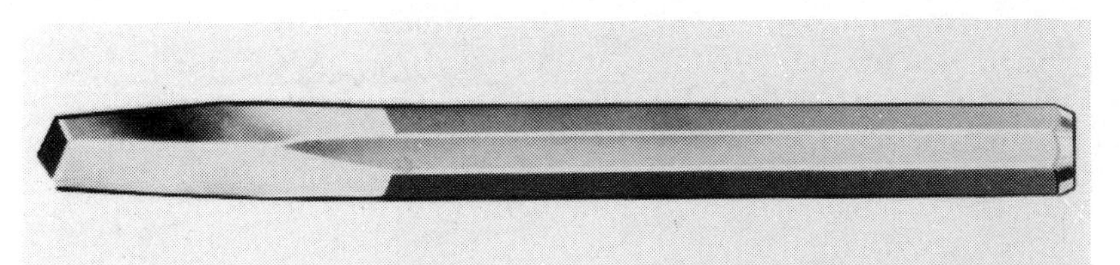

Courtesy Millers Falls Company

Fig. 2-12. Diamond-point chisel.

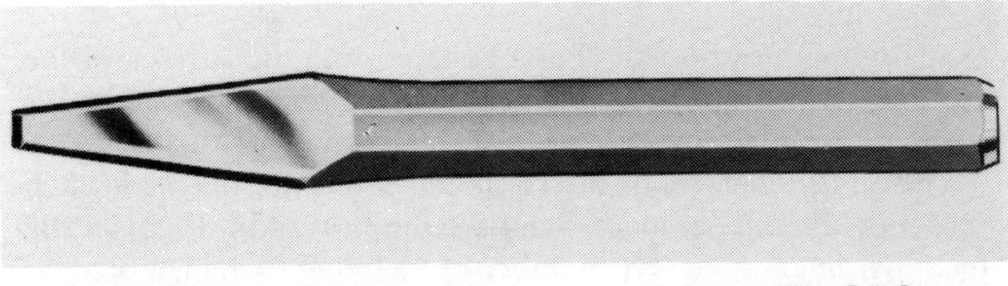

Courtesy Millers Falls Company

Fig. 2-13. Round-nose cape chisel or round-nose chisel.

concentricity with the drilling circles. They are also used on large round-bottomed channels, and for cutting channels such as oil grooves.

The stock generally used for all the aformentioned forms of chisels is octagonal and of a good grade of tool steel, carefully forged, hardened, and tempered.

SAWING

Hacksaws

The sawing of metal is one of the most common benchwork operations. Hand hacksaws are available with either a fixed frame or an adjustable frame. The adjustable frame hacksaw (Fig. 2-14) can be changed to hold 8-inch, 10-inch, and 12-inch blades. Most blades are ½ inch wide and 0.025 inch thick.

The workpiece must be held securely in a vise. The workpiece should be sawed near the vise jaws to prevent chattering. To hold nonrectangular shaped pieces (Fig. 2-15) clamp the work to allow

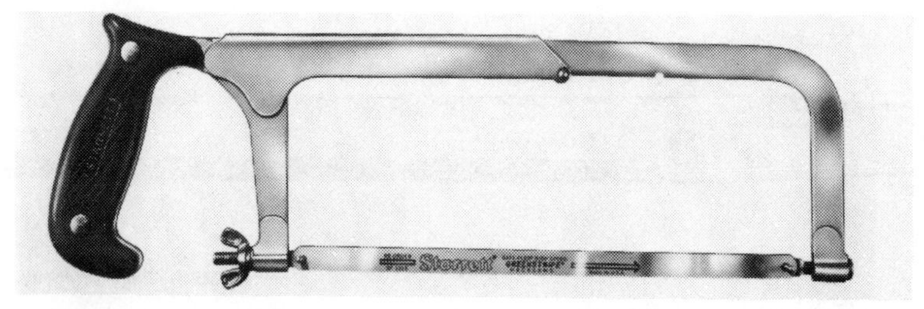

Fig. 2-14. Adjustable frame hacksaw.

Courtesy L. S. Starrett Company

as many teeth as possible be in contact with the surface of the workpiece. Polished work surfaces should be protected from the steel vise jaws by covering them with soft metal jaw caps.

The type of metal to be sawed should determine the blade pitch (number of teeth per inch). Standard pitches are 14, 18, 24 and 32 teeth per linear inch. The number of teeth per inch on a blade is

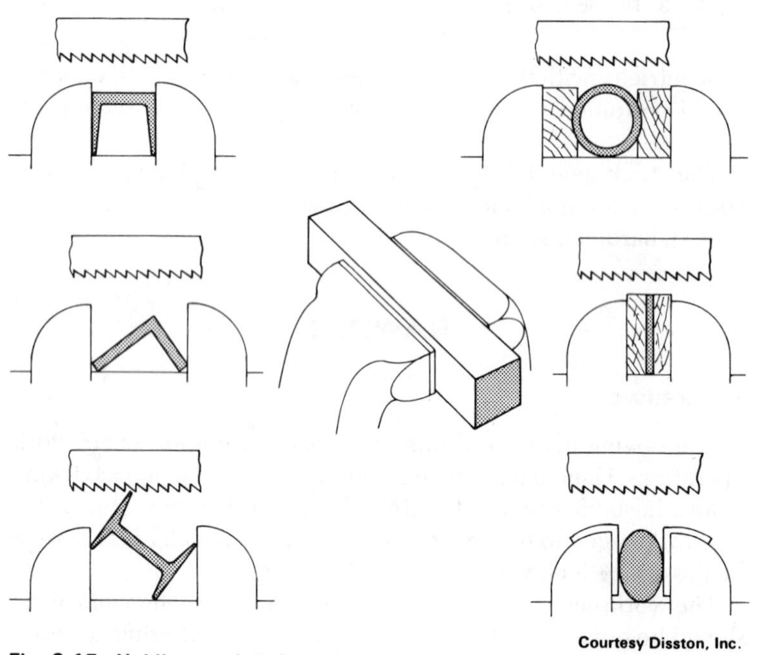

Fig. 2-15. Holding work to be cut

Courtesy Disston, Inc.

important because at least two teeth should be in contact with the work at all times (Fig. 2-16).

Cutting With a Hacksaw

To start a hacksaw cut, it is a good practice to guide the blade until the cut is well established. To start an accurate cut, use the thumb (Fig. 2-17) as a guide and saw slowly with short strokes. As the cut deepens, grip the front end of the frame firmly and take a full-length stroke.

When sawing, stand facing the work with one foot in front of the other and approximately 12 inches apart as shown in Fig. 2-18. Pressure should be applied on the forward stroke and released on the return stroke because the blade only cuts on the forward stroke. Do not permit the teeth to slip over the metal as this dulls the teeth and may cause blade breakage. Once the kerf, the slot made by the blade, is established the hacksaw should be moved at about 40 strokes per minute.

FILING

Filing is a difficult operation for the beginner because it depends on the motion of the hands, without a means of guiding the tool, so that it will move over the work in the correct direction. Proficiency is obtained by practice only when the proper methods are followed.

How to File

The correct position and method of holding the file are important. The work should be at the proper height—about level with the elbows on light work, and a little lower on heavy work (Fig. 2-19). The feet should be about eight inches apart and at right angles to each other, the left foot being parallel with the file. Hold the file handle with the right hand—thumb on top and fingers below the handle (Fig. 2-19).

In filing, pressure should be exerted on the forward stroke only because the teeth or cutting edges are pointed toward the end of the file. Pressure on the return stroke produces no cutting action, but tends only to dull the teeth. Fig. 2-20 shows an incorrect position of the body in filing.

Drawfiling

When the file is grasped by the ends and moved sidewise across the work, the action is known as drawfiling (Fig. 2-21). This produces a smooth finish on narrow surfaces and edges and is used on turned work to remove any tool marks. Drawfiling is light filing—for producing a smooth surface (Fig. 2-22). A second cut or smooth file should be used; a single-cut file is better than a double-

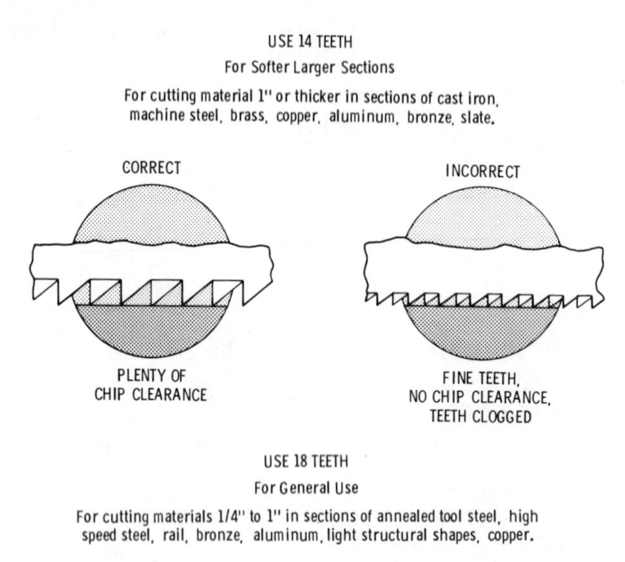

USE 14 TEETH
For Softer Larger Sections
For cutting material 1" or thicker in sections of cast iron, machine steel, brass, copper, aluminum, bronze, slate.
CORRECT
INCORRECT
PLENTY OF CHIP CLEARANCE
FINE TEETH, NO CHIP CLEARANCE, TEETH CLOGGED
USE 18 TEETH
For General Use
For cutting materials 1/4" to 1" in sections of annealed tool steel, high speed steel, rail, bronze, aluminum, light structural shapes, copper.

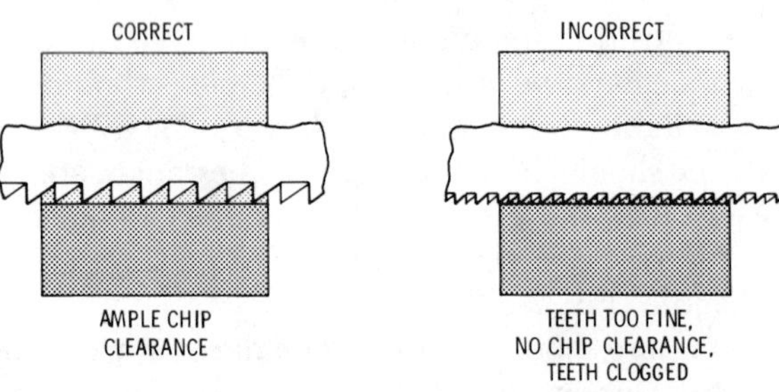

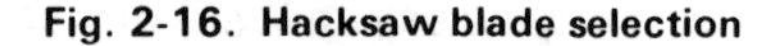
Fig. 2-16. Hacksaw blade selection

cut file because the single-cut is less likely to scratch the surface of the work.

For most filing operations, begin with a coarse file and continue using successively finer grades of file, finishing with a smooth or dead-smooth file, according to the degree of finish desired (Fig. 2-23).

Particles of metal, or pins, often remain in the teeth of the file, and they either reduce its cutting qualities or scratch the work. These particles can be removed by using either a stiff brush (Fig. 2-24A) or a file card (Fig. 2-24B) frequently for cleaning them from the file.

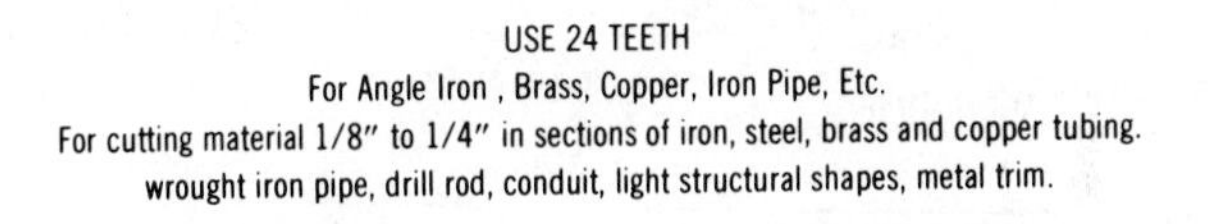

USE 24 TEETH
For Angle Iron, Brass, Copper, Iron Pipe, Etc.
For cutting material 1/8″ to 1/4″ in sections of iron, steel, brass and copper tubing.
wrought iron pipe, drill rod, conduit, light structural shapes, metal trim.

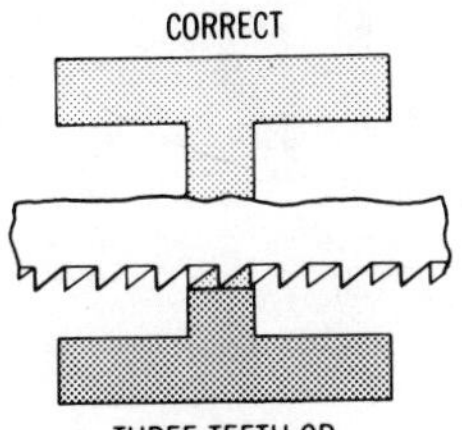

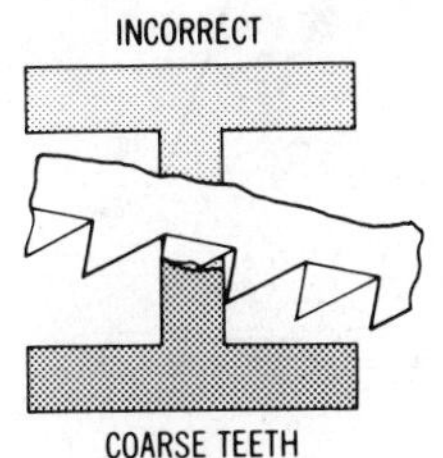

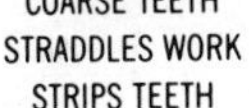

USE 32 TEETH
For Conduit and Other Thin Tubing, Sheet Metal Work
For cutting material similar to recommendations
for 24 tooth blades for 1/8″ and thinner.

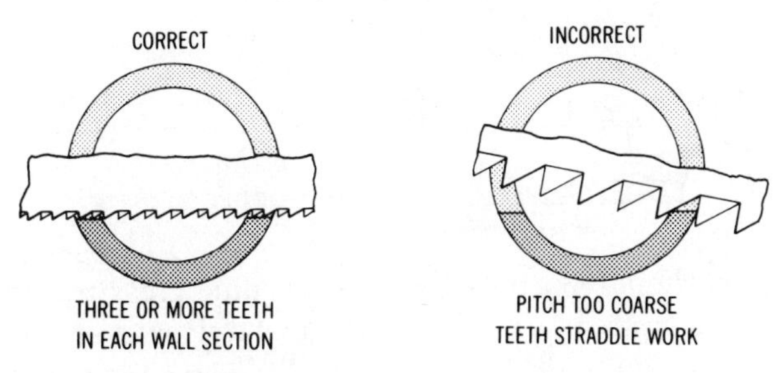

Courtesy Disston, Inc.

for various cutting operations.

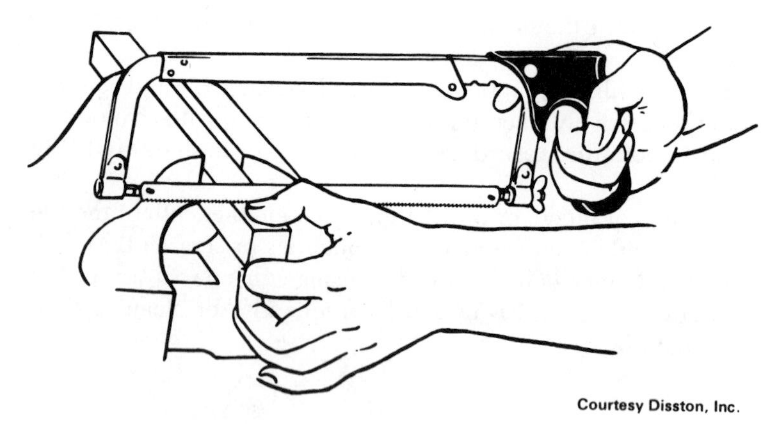

Courtesy Disston, Inc.

Fig. 2-17. Starting a cut.

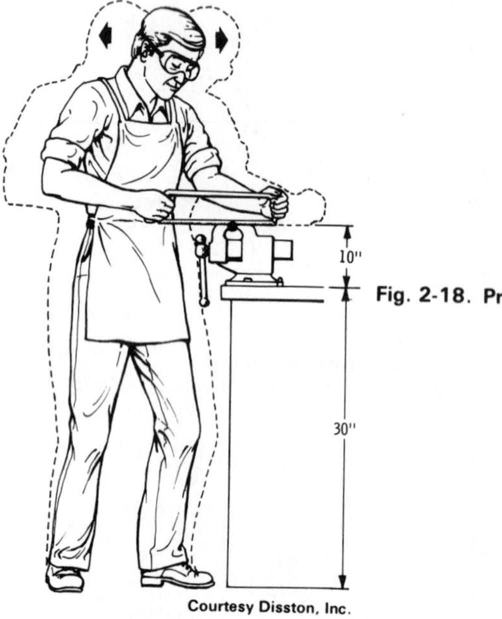

Courtesy Disston, Inc.

Fig. 2-18. Proper stance for cutting.

Cast iron should not be allowed to become greasy, as the file tends to slide without cutting into the metal. However, frequent pinning (clogging of the teeth with small slivers of metal) can be prevented by the use of oil when filing steel.

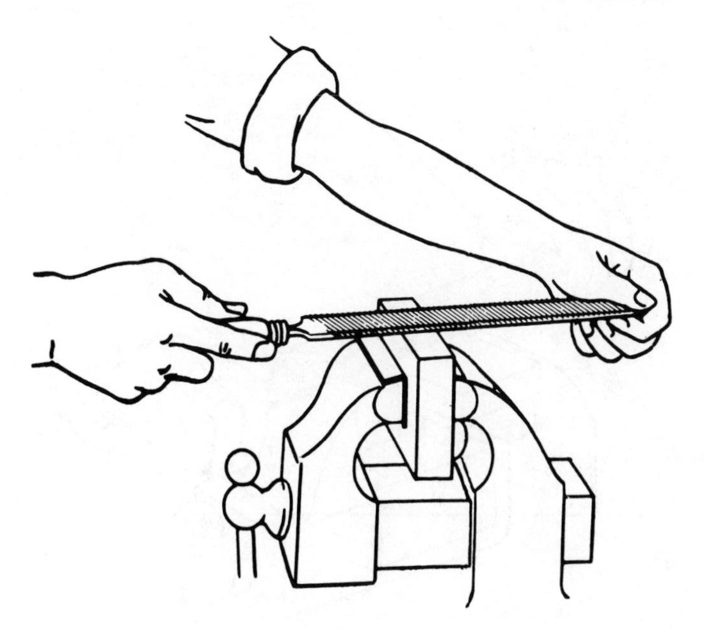

Courtesy Disston, Inc.

Fig. 2-19. Correct position of hands and arms in filing.

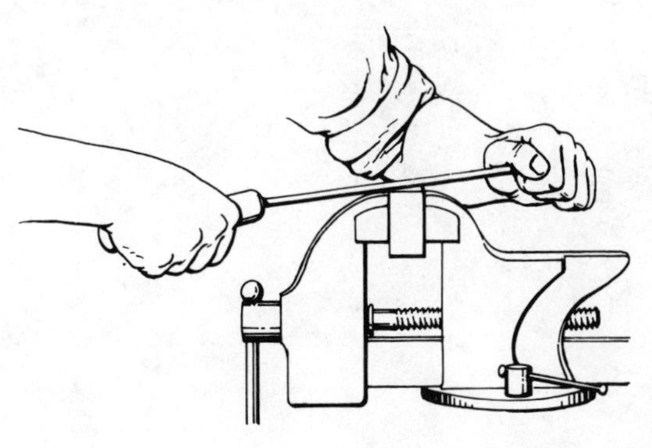

Courtesy Nicholson File Co.

Fig. 2-20. Incorrect position of body in filing.

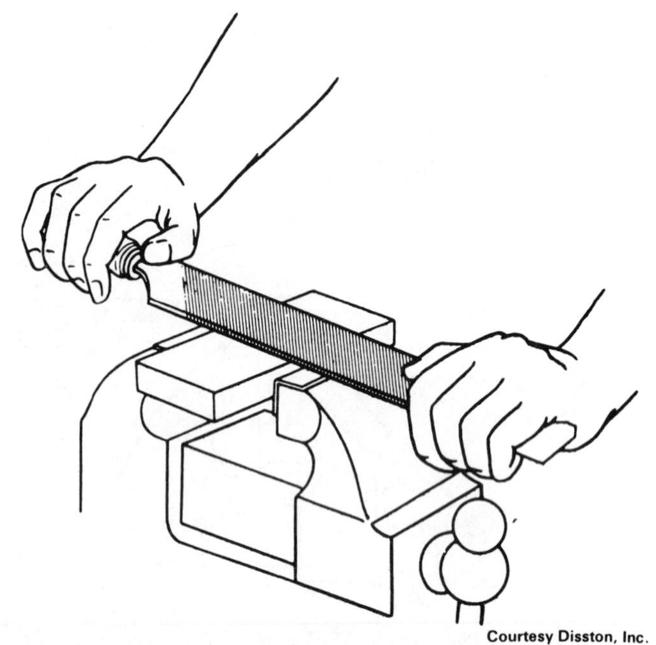

Courtesy Disston, Inc.

Fig. 2-21. Draw-filing for producing a smooth surface.

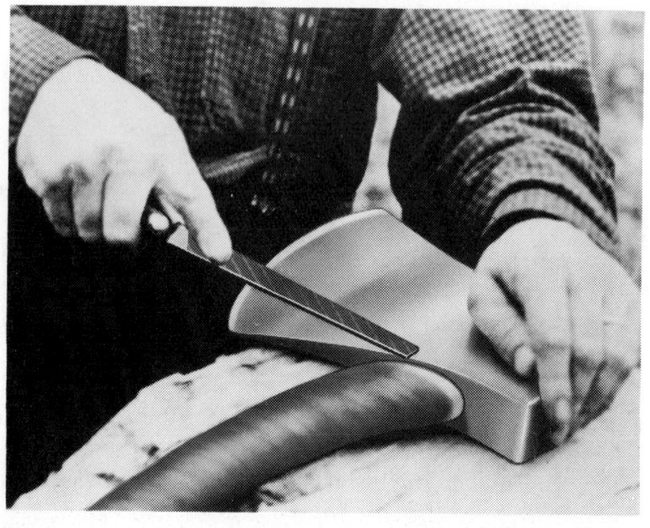

Courtesy Nicholson File Co.

Fig. 2-22. Using one hand to do light filing.

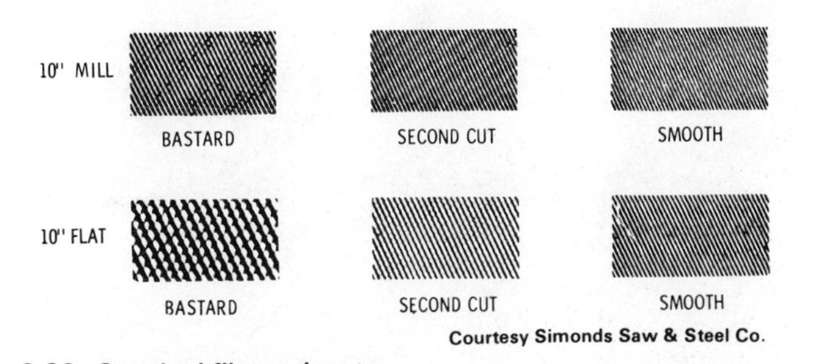

Courtesy Simonds Saw & Steel Co.

Fig. 2-23. Standard file tooth cuts.

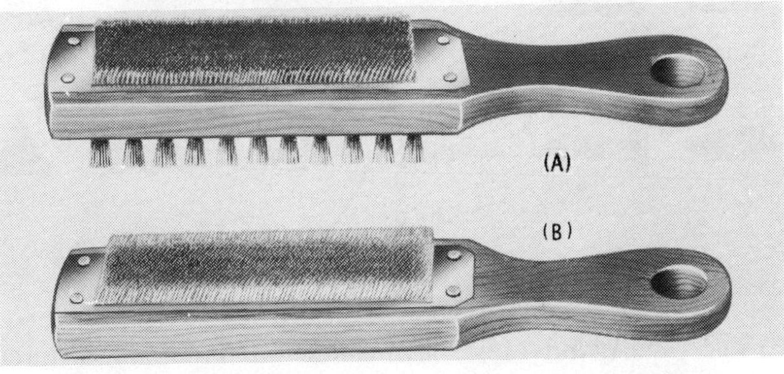

Courtesy Nicholson File Co.

Fig. 2-24. File cleaners: (A) File brush, (B) File card.

Files

A file differs from a chisel in that it has a large number of cutting points, instead of a single cutting edge; and the file is driven by hand, rather than by a hammer. When a file is applied to a metal surface with a reciprocating motion, the teeth act as small chisels, each removing small chips.

Files (Fig. 2-25) have three distinguishing characteristics: (1) length, which is always measured from the heel to the point, the tang not being included; (2) kind, which refers to the shape or style; and (3) cut, which refers to both the character and the relative degrees of coarseness of the teeth.

Length—File lengths vary from 3 inches to 20 inches. Most machinists' files are from 4 to 6 inches in length (Fig. 2-25A).

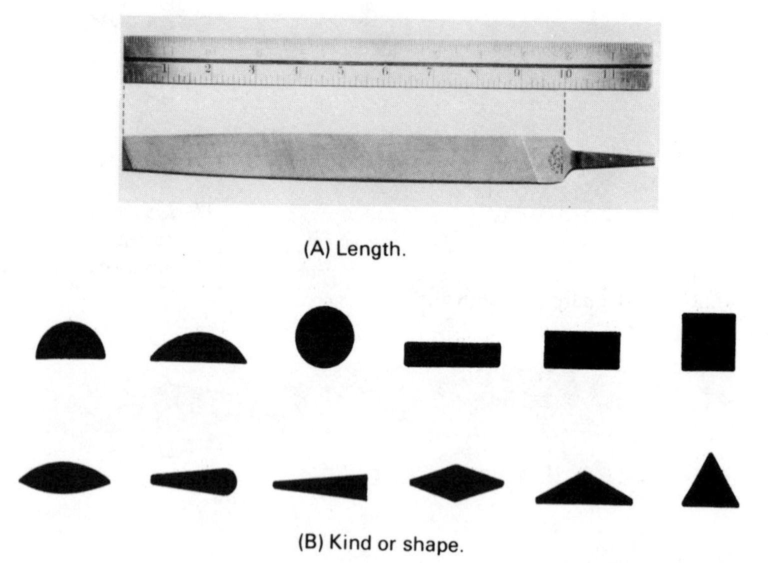

(A) Length.

(B) Kind or shape.

Fig. 2-25. File

Kind—Many kinds of files are manufactured for many different purposes. Shapes of files in common use are mill, flat, hand, square, three-square, half-round, and round files (Fig. 2-25B).

Cut—The teeth on a file are shaped to form a cutting edge similar to that of a tool bit; and they have both rake and a clearance angle. *Single-cut* files (Fig. 2-25C) are made with a single set of teeth cut at an angle of 65° to 85°. They are usually used with light pressure to produce a smooth finish on a surface, or to produce a keen edge on a knife or other cutting implement. *Double-cut* files (Fig. 2-25C) are made with two sets of teeth that cross each other. One set is cut at approximately 45° and the other set at 70° to 80°.

Machinists' files (Fig. 2-26) are used throughout the industry wherever metal must be removed rapidly and finish is of secondary importance. They include flat, hand, round, half-round, square, pillar, three-square, warding, knife, and several less commonly known kinds of files. Most machinists' files are double cut (Fig. 2-27).

The cut (coarseness) of small files is usually designated by numbers as: 00, 0, 1, 2, 3, 4, 5, 6, 7, 8. The coarsest cut is 00, and 8 is

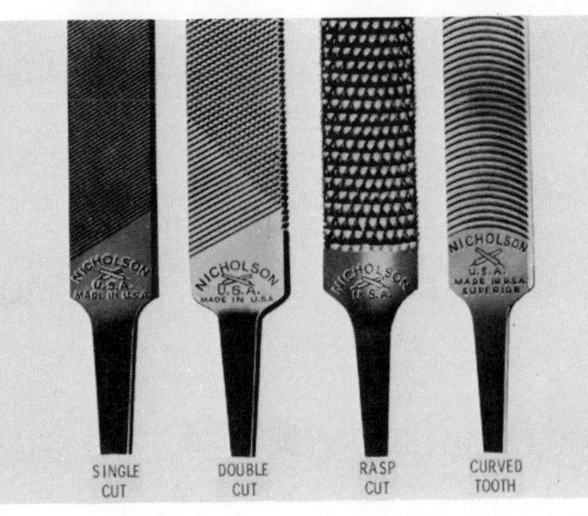

Courtesy Nicholson File Co.

(C) Cut.

characteristics.

the finest cut (Fig. 2-28). The cut or coarseness in larger files is designated as: rough, coarse, bastard, second-cut, smooth, and dead smooth. These designations are relative and depend on the length of a file. A 14-inch bastard file is much coarser than a 6-inch bastard file (Fig. 2-29).

SCRAPING

Scraping is the operation of correcting the irregularities of machined surfaces by means of scrapers (Fig. 2-30) so that the finished surface is a plane surface. Although it is impossible to produce a true plane surface, scrapers are used to approach a plane surface, or for truing up a plane surface. Scrapers are also used for truing up circular surfaces such as bearings.

How to Use a Scraper

In scraping operations (Fig. 2-31), a surface plate is used to indicate irregularities or high spots. Any dust or grit should be

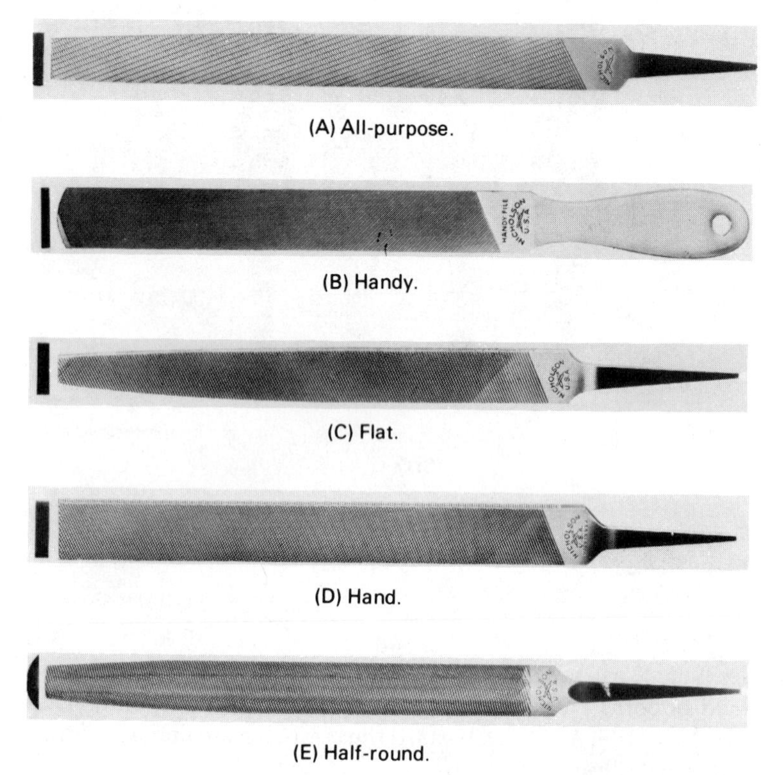

(A) All-purpose.

(B) Handy.

(C) Flat.

(D) Hand.

(E) Half-round.

Fig. 2-26. Machinists'

wiped off the surface, and any burrs on the metal should be removed with a very fine file.

After thoroughly cleaning the surface plate, coat it with a marking material and rub the work over the surface plate a few times. High spots on the work will be indicated by the marking material that has been rubbed off. These high spots are removed by scraping. Continuing the process will bring up more high spots. After repeated scraping and testing with the surface plate, the entire work surface will be covered with marking material, which indicates that the work is finished.

The correct use of the scraper is important. When a flat scraper is used, cutting is done on the forward stroke. Cutting is done on

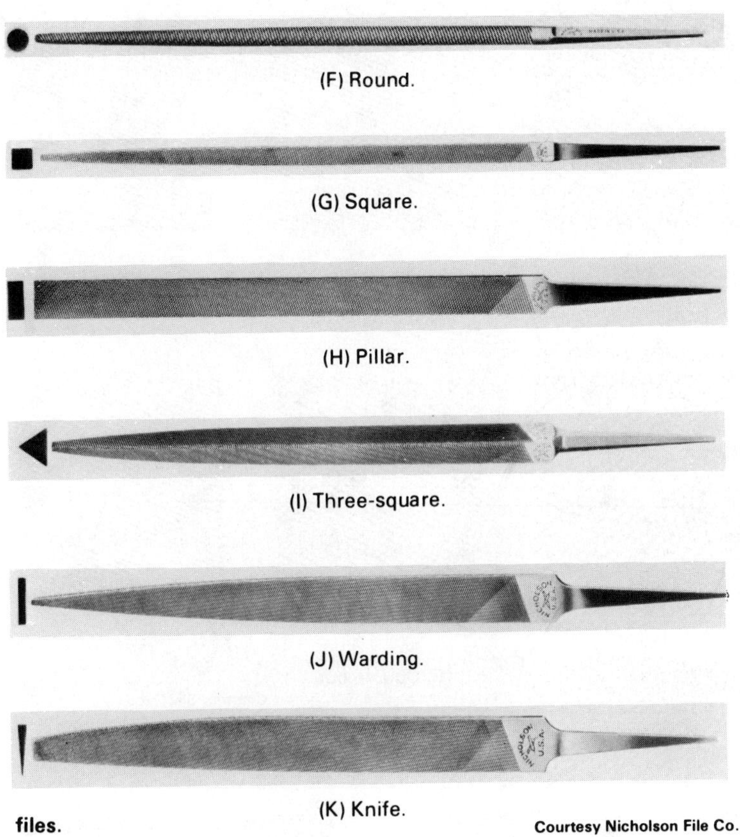

(F) Round.

(G) Square.

(H) Pillar.

(I) Three-square.

(J) Warding.

(K) Knife.

files.

Courtesy Nicholson File Co.

the return stroke when a hook scraper is used. Scraping requires a delicate touch. Less metal is removed by a scraper than by a file. The cutting operation, as done with scrapers, should be perfectly smooth and free from scratches.

The cutting edge of a scraper should be at 3/32 of an inch in thickness and 1½ inches in width. The scraper should be ground on a grinding wheel and carefully honed on an oilstone. Scrapers are sometimes made from discarded files.

Scrapers

Various forms of scrapers are used. The nature of the scraping

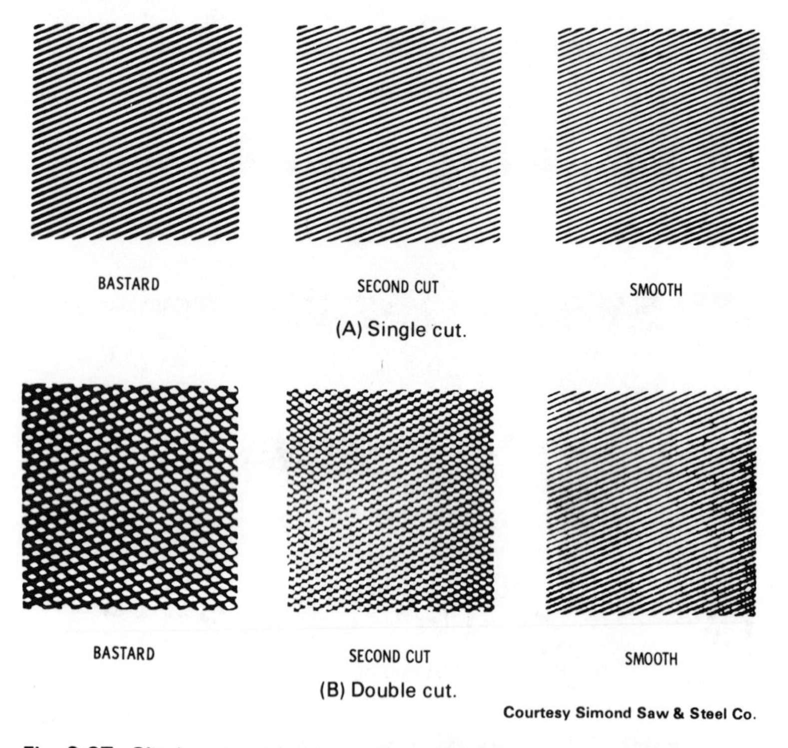

Fig. 2-27. Single cut and double cut files. Each type has its own application.

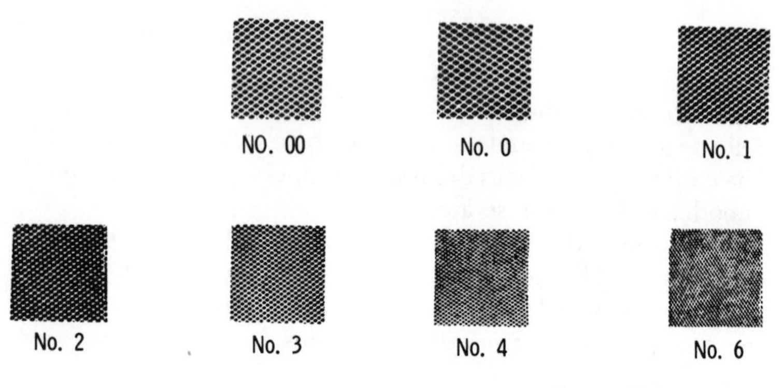

Fig. 2-28. Small files are designated by numbers from 00 to No. 8.

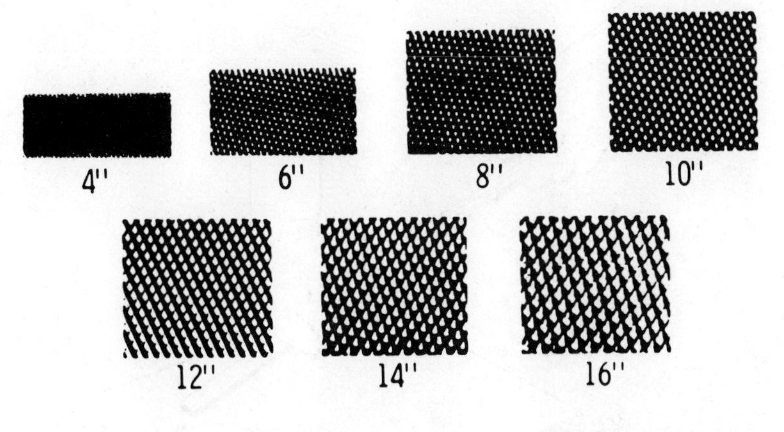

Courtesy Nicholson File Co.

Fig. 2-29. The coarseness varies for flat bastard files used by machinists. Length of the file determines coarseness.

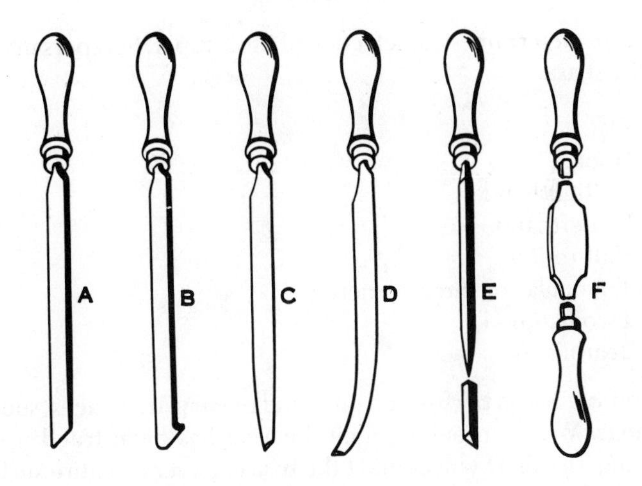

Fig. 2-30. Typical scrapers: (A) flat or straight, (B) hook, (C) straight half round, (D) curved half round, (E) three cornered or triangular, (F) double handle.

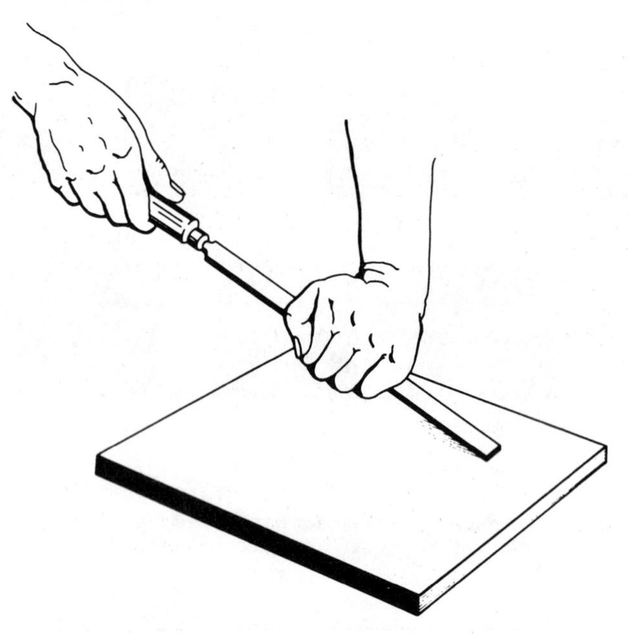

Fig. 2-31. Correct method of holding a scraper.

operation determines the selection of the scraper. Scrapers may be classified as:

1. Flat.
2. Hook.
   a. Righthand.
   b. Lefthand.
3. Half round.
4. Triangular or three cornered.
5. Two handled.
6. Bearing.

Scraping is also performed on round or curved surfaces, such as bearings. When an engine main bearing has been trued up by scraping, the shaft will contact the bearing over its entire surface instead of making contact only at the high spots. Consequently, the bearing surface is presented, and the pressure is distributed over the entire bearing instead of being concentrated on the high spots.

SUMMARY

Benchwork is activity performed by the mechanic at the machinists' bench with hand tools and a vise. Hand tools consist of a vise, hammers, chisels, hacksaws, files, and scrapers. There are various types of vises, such as pipe, combination, machinist, and blacksmith. Peen hammers are classified as ball, straight, and cross.

Various types of chisels, such as cold, cape, diamond-point, and round-nose, are used in benchwork. A cold chisel is usually used for all general chipping operations. The chisel should be held firmly with the second and third fingers, and the little finger should be used to guide the chisel as required. Reset the chisel to its proper position after each blow of the hammer.

Hacksaws cut different kinds and shapes of materials. The most common hacksaws are adjustable and use blades 8 to 12 inches long with various numbers of teeth per inch. Blades should be chosen based on the material to be sawed.

REVIEW QUESTIONS

1. Name any five of the ten most popular machinists' files.
2. How many various forms of scrapers are used? Name them.
3. Name the three types of peen hammers and the four types of chisels.
4. Name the four types of bench vises.
5. Name a few bench tools.
6. Name four hacksaw blade pitches.
7. How many hacksaw blade teeth should be in contact with the work piece?
8. What is the meaning of benchwork?
9. What are four operations that can be performed at the bench?
10. Name four types of chipping chisels.
11. What is drawfiling?
12. How does a file differ from a chisel?
13. What are the three distinguishing characteristics of files?
14. Why is scraping used on bearings?

CHAPTER 3

Layout and Measurement

Layout means the marking or scribing of lines on material to indicate the exact location of cutting or forming operations. The machinist gets the necessary information for layout from various types of blueprints.

The subject of shop measurements is an important one, especially for the machinist, because his work must be done with a much greater degree of precision than for many other lines of work.

For the machinist, measuring means the act of determining the dimension, or dimensions, of an object. The machinist has precision measuring tools available for taking measurements with great accuracy.

The machinist must be skilled in handling measuring tools when attempting layout work. It is essential that measuring tools be kept clean and in order.

CARE OF MEASURING TOOLS

While a carpenter may be known by his chips, an index of the efficiency of a machinist is the condition of his measuring tools. A battered steel rule, or scale, that is difficult to read may cause a workman to ruin a piece of work, thereby wasting considerable time and effort. Dividers that are either dull or have loose joints cannot be used effectively. Calipers that are either bent or out of shape may cause difficulty. Likewise, a strained micrometer may give inaccurate measurements.

All those items mentioned, along with many others that could be mentioned, indicate the importance of keeping measuring tools in excellent condition. Precision tools should be kept in special boxes and treated with the same care that a draftsman gives his drawing instruments.

MEASURING TOOLS

To obtain accuracy, special measuring tools are used. A great variety of such tools, which are adapted to all kinds of measurements, is available. The measuring tools most commonly used are:

1. Rule.
2. Square.
3. Calipers.
4. Dividers.
5. Protractors.
6. Bevels.

There are many variations of each of the measuring tools listed above. Measuring tools may be classified as:

1. Linear.
2. Angular.

RULES

The different kinds of rules used by machinists and other mechanics can usually be classified as follows:

1. Flat.
2. Flexible.
3. Hook.
4. Holder.
5. Caliper.
6. Key seat.
7. Shrink.

All the rules listed above are made of tempered steel, and the graduations are cut with great accuracy. There is a great variety of graduations to choose from, and the rules are made in various lengths. The thickness of steel rules varies from 1/64- to 1/20-inch, depending on their length.

A plain flat 6-inch steel rule is a type commonly used. The inches are graduated into 8ths and 16ths of an inch. The appearance of smaller graduations is shown in Fig. 3-1. Note the figures used. On the 64th of an inch side, every 8th graduation is numbered.

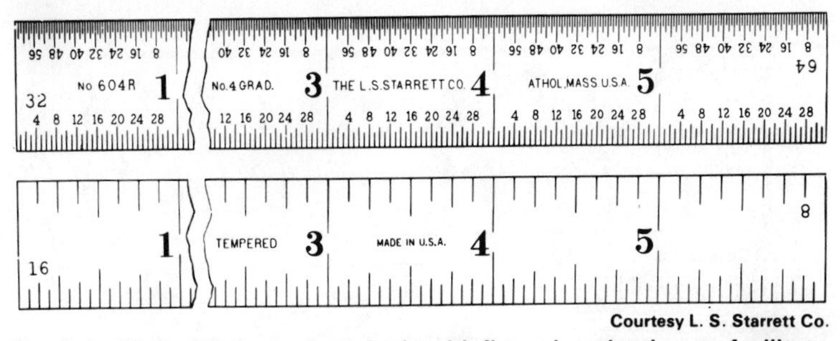

Courtesy L. S. Starrett Co.

Fig. 3-1. Six-inch tempered steel rule with figured graduations to facilitate readings in fine dimensions. Front and reverse sides are shown.

In taking measurements with the steel rule, place the 1-inch index mark, instead of the end of the rule, at one edge of the piece to be measured. The steel rule should be laid parallel to the edge of the piece (Fig. 3-2). Remember to deduct 1 inch from the reading. Precaution should be used in setting the 1-inch mark at the edge of the piece. The edge of the piece should coincide with the center of the index mark on the steel rule (Fig. 3-3). Even though the graduation lines on machinists' rules are much finer than those on nonprecision rules, it is impossible to make them without width. Width, of course, is necessary for the lines to be seen without difficulty.

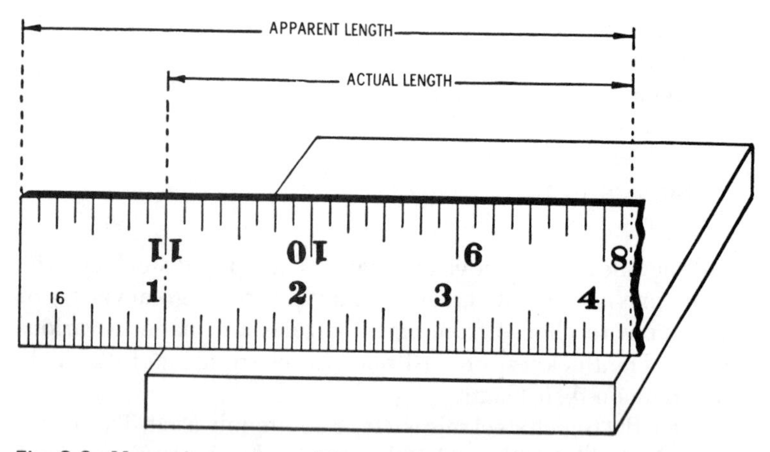

Fig. 3-2. Measuring a piece of flat stock with the machinists rule. Some mechanics prefer counting the inch marks on the rule to subtracting one inch from the reading on the rule.

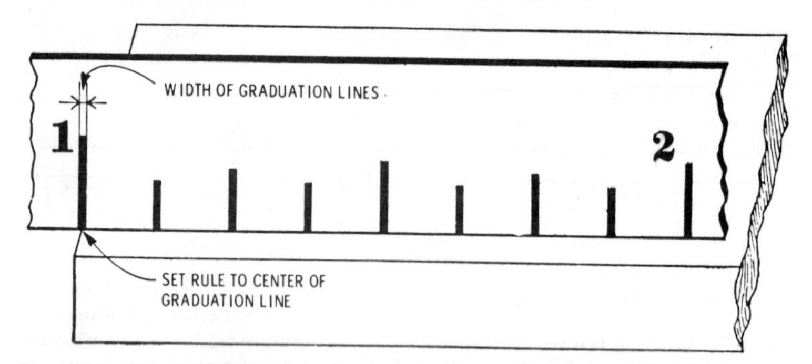

Fig. 3-3. An enlarged one-inch division on the machinists' rule. Note the width of the graduation marks and the necessity for placing the rule correctly to avoid errors in measurement.

An internal diameter, such as the diameter of a cylinder, may be measured with the steel rule (Fig. 3-4). To avoid an incorrect reading, do not place the steel rule too far inside the cylinder. By moving the rule back and forth through a small arc, various readings can be obtained. The maximum reading obtained will give the correct diameter of the cylinder.

To determine the external diameter of a piece of round stock, place the rule directly across the end of the stock (Fig. 3-5). Again,

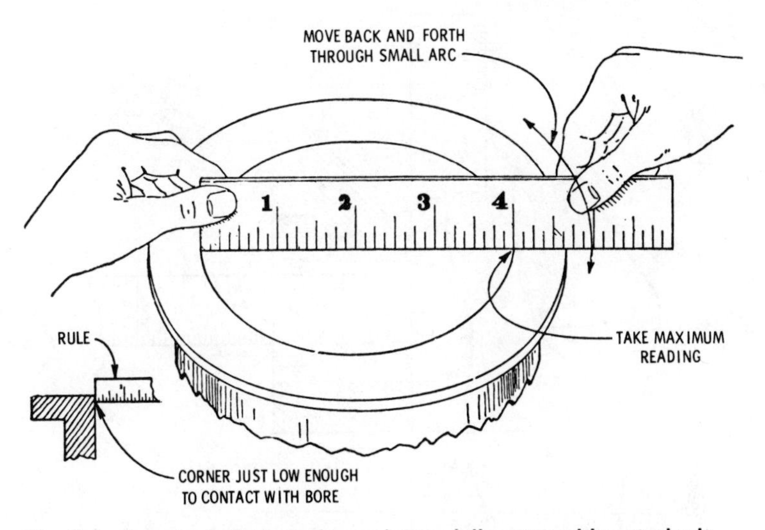

Fig. 3-4. A method of measuring an internal diameter with a steel rule.

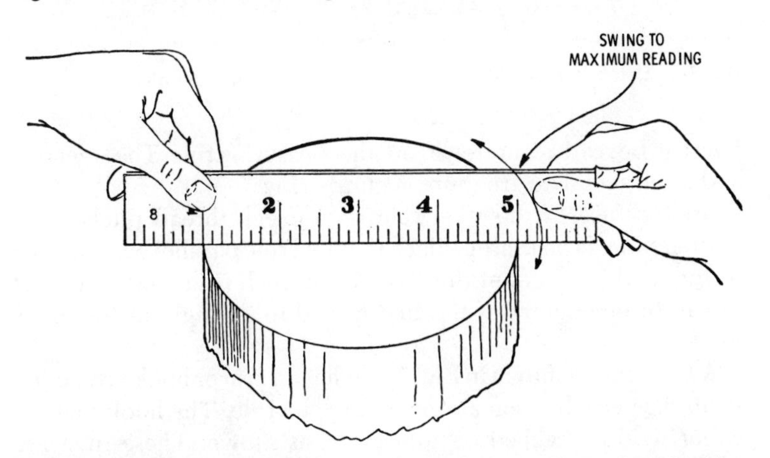

Fig. 3-5. A method of measuring the external diameter of round stock with a steel rule.

use any index mark of the scale, and move the rule back and forth through small arcs to obtain the maximum diameter reading.

To measure the depth of a hole, the use of two steel rules is advantageous (Fig. 3-6). One steel rule, or a straightedge, is placed across the end of the piece with the index mark of the measuring rule registering accurately with the straightedge. The depth of the

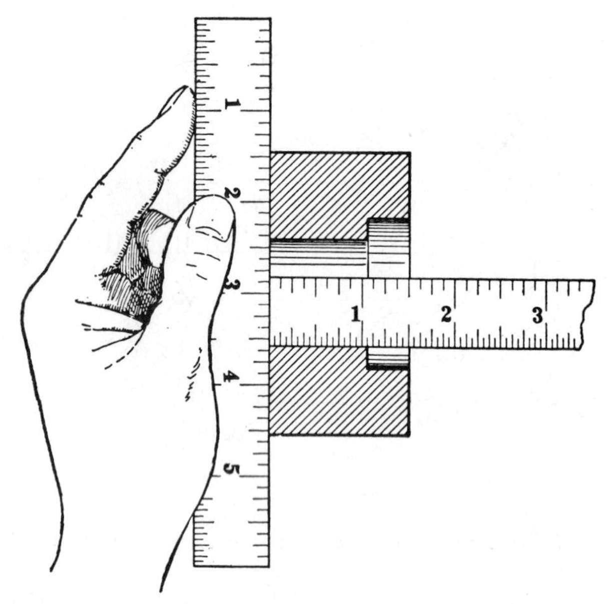

Fig. 3-6. Using two steel rules to measure the depth of a hole.

hole can be read accurately from the measuring rule. The steel rule and a surface plate are more accurate (Fig. 3-7).

Another use of the steel rule is to check thread pitch when cutting threads on a lathe. Place the steel rule parallel with the axis of the work, and count the threads per inch (Fig. 3-8). The first thread to be counted is the first thread to the right of the index mark.

A *hook* rule is shown in Fig. 3-9. It has an arm or hook projecting from the zero division at one end of the rule. The hook may be either rigidly attached or adjustable, as shown. These rules are convenient for taking measurements from points where a person cannot see whether the rule is even with the measuring edge, for measuring from round corners, for measuring through hubs of pulleys, and for setting inside calipers.

A hook rule is preferable to the steel rule in some instances (Fig. 3-10). The hook eliminates error in placing the rule, as it permits an exact setting of the rule. Thus, the actual dimension is read from the rule. The hook rule, along with a steel rule or straightedge, can

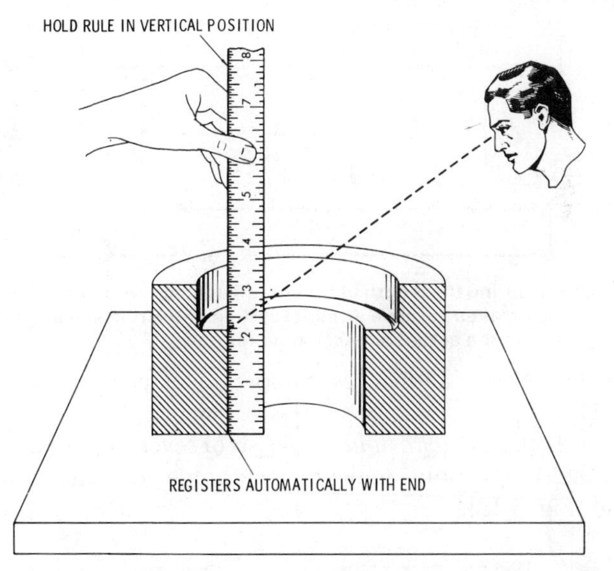

Fig. 3-7. Measuring the depth of a hole with a steel rule and a surface plate.

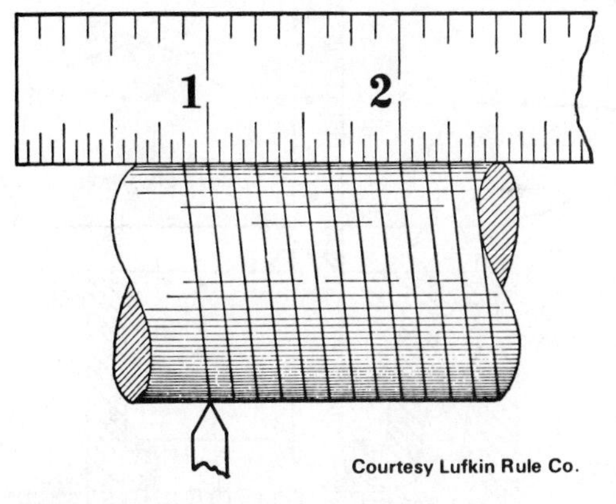

Courtesy Lufkin Rule Co.

Fig. 3-8. Measuring thread pitch with a steel rule.

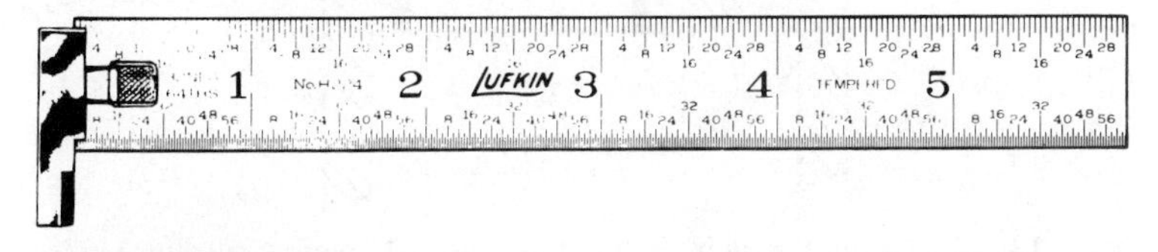

Fig. 3-9. A hook rule.

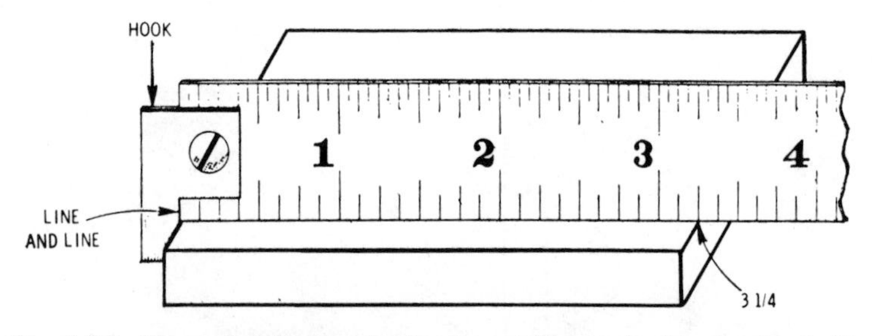

Fig. 3-10. Measuring the length of a piece of flat stock with a hook rule. The hook makes possible an exact setting of the rule and enables one to read the actual dimension on the rule.

be used to measure an inaccessible part, such as an engine part (Fig. 3-11).

A set of *small rules with holder* consists of several short rules that may be held at various angles by the holder designed for the purpose (Fig. 3-12).

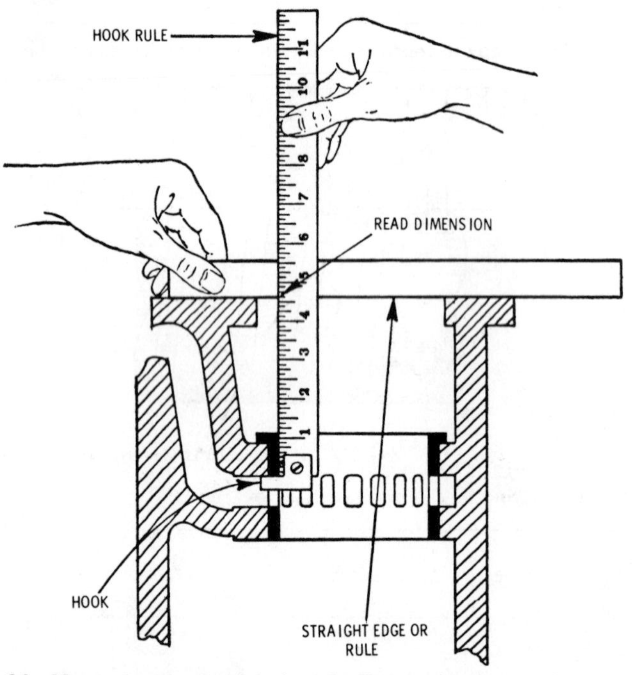

Fig. 3-11. Measuring the depth below the flange of a piston valve part with the aid of a hook rule and a straightedge.

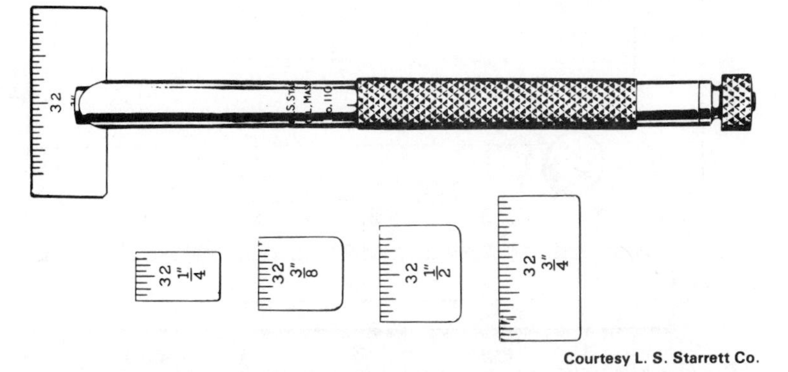

Courtesy L. S. Starrett Co.

Fig. 3-12. A set of small rules with their holder. The rules are in various lengths with 32nd and 64th graduations. The rules are held in a split chuck adjusted by a knurled nut at the top of the holder and can be set at various angles. The rules are of tempered steel, graduated on both sides. These rules are used for general tool and die work or in measuring a recess or keyway where an ordinary rule cannot be used.

Whenever quick measurements are desired on small rods, tubing, sheet stock, etc., it is convenient to use an instrument that measures the object as it is held between two contacts. The *caliper rule* (sometimes called slide-caliper rule) is used for this purpose, as shown in Fig. 3-13. A *pocket-slide caliper rule* is shown in Fig. 3-14.

A *shrink* or *shrinkage rule* (Fig. 3-15) is essential for pattern work as it is a rule designed to allow for shrinkage of castings. The

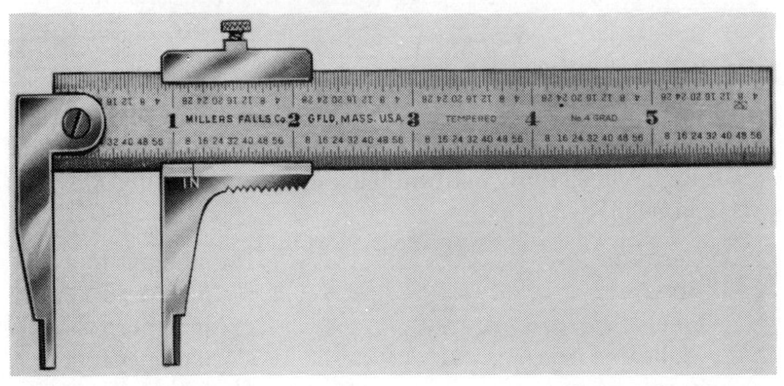

Courtesy Millers Falls Co

Fig. 3-13. A slide-caliper rule. Used for quick measurements.

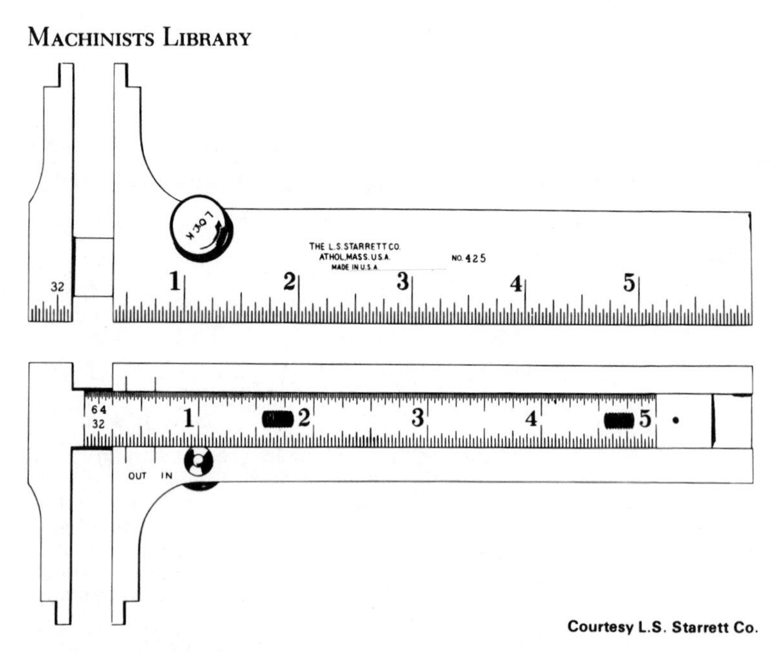

Courtesy L.S. Starrett Co.

Fig. 3-14. A pocket-slide caliper rule. The nibs of the jaws can be inserted into holes as small as $\frac{1}{8}$ inch. The clamp nut locks the slide and holds it set for any particular measurement. The button on the slide aids in operating and closing the jaws.

SHRINK 1-16 TO FOOT No. C 84 A LUFKIN

Courtesy Lufkin Rule Co.

Fig. 3-15. A shrink rule. Note the marking "shrink $\frac{1}{16}$ to foot."

allowance necessary for shrinkage of castings varies for different metals and the different conditions under which they are cast. For castings having a fairly uniform thickness, the following allowances can be made:

Cast iron	$\frac{1}{8}$ inch per foot
Brass	$\frac{3}{16}$ inch per foot
Steel	$\frac{1}{4}$ inch per foot
Malleable iron	$\frac{1}{8}$ inch per foot
Zinc	$\frac{5}{16}$ inch per foot
Tin	$\frac{1}{32}$ inch per foot

Aluminum .... 3/16 inch per foot
Britannia .... 1/32 inch per foot

The above shrinkages are standard shrinkages for the different metals, but some consideration must be given to the size and the shape of the casting. The thicker castings shrink less under the same conditions, and the thinner castings shrink more than standard. The quality of the material and the manner of molding and cooling also make differences in shrinkages.

When a sand mold made from a wooden pattern is filled with molten metal, its temperature is high. As the metal cools and solidifies, it contracts or shrinks. To compensate for this shrinkage, the patternmaker must add to the size of the pattern. Therefore, all dimensions are measured with a shrinkage rule. The shrink rule "shrink 1/8" to foot" is used for cast iron (actual measurement is 12 1/8 inches).

Since the shrinkage of different metals varies greatly, rules for the various shrinkages can be obtained. For example, a patternmaker working with steel (1/4 inch per foot) would use a "shrink 1/4" to foot" shrinkage rule.

A *rule clamp* is used to clamp two steel rules together to measure lengths longer than either of the rules (Fig. 3-16).

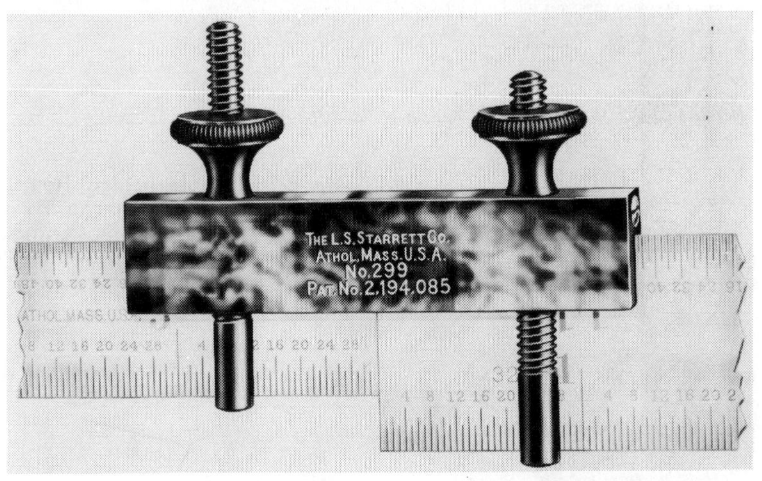

Courtesy L.S. Starrett Co.

Fig. 3-16. A rule clamp. Used to clamp two steel rules together.

Key-seat clamps are used to convert straightedges, steel rules, and combination square blades into key-seat rules for laying out keyways and for scribing parallel lines on shafts or round work (Fig. 3-17).

A *rule holder* is used in an upright position for transferring measurements with surface gages, etc. (Fig. 3-18).

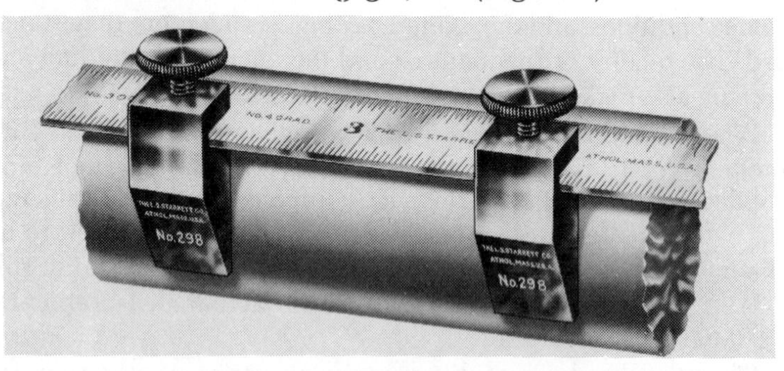

Courtesy L.S. Starrett Co.

Fig. 3-17. Key-seat clamps. Designed to transform any common steel rule or straightedge into a key-seat rule.

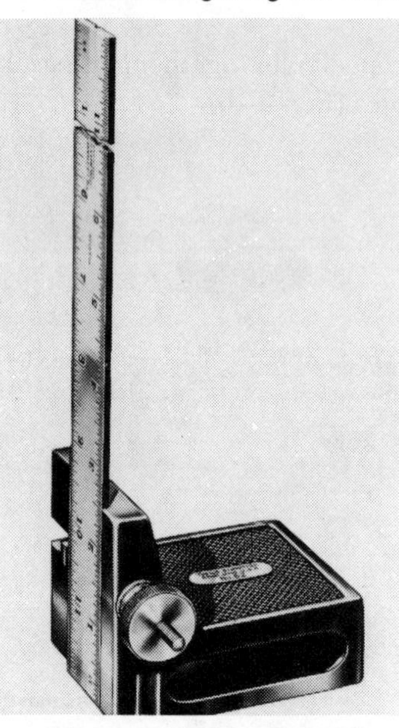

Courtesy L.S. Starrett Co.

Fig. 3-18. A rule holder. Holds rules in upright position for use in transferring measurements with surface gages, etc.

SQUARES

Many kinds of squares are used by machinists and toolmakers. The *try square* consists of a thick beam, sometimes called stock, and a thin blade set at a 90° angle (Fig. 3-19). The stock is made thicker than the blade so that the square will stand on a flat surface, and will present a bearing when it is pressed against the edge of the work.

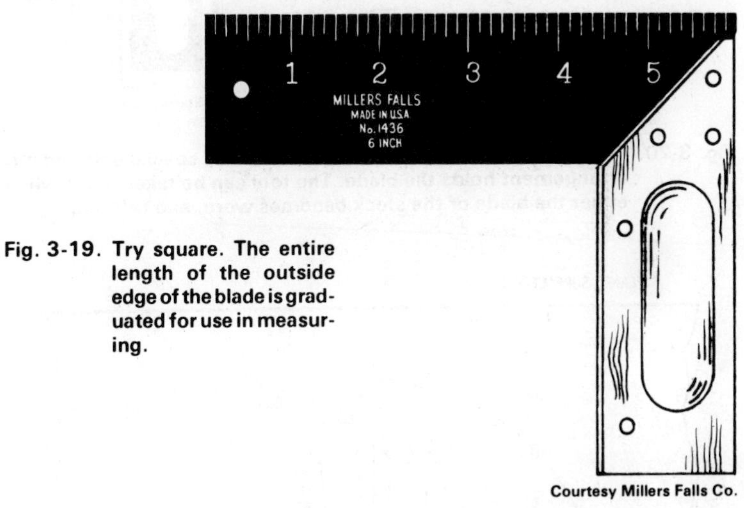

Fig. 3-19. Try square. The entire length of the outside edge of the blade is graduated for use in measuring.

Courtesy Millers Falls Co.

A good steel try square is a precision instrument and should be handled with care. Its accuracy is affected by temperature changes. Both the beam and the edge are hardened and accurately ground to insure parallelism and straightness. The sides of the blades are not working surfaces, and they are not held precisely at the right angles with the bottoms of the beams.

A try square is commonly employed by machinists to test or "try" the accuracy of work. This definition would indicate that any square used to test squareness of work is a try square. However, if the try square were used only for that purpose, graduations on the blade would not be required.

Some try squares are designed so that they may be dismantled (Fig. 3-20). A good steel square is constructed with the blade and beam at exactly 90° (Fig. 3-21). Its accuracy is affected greatly by

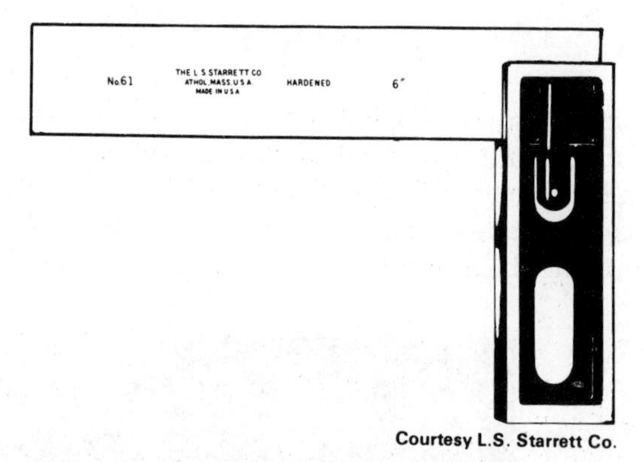

Courtesy L.S. Starrett Co.

Fig. 3-20. Try square with the blade not graduated. A special bolt and nut arrangement holds the blade. The tool can be taken apart when either the blade or the stock becomes worn, and refitted.

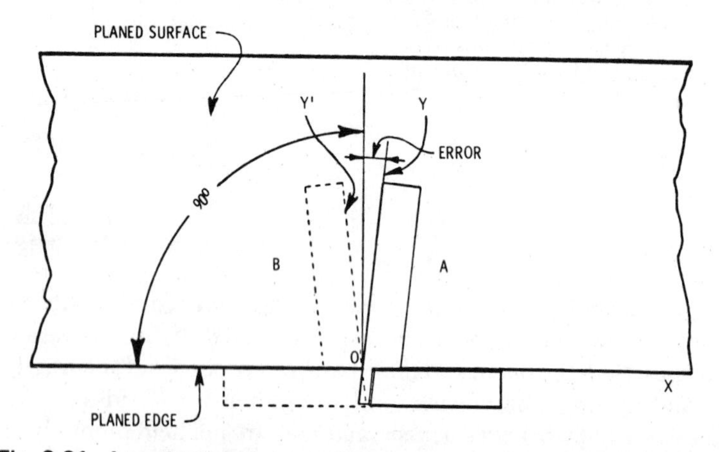

Fig. 3-21. A method used to test the squareness of a try square. Place the square in position A, and scribe line Y along the blade. Reverse the square to position B, as shown by the broken lines. If the blade does not register with the scribed line, but assumes the other position (Y'), the square is out of true. The error is half the angle between the two lines, or ½ YOY'

abuse. To test two machined surfaces for squareness, place the beam of the try square against one surface, and move the square until the blade contacts the other surface. If the blade does not register squarely with the surface, the work is "out of square."

Combination Square

This instrument combines a try square, a miter square, and a level. It is constructed in such a manner that the head slides along the blade and clamps at any desired place (Fig. 3-22). The head slides along a central groove in which a guide located in the head of the square travels. In all blades the groove is concave to eliminate congestion of dirt and to permit a free and easy slide. The scale can be pulled out and used solely as a rule. The head may be used as an ordinary level. A center head may be used on the scale to find the centers of shafts and other cylindrical pieces.

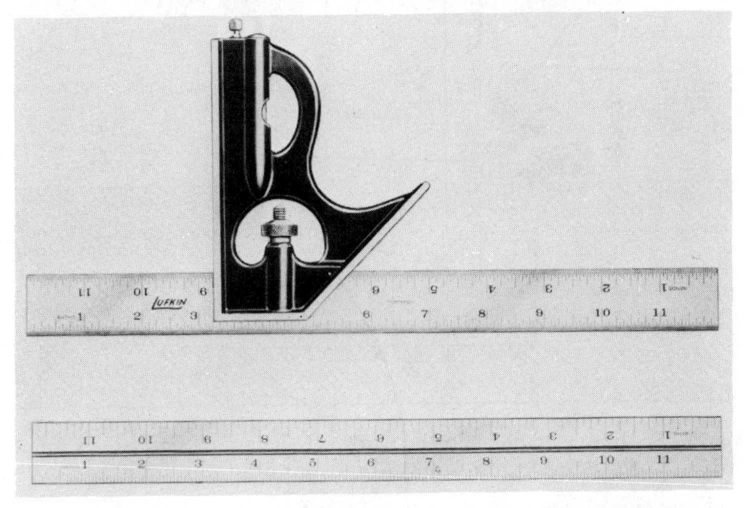

Courtesy Lufkin Rule Co.

Fig. 3-22. Combination square.

The combination square is used for checking either 45° or 90° angles, for scribing lines at right angles to a surface, or as a depth gage. It may also be used as a height gage by setting the blade flush with the head. The combination square is also conveniently used to square a piece with a surface and to indicate levelness or plumbness at the same time.

This is a very useful tool for the machinist, and many different measurements can be made with it. The combination square should be kept in proper condition, or the blade will not align with the slot.

To measure roughly the depth of a slot, place the square in position, and adjust the blade to the depth of a slot (Fig. 3-23). For a precision measurement of the depth of a slot, place a block gage on the machined surface of the work, and set the square on the top of the block (Fig. 3-24A). Place a piece of paper under the end of the blade of the square, and adjust the blade until the paper will barely slide back and forth. Remove the block and place the square on the casting (Fig. 3-24B).

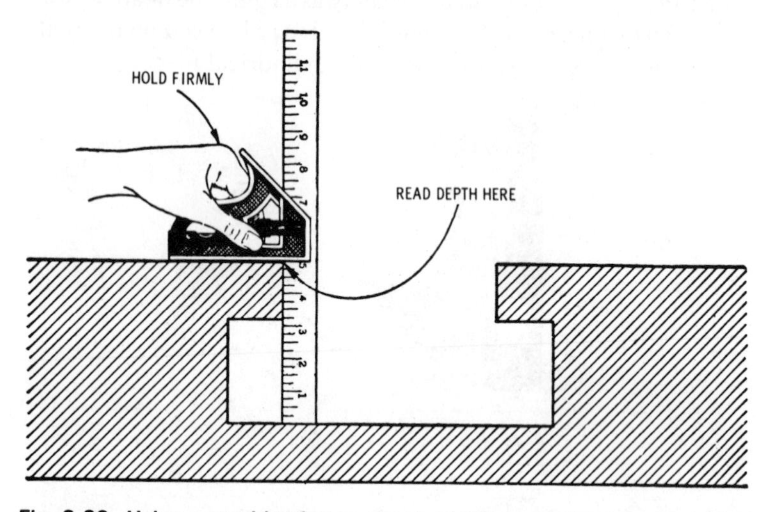

Fig. 3-23. Using a combination square to obtain a rough measurement of the depth of a slot.

The center head is used with the combination square to measure the diameter of round stock (Fig. 3-25). The center head is pressed firmly against the piece. This aligns the blade across the center of the work, and permits an accurate measurement of the diameter of the round stock.

The *diemakers' square* is an instrument used for measuring die clearances (Fig. 3-26). The tool is also useful to the patternmaker for determining drafts on patterns. The blade is locked in position by a small clamp screw and can be set for any angle up to eight degrees on either side of zero. The graduations indicate the setting in degrees. The clamp screw on the back of the square locks the setting. The offset blade is used in positions where it is impossible to read the straight blade. A toolmaker's square (Fig. 3-27) has

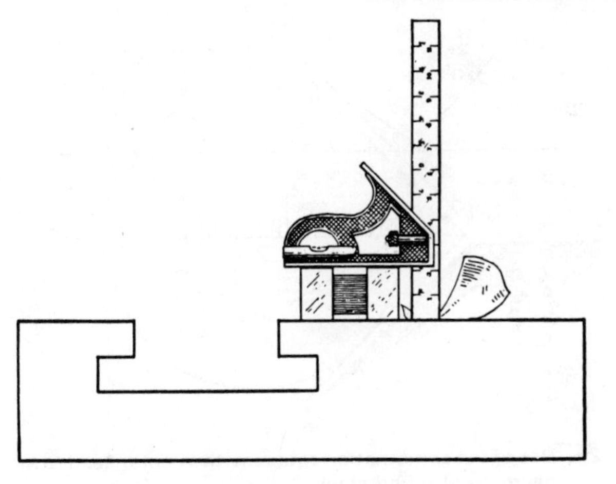

(A) Set the depth on the block gage.

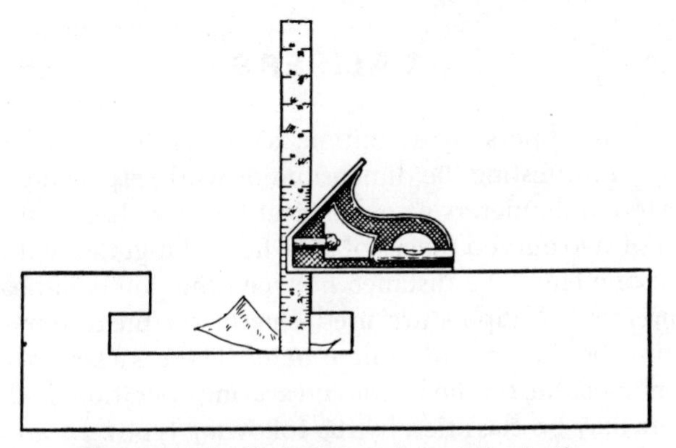

(B) Remove the block gage and check the depth.

Fig. 3-24. Using a combination square and a block gage to obtain a precision measurement of the depth of a slot.

three blades. All are interchangeable and lock securely by tightening the thumbnut in the beam.

Magnetic blocks are used by the toolmaker for holding pieces at right angles, such as holding a steel rule vertically on a surface plate (Fig. 3-28). These blocks have three adjacent magnetic sides that are square and parallel. The outer shell on the other three sides is nonmagnetic aluminum.

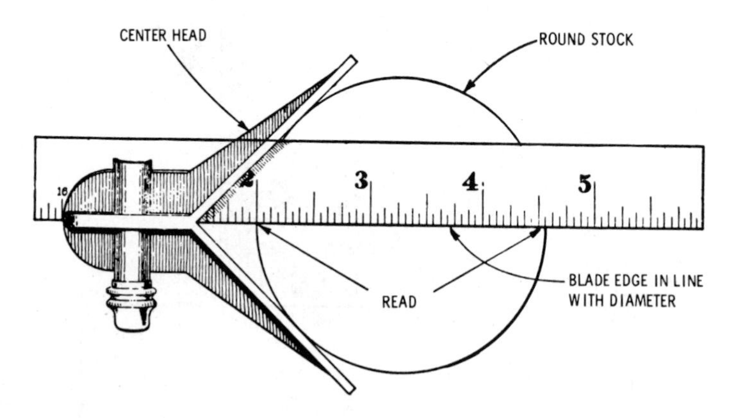

Fig. 3-25. Use of the combination square with a center head to measure the diameter of round stock.

CALIPERS

The term calipers, by definition, is used to designate an instrument used for testing the dimensions of work, especially internal and external diameters of cylindrical pieces. These instruments consist of two curved pieces of steel hinged together with a tight joint at one end. The distance between the points indicates the measurement. Calipers are used chiefly for measurements not requiring the accuracy of a micrometer. There is a large variety of calipers available for the various measuring operations in the shop. Calipers may be classified by the following types:

1. Firm joint.
2. Spring.
3. Transfer or lock joint.
4. Hermaphrodite.

Calipers should never be used on revolving work, either in a lathe or in other machines, because if one contact of the caliper is placed against the work, the other contact is likely to be drawn over the work by the friction of the moving surfaces. Only a slight force is needed to spring the legs of calipers. Measurements taken from moving pieces are inaccurate.

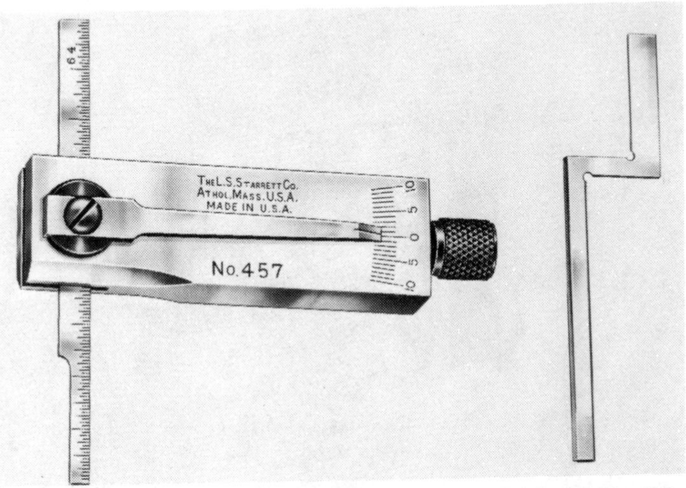

Courtesy L.S. Starrett Co.

Fig. 3-26. A diemakers' square. Used for measuring die clearances. The blade is locked in position by a small clamp screw and can be set for any angle up to 8°. The offset blade is used where it is impossible to sight the straight blade.

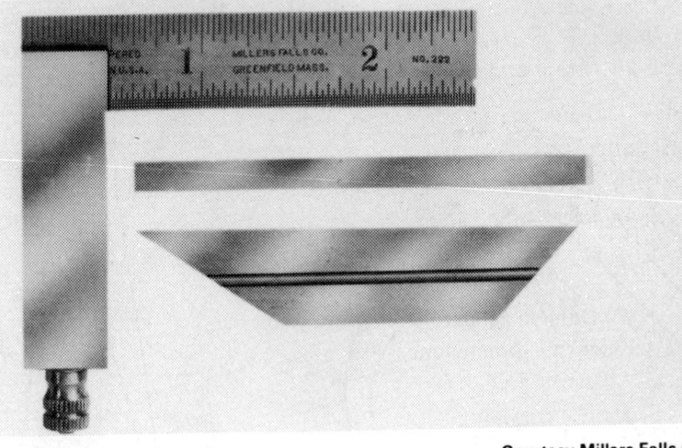

Courtesy Millers Falls Co.

Fig. 3-27. Toolmakers' square.

The terms *outside* and *inside* mean that these calipers are used for outside and inside dimensions, respectively (Fig. 3-29).

Firm-joint calipers (Fig. 3-30) have a friction joint connecting the legs, whereas *spring calipers* (Fig. 3-31) are provided with a

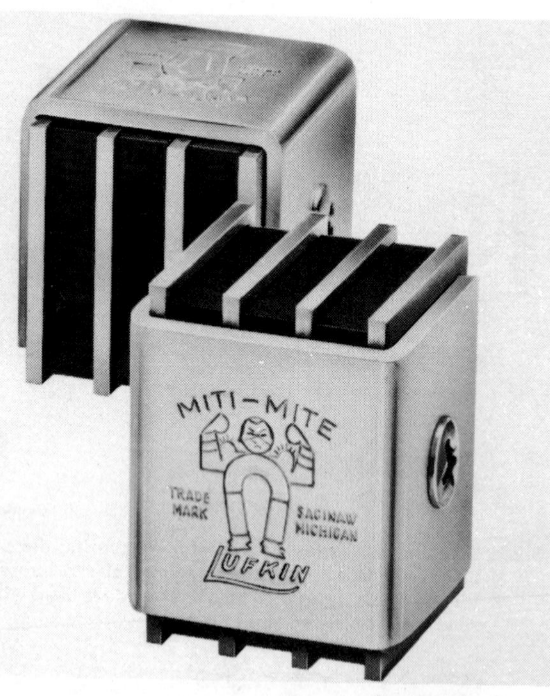

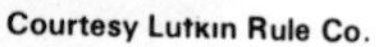
Courtesy Lufkin Rule Co.

Fig. 3-28. Magnetic blocks.

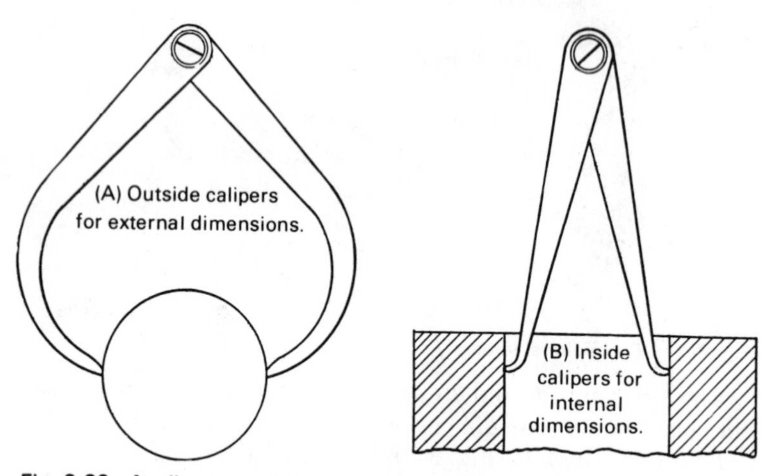

Fig. 3-29. Applications of the firm-joint calipers.

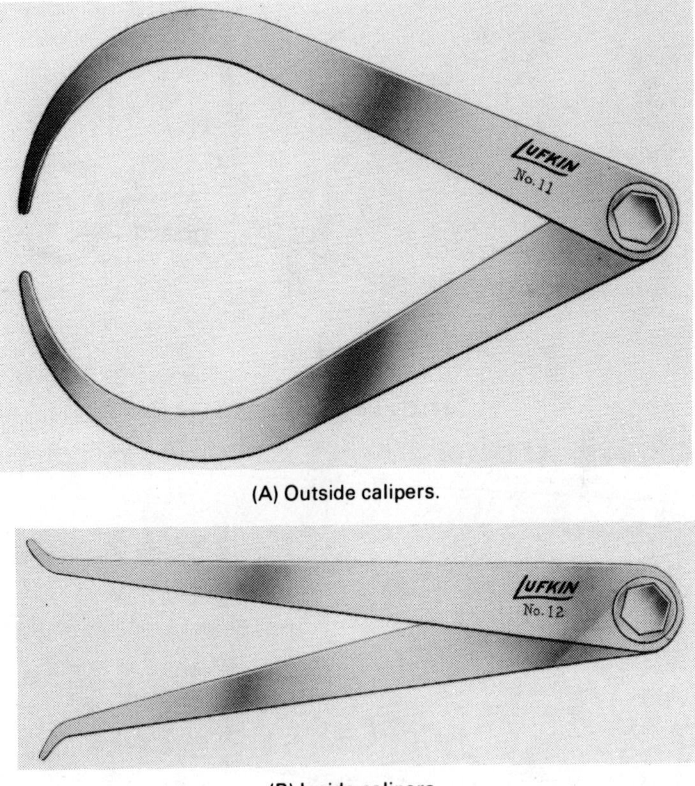

(A) Outside calipers.

(B) Inside calipers.

Courtesy Lufkin Rule Co.

Fig. 3-30. Firm-joint calipers.

spring that connects the two legs, along with an adjusting screw and nut, the two legs being forced together against the tension of the curved spring.

Both outside and inside firm-joint calipers are set to the approximate dimension, and then closed or opened to the desired dimension by gently tapping one of the legs against a convenient object (Fig. 3-32). This method requires practice. Too much movement is usually the result at first because the beginner tends to tap too hard.

Spring calipers are not made in the large sizes as are the friction-joint type. The steel rule is usually used to set calipers to a given

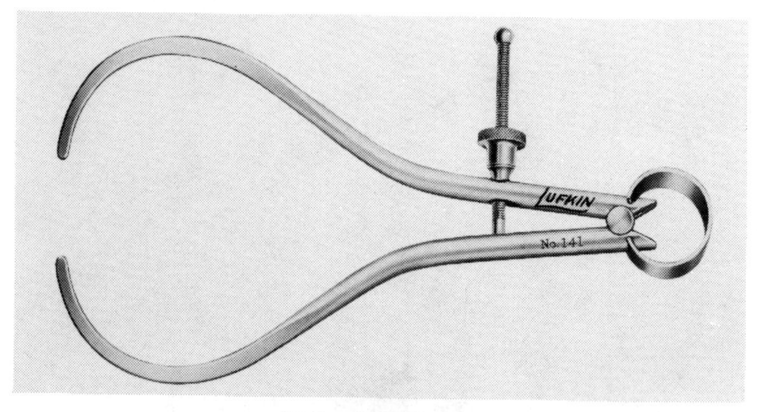

(A) Outside calipers.

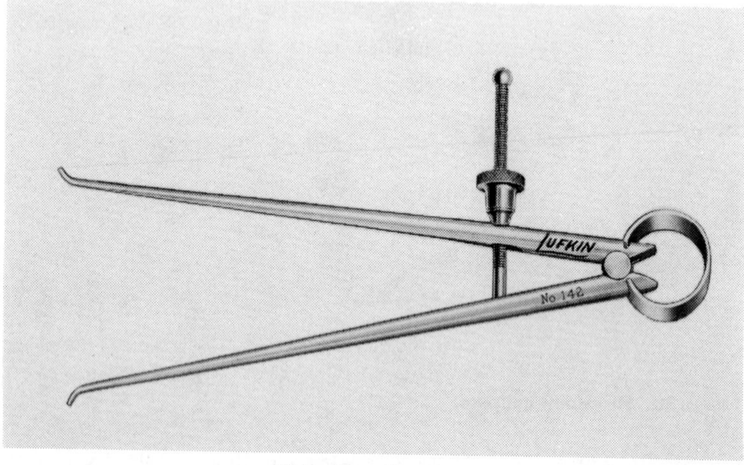

(B) Inside calipers.

Courtesy Lufkin Rule Co.

Fig. 3-31. Spring-joint calipers.

dimension. In setting outside spring calipers, one finger of the left hand supports the end of the rule and one leg of the calipers. The adjustment nut of the calipers is turned by the thumb and forefinger of the right hand (Fig. 3-33).

Outside calipers can be used to measure the diameter of round stock. Skilled machinists have a highly developed sense of touch, or "feel," which determines the accuracy of their work with the

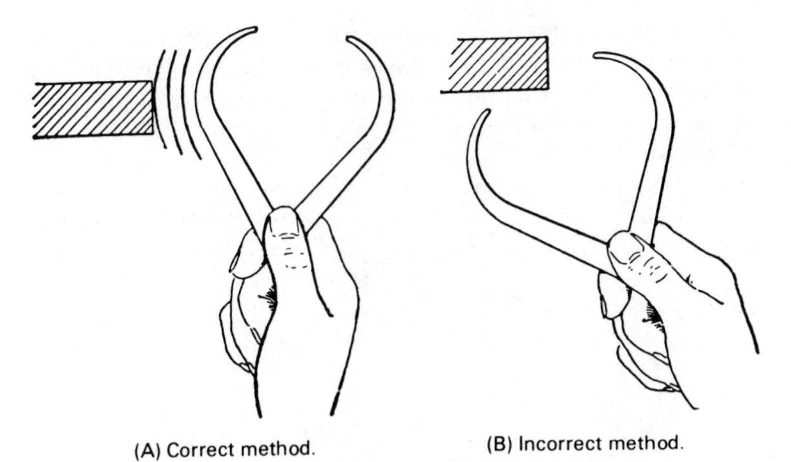

(A) Correct method. (B) Incorrect method.

Fig. 3-32. Correct and incorrect methods of setting firm-joint calipers by tapping a leg of the calipers.

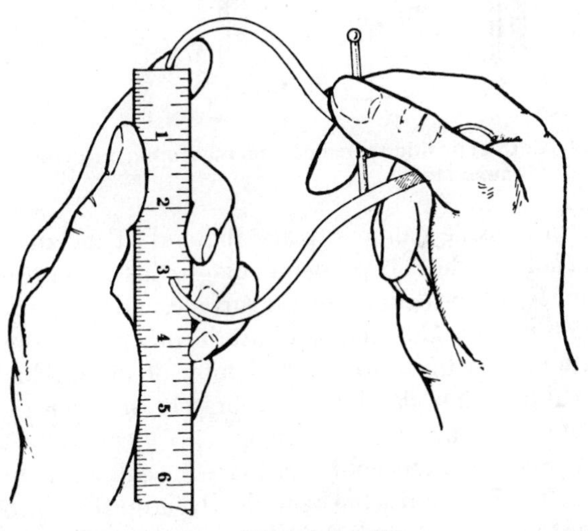

Fig. 3-33. Method of setting outside spring calipers.

calipers. Only the tips of the fingers should be used to hold the calipers. Calipers should not be gripped too tightly. Relaxation of the fingers increases their sensitivity to touch or feel. A skilled mechanic can detect differences in diameter as small as one-thousandth of an inch.

When the calipers measure properly, the weight of the calipers themselves should just force the calipers over the shaft. The true diameter of the work is measured at right angles to the center line of the work, where the calipers come in contact with the work (Fig. 3-34).

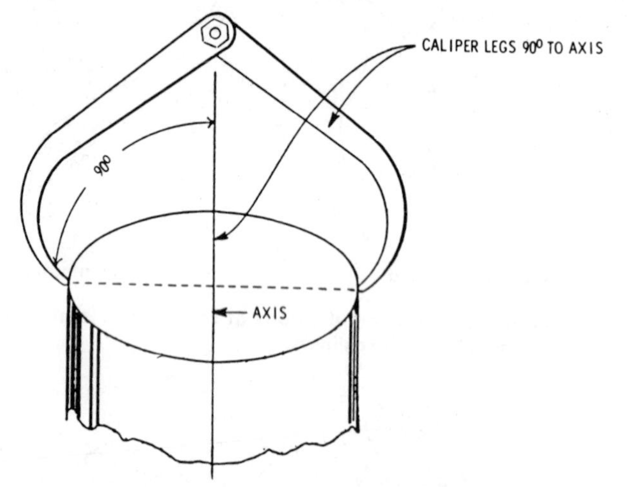

Fig. 3-34. Correct position of calipers for measuring the diameter of a piece of round stock.

In setting inside calipers, place the end of the rule on a flat surface, holding the rule perpendicular to the supporting surface. Place one leg of the calipers on the surface, and use the thumb and forefinger of the right hand to adjust the other leg to the required graduation. Be careful to set the center of the graduation line. Inside calipers can also be set accurately between the contact points of outside micrometer calipers.

To measure diameters of holes, place the calipers in the hole, as shown in Fig. 3-35. Raise the hand slowly, adjusting the calipers at the same time. Do not force the calipers, and be sure that the points of the calipers are placed across the diameter of the hole being measured.

To transfer settings from outside to inside calipers, the point of one leg of the inside calipers should rest on a similar point of the outside calipers (Fig. 3-36). While using this contact point as a pivot, move the inside calipers along the dotted line shown in the

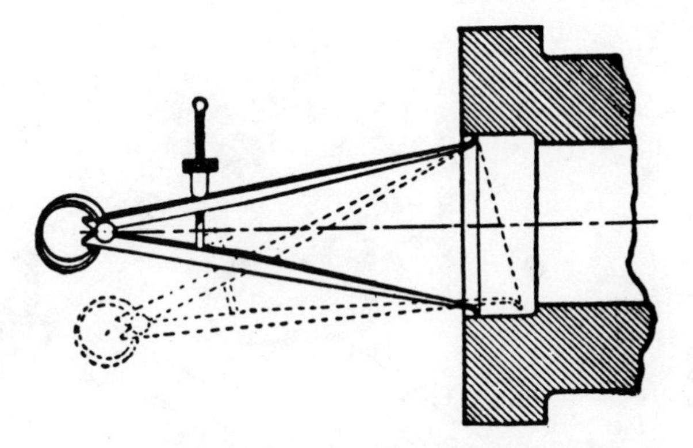

Fig. 3-35. Method of obtaining a measurement with inside calipers.

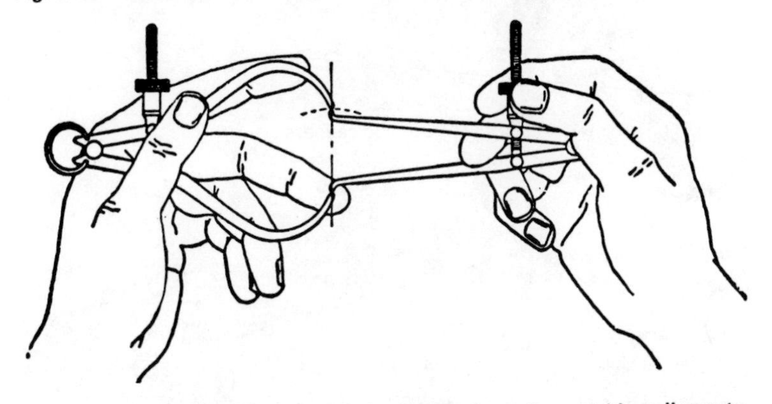

Fig. 3-36. Method of transferring the setting from the outside calipers to the inside calipers.

illustration, and move the adjustment nut until the "feel" indicates that the setting has been transferred correctly.

The degree of accuracy attained in using the calipers for testing sizes is dependent on the skill, judgment, and experience of the machinist using the calipers. Some machinists and toolmakers work within very close limits, while other may lack the required delicate sense of touch.

Transfer or lock-joint calipers can be used advantageously in semiaccessible places, such as behind shoulders or chamfered cavities where other calipers could not be removed after setting (Fig. 3-37). These calipers are so constructed that they can be set,

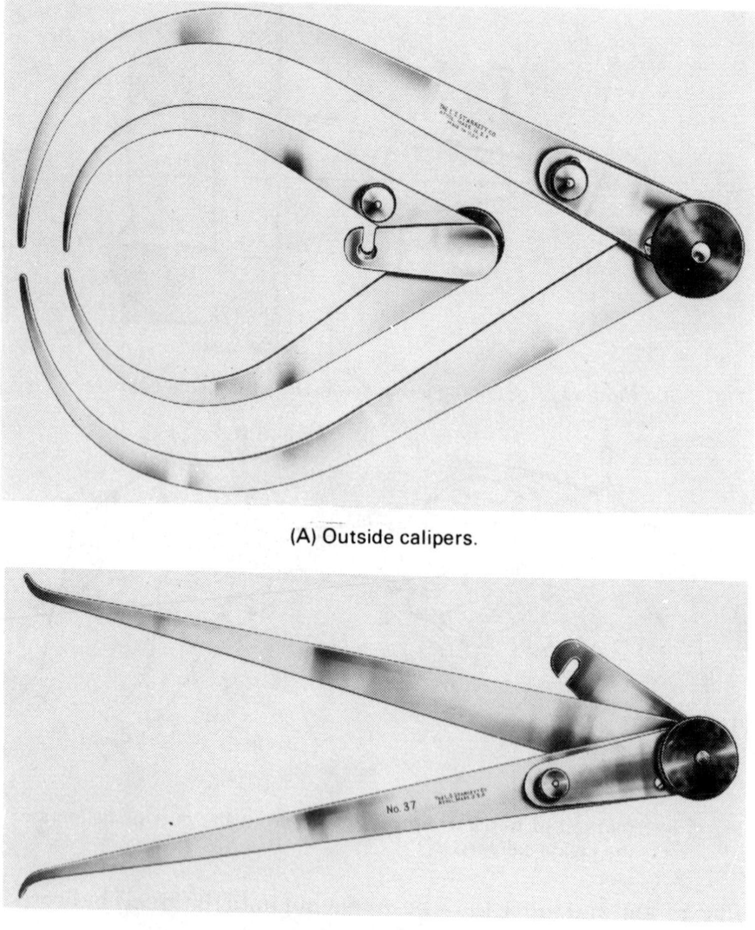

(A) Outside calipers.

(B) Inside calipers.

Courtesy L.S. Starrett Co.

Fig. 3-37. Firm-joint transfer or lock-joint calipers.

closed to clear the obstruction, and reset accurately in the original setting.

Hermaphrodite calipers are calipers having a short divider leg and a caliper leg (Fig. 3-38). These calipers may be used to scribe lines on a shaft, or to make lines on a surface parallel to an edge.

The pointed leg, or scriber, should be adjusted so that it is

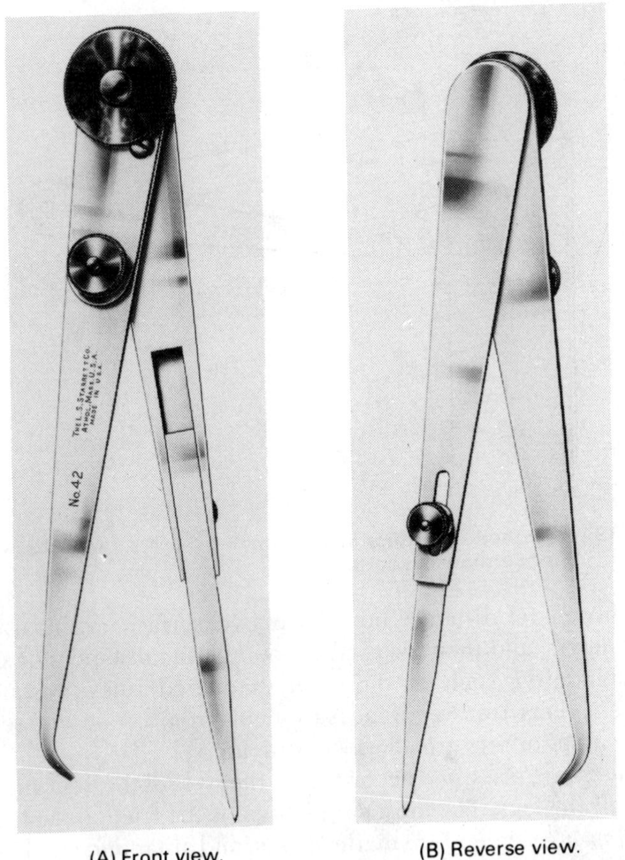

(A) Front view.

(B) Reverse view.

Courtesy L.S. Starrett Co.

Fig. 3-38. Hermaphrodite calipers with offset leg and adjustable round point.

slightly shorter than the bent leg. Place the bent leg against the end of the rule and adjust the scriber leg to the desired graduation mark on the rule (Fig. 3-39).

DIVIDERS

This instrument is similar in construction to calipers, except that the legs are round in cross section and they terminate in a sharp

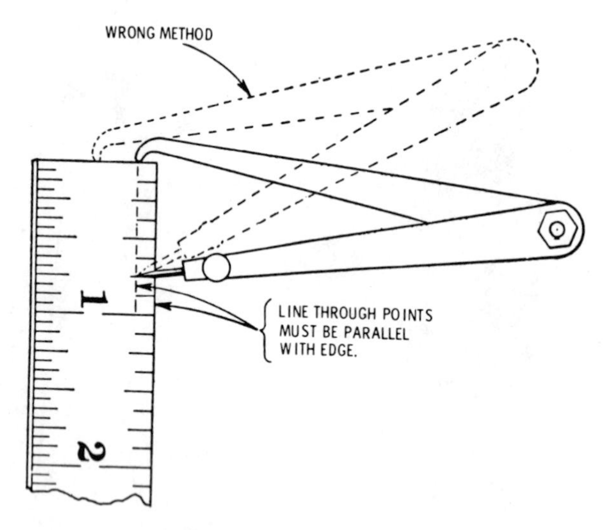

Fig. 3-39. A method of setting hermaphrodite calipers. Note the wrong method shown in dotted lines.

point. In general, distances between or over surfaces are measured with calipers; and they are used for comparing distances or sizes with standards, such as those on graduated rules. Distances between points, transfer of measurements from a scale, and scribing of circles or arcs are all uses for dividers.

Dividers are classified according to the style of the joint and the length of the legs. The most simple joint is the friction joint. The legs of *spring dividers* are made of steel and are round and highly polished; the points come together evenly (Fig. 3-40).

Lock-joint dividers can be moved freely to the approximate correct position, the joint locked, and the adjusting screw used for the final setting. In the *wing dividers*, the construction is similar to the lock-joint type, except that the setting is fastened on the wing instead of at the pivot (Fig. 3-41).

The points on all dividers must be kept sharp. They should be ground to an included angle of about 25°. A greater angle makes them more difficult to set to a given point or line; a smaller angle allows the points to wear too rapidly, which necessitates frequent sharpening. The points of dividers should be honed on an oilstone (Fig. 3-42). When dividers receive constant use, frequent honing

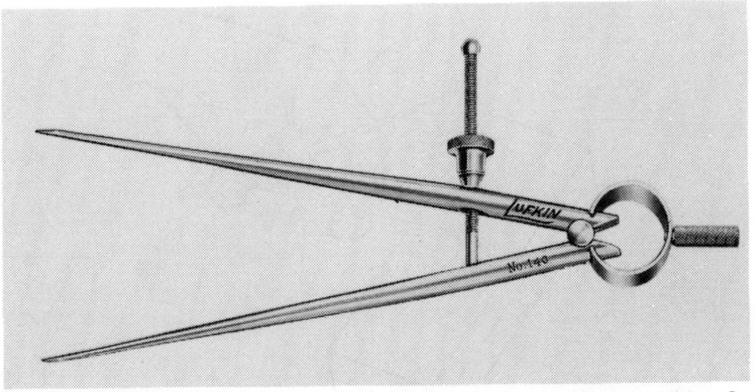

Courtesy Lufkin Rule Co.

Fig. 3-40. Toolmakers' spring dividers. The legs are of steel, round, and highly polished; the measuring points come together evenly.

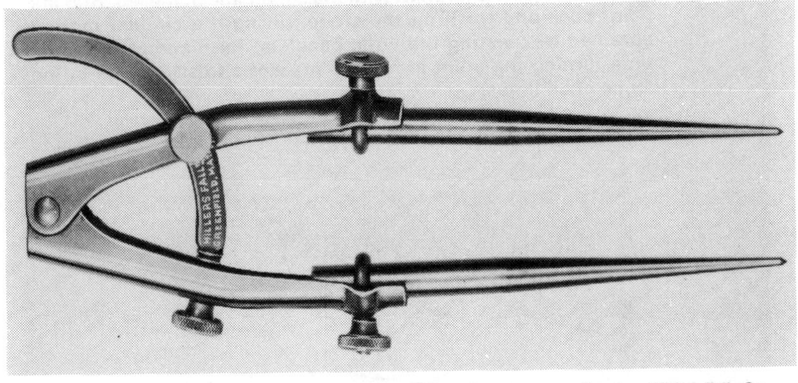

Courtesy Millers Falls Co.

Fig. 3-41. Wing dividers.

avoids the necessity for grinding the points on the face of a grinding wheel.

The points of dividers are delicate and sharp. Care should be taken not to injure them. Some mechanics hold the dividers perpendicular to the rule in setting, but this is liable to injure the points and scratch the rule.

In setting the dividers, place the rule on a flat surface, and hold the dividers in the manner shown in Fig. 3-43. Place one of the divider points on an index mark on the rule. Place the other point

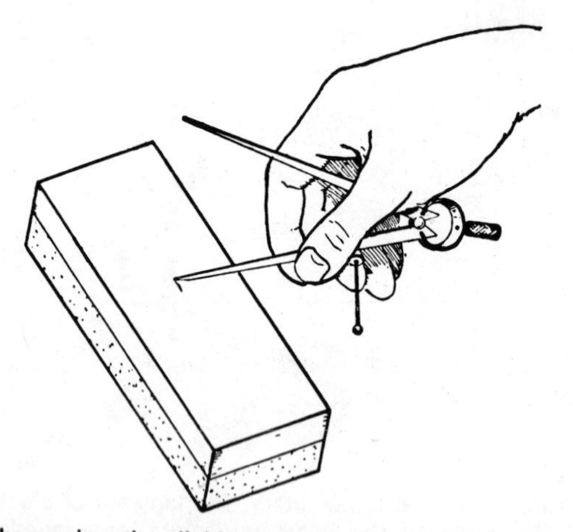

Fig. 3-42. Sharpening the dividers on an oilstone. The point should be ground so that it is straight with the center line of the leg. Rub the point back and forth on the stone, giving it a circular motion obtained by twisting the wrist back and forth and at the same time turning the point around to present a fresh surface to the stone at all times.

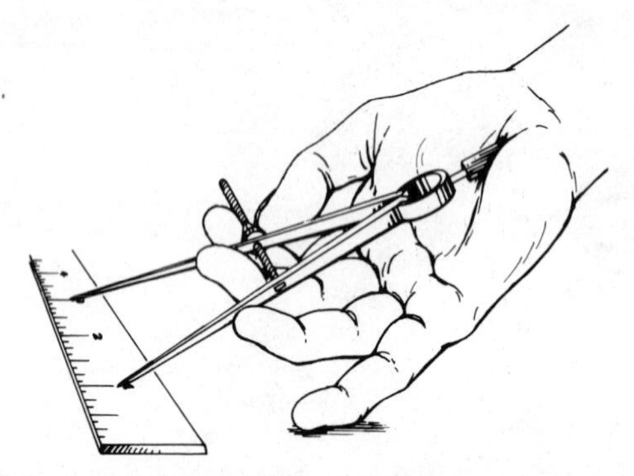

Fig. 3-43. A method of setting spring dividers. Set the points of the dividers to the centers of the index marks on the rule. It is important that the points of the dividers are equally distant from the edge of the rule.

on the rule, and turn the adjusting screw with thumb and forefinger to set the dividers to the desired dimension on the rule. Set the points of the dividers to the centers of the index marks on the rule. The dividers should be held so that a line drawn between the two points will be parallel with the edge of the rule.

Firm-joint dividers are set by first setting them to the approximate dimension and then tapping them gently. They are set in a manner similar to the method used for setting firm-joint calipers.

PROTRACTORS

A protractor is an instrument used to measure or to construct angles. In its basic form, it is a semicircle graduated to 180° (Fig. 3-44). A hole or notch is provided at the center of its straight base; this is placed on the starting point of the desired line, the degrees in the angle being read from the circumference of the protractor. Higher quality and more expensive protractors are fitted with verniers and pointers so that minutes or even finer graduations may be read.

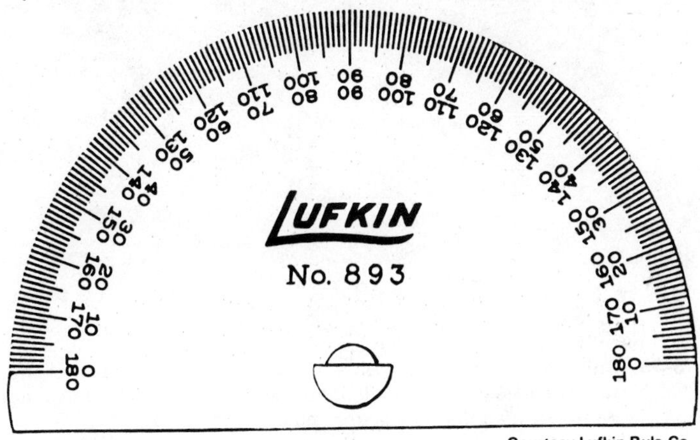

Courtesy Lufkin Rule Co.

Fig. 3-44. Steel protractor used for measuring angles.

BEVELS

A bevel may somewhat resemble a try square, but the blade is hinged on the stock so that it may be moved and set at any angle in its own plane.

A *universal bevel* has an offset in the blade to increase its capacity and usefulness for bevel gear work, etc., so that any angle may be obtained (Fig. 3-45). One edge of the case is solid, which forms a rest directly under the blade where thin templates may be placed and fitted accurately. The universal bevel is useful in working the draft on patterns and in turning angles on the lathe that cannot be reached with an ordinary bevel.

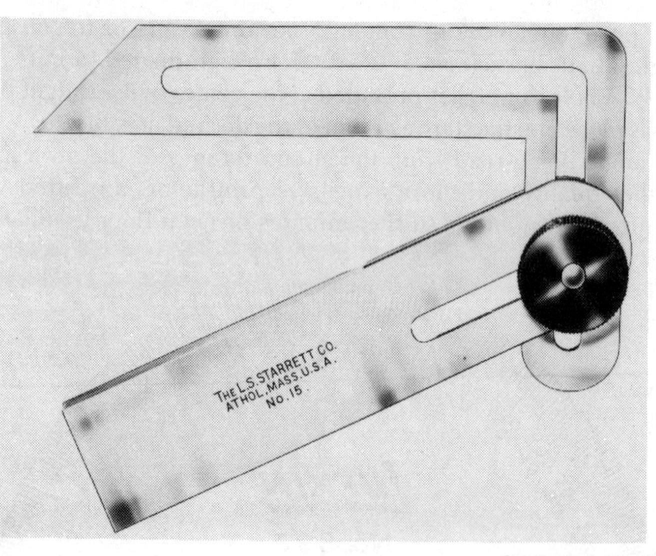

Courtesy L.S. Starrett Co.

Fig. 3-45. Universal bevel. Used with a protractor to form a bevel protractor. The offset in the blade increases its capacity for bevel gear work, etc., so that any slight angle may be obtained.

A *combination bevel* has a stud riveted in the straightedge stock or head, on which its straight blade is hinged so that it can swing over the stock to be clamped at any angle (Fig. 3-46). The slotted auxiliary blade with the clamp bolt may be slipped onto the split blade, clamped at any desired angle, and used, in combination with the stock of the other blade, for layout work, measuring, etc.; and when so combined, the instrument will lie flat on the work. The many uses for which combination bevels are adapted are shown in Fig. 3-47.

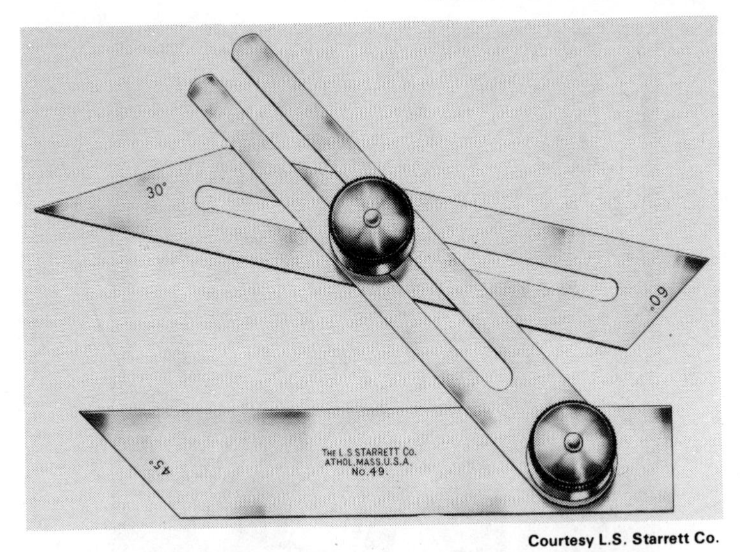

Courtesy L.S. Starrett Co.

Fig. 3-46. Combination bevel. It is constructed so as to swing over the stock and be clamped at any angle for laying out, measuring, or showing any angle desired.

SUMMARY

For a machinist to make accurate measurements, it is essential that measuring tools be kept clean and in working order. When making measurements with a steel rule, place the 1-inch index mark, instead of the end of the rule, at one edge of the piece to be measured. Precaution should be used in setting the 1-inch mark at the edge of the piece. Remember to always deduct 1 inch from the reading. A hook rule is preferable to the steel rule in some instances because the hook eliminates error in placing the rule on the piece to be measured.

A very useful tool for the machinist is the combination square. To measure roughly the depth of a slot, place the square in position and adjust the blade to the depth of the slot.

When extreme accuracy in making contact measurements is not required, spring calipers are generally used. Outside spring calipers can be set by means of a steel rule. Outside calipers can be used to measure the diameter of round stock. Calipers should not be

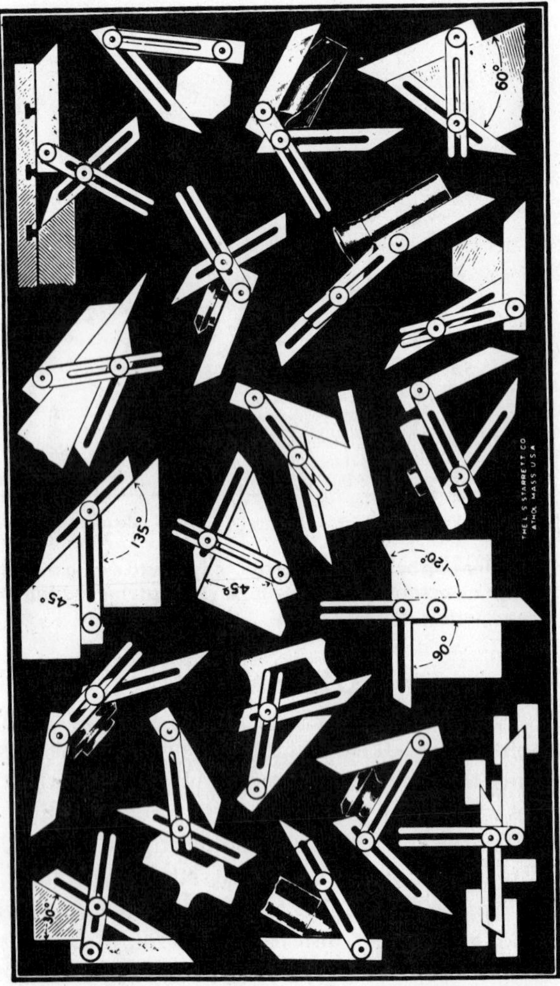

Courtesy L.S. Starrett Co.

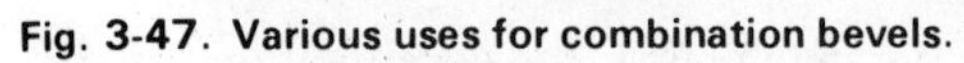

Fig. 3-47. Various uses for combination bevels.

gripped too tightly. Relaxation of the fingers increases their sensitivity to touch or feel.

REVIEW QUESTIONS

1. Why is the 1-inch index mark used when measuring with a steel rule?

2. How is the combination square used in machinist measurements?
3. What are the advantages in using a hook steel rule?
4. What is the correct way to hold or use calipers?
5. What shop measuring tools are used for angular measurements?
6. List at least five measuring tools.
7. How are measuring tools classified?
8. List at least seven types of rules.
9. What is the allowance necessary for shrinkage of cast iron?
10. How is a rule clamp used?
11. What is a try square?
12. What is a combination square used for?
13. What is the diemakers' square?
14. How do machinists use magnetic blocks?
15. What is a hermaphrodite caliper?
16. What is another name used to identify a lock-joint caliper?
17. How are dividers used for measurement?
18. How do wing dividers differ from lock-joint dividers?
19. How is a protractor used for measurement?
20. What is a universal bevel?

CHAPTER 4

Precision Measurement and Gaging

The worker in the machine shop uses many tools, instruments, and gages to produce accurate measurements. Precision measurements are generally written in decimals and are read in thousandths (0.001) and ten-thousandths (0.0001) of an inch.

MICROMETER CALIPERS

The word "micrometer" indicates a precision instrument for small measurements, which are usually made by rotating a screw with a fine pitch. Micrometer calipers have a U-shaped frame with a hardened anvil at one end and an indicating thimble at the other end. The micrometer is the precision tool widely used for measurements in thousandths or even ten-thousandths of an inch (Fig. 4-1). Micrometers are made in many styles and sizes for outside, inside, and depth measurements. The pitch of the screw thread on the spindle is $1/40$ inch (40 threads per inch). Therefore, one complete revolution of the thimble moves the spindle $1/40$ inch, or 0.025 inch, either toward or away from the anvil space. The longitudinal

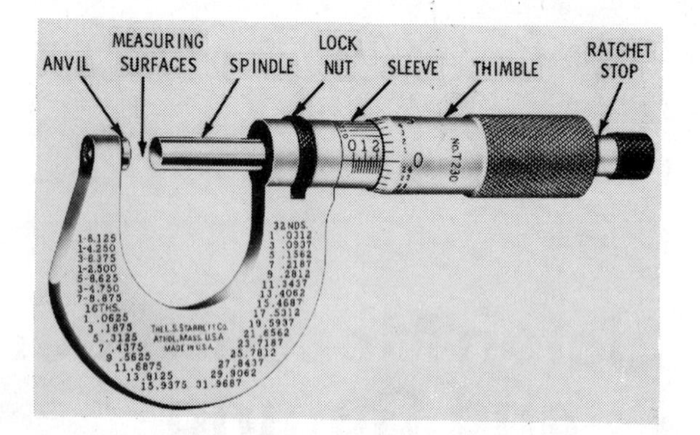

(A) Parts of the micrometer.

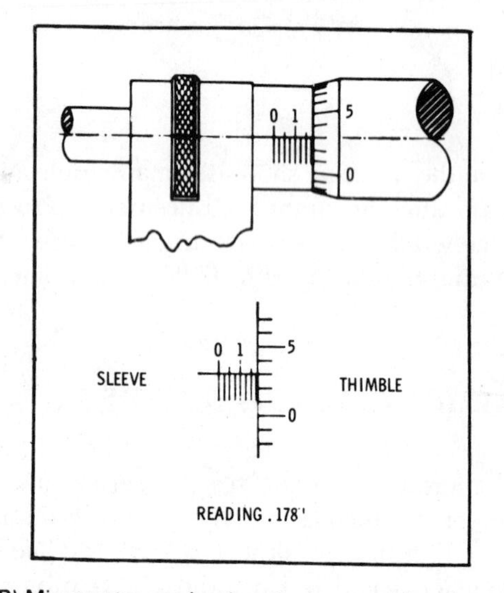

(B) Micrometer graduations in thousandths of an inch.

Courtesy L. S. Starrett Co.

Fig. 4-1. Micrometer calipers graduated in thousandths of an inch.

line on the sleeve is divided into 40 equal parts by vertical lines corresponding to the number of the threads on the spindle. Therefore, each vertical line designates 1/40 inch, or 0.025 inch; every fourth line indicates hundreds of thousandths. The line marked "1" indicates 0.100 inch; the line marked "2" indicates 0.200 inch, etc.

The beveled edge of the thimble is divided into 25 equal parts, each line representing 0.001 inch. To read the micrometer in thousandths, multiply the number of vertical divisions visible on the sleeve by 0.025 inch, and add the number of thousandths indicated by the line on the thimble, which coincides with the longitudinal line on the sleeve.

Example: The "1" line on the sleeve is visible, representing 0.100 inch. Three additional lines are visible, each representing 0.025 inch; 3 times 0.025 inch = 0.075 inch. Line "3" on the thimble coincides with the longitudinal line on the sleeve, each representing 0.001 inch; 3 times 0.001 inch = 0.003 inch. The micrometer reading is the total (0.100 + 0.075 + 0.003 = 0.178 inch or 178 thousandths of an inch).

Micrometer calipers can be adjusted to compensate for wear. This is done by adjustment of the friction sleeve, as follows: Take up the wear of the screw and nut; insert the spanner wrench in the slot of the adjusting nut, and tighten just enough to eliminate play; then, carefully bring the anvil and spindle together, and insert the spanner wrench in the small slot of the sleeve. Turn the sleeve until the line on the sleeve coincides with the zero line on the thimble (Fig. 4-2).

In using a 1-inch micrometer for small work, hold the tool in one hand, turning the thimble with the thumb and forefinger (Fig. 4-3A). This permits freedom for holding the work with the other hand. On some types of work, such as measuring over two flat surfaces, the micrometer is held in the left hand, and the thumb and forefinger are used to turn the sleeve to adjust to the dimension. Round stock may be held in the left hand, and the micrometer held in the right hand. The thumb and forefinger are used to turn the sleeve until the correct setting is indicated by "feel" (Fig. 4-3B). In larger work, or in stationary work, the frame should be held securely in one hand while the other hand turns the thimble (Fig. 4-4).

Some mechanics change a micrometer setting quickly by hold-

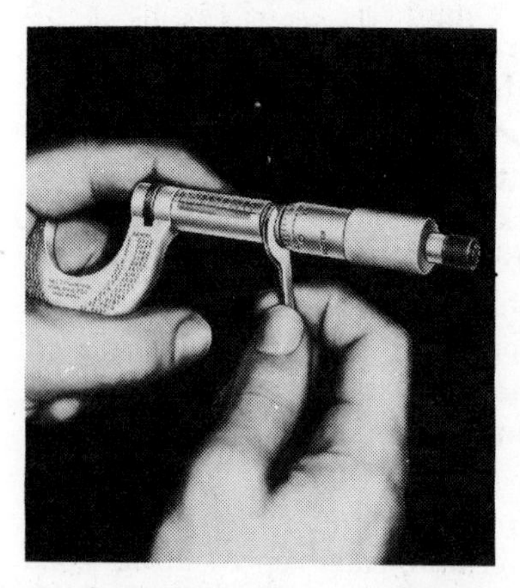

(A) Back off thimble, and tighten the adjusting nut to eliminate play in spindle nut.

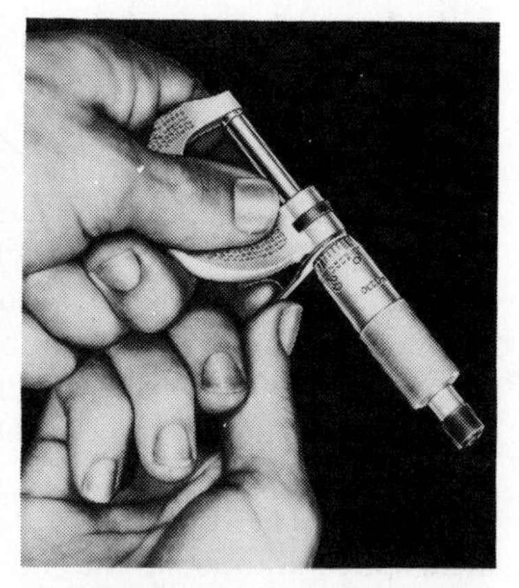

(B) With anvil and spindle in contact, adjust sleeve, so that line on sleeve coincides with zero line on thimble.

Fig. 4-2. Adjusting micrometer calipers. Courtesy L. S. Starrett Co.

ing the sleeve and swinging the frame around several times. This practice should be avoided because the centrifugal force generated by the whirling frame unduly wears the threads on the spindle, rendering the instrument inaccurate.

The inside micrometer is used for obtaining precision measurements of internal diameters of cylinders, holes, etc. (Fig. 4-5).

VERNIER MICROMETER CALIPERS

Micrometers graduated in ten-thousandths of an inch are used in the same manner as the micrometers graduated in thousandths of an inch (Fig. 4-6), except that an additional reading in ten-thousandths is obtained from a vernier and is added to the thousandth reading.

The vernier has ten divisions on the sleeve (Fig. 4-6B), which occupy the same amount of space as nine divisions on the thimble. Therefore, the difference between the width of one of the ten spaces on the vernier and one of the nine spaces on the thimble is one-tenth of a division on the thimble, or one-tenth of one-thousandth, which is one ten-thousandth of an inch.

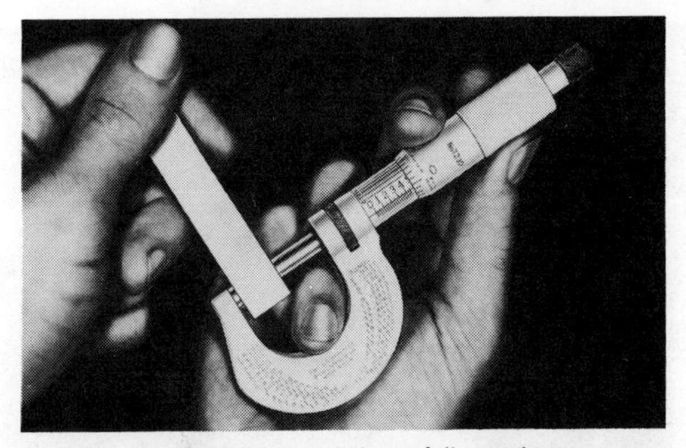

(A) Measuring a piece of die steel.

(B) Measuring tubular work.

Courtesy L. S. Starrett Co.

Fig. 4-3. Using the micrometer calipers.

To read the ten-thousandths micrometer, obtain the thousandths reading and note which of the lines on the vernier coincides with a line on the thimble. If line "1" on the vernier coincides, add one ten-thousandth; if line "2" coincides, add two ten-thousandths, etc.

VERNIER CALIPERS

By definition, a *vernier caliper* has a graduated blade and an adjustable tongue (Fig. 4-7). The blade has graduations and carries

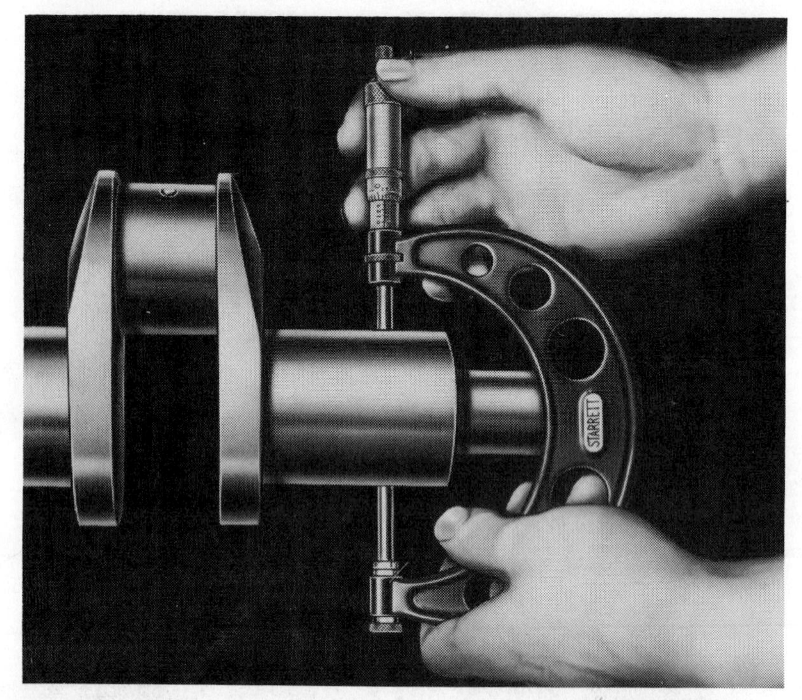

Courtesy L. S. Starrett Co.

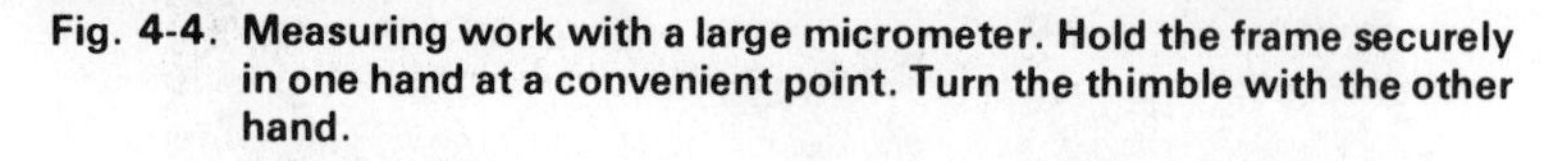

Fig. 4-4. Measuring work with a large micrometer. Hold the frame securely in one hand at a convenient point. Turn the thimble with the other hand.

two crossheads, one of which is slightly adjustable by a nut, the other being movable along the blade. The crossheads are adapted to the measurement of interior diameters or sizes, and the other side is adapted to external measurements.

The bar of the calipers is graduated in 40ths, or 0.025 of an inch, every fourth graduation being numbered, representing a tenth of an inch (Fig. 4-8). The vernier scale is divided into 25 divisions, numbered 0, 5, 10, 20, and 25. The 25 divisions on the vernier scale occupy the same space as 24 divisions on the bar.

Since one division on the bar is equal to 0.025 inch, 24 divisions equal 24×0.025 inch, or 0.600 inch, and 25 divisions on the vernier scale also equal 0.600 inch. Therefore, each division on the vernier is equal to $\frac{1}{25} \times 0.600$ inch, or 0.024 inch. The difference

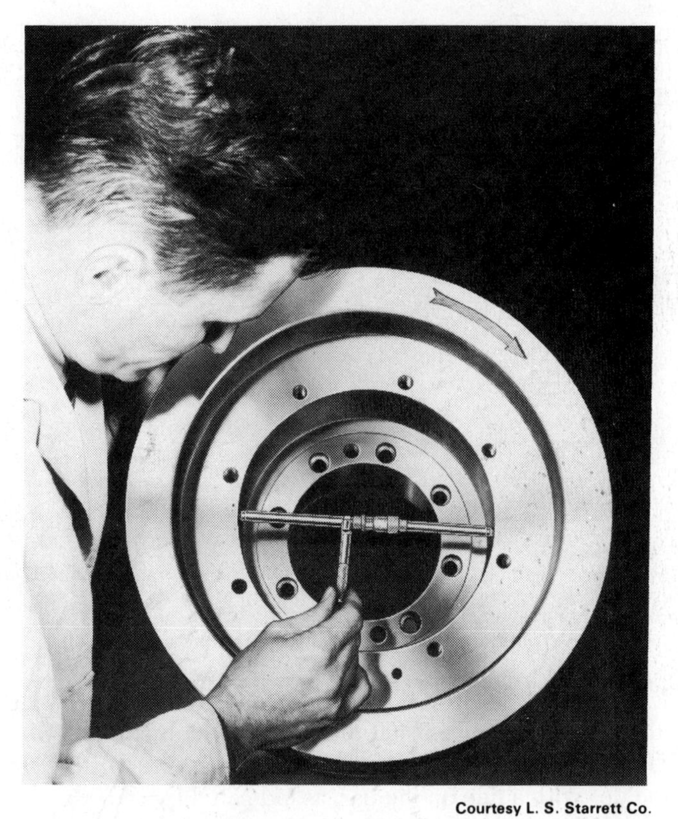

Courtesy L. S. Starrett Co.

Fig. 4-5. Measuring an inside diameter with an inside micrometer calipers.

between one bar division (0.025 inch) and one vernier division (0.024 inch) equals 0.025 inch less 0.024 inch, or 0.001 inch. If the zero line on the vernier coincides with the zero line on the bar, the line to the right of the zero on the vernier will differ from the line to the right of the zero on the bar by 0.001 inch; the second line by 0.002 inch, etc. The difference increases by 0.001 inch for each division until the "25" on the vernier coincides with the "24" on the bar. In reading the calipers, note the number of inches, tenths (or 0.100 inch), and fortieths (or 0.025 inch) the zero line on the vernier is from the zero mark on the bar; add the number of thousandths indicated by the line on the vernier that coincides with the line on the bar.

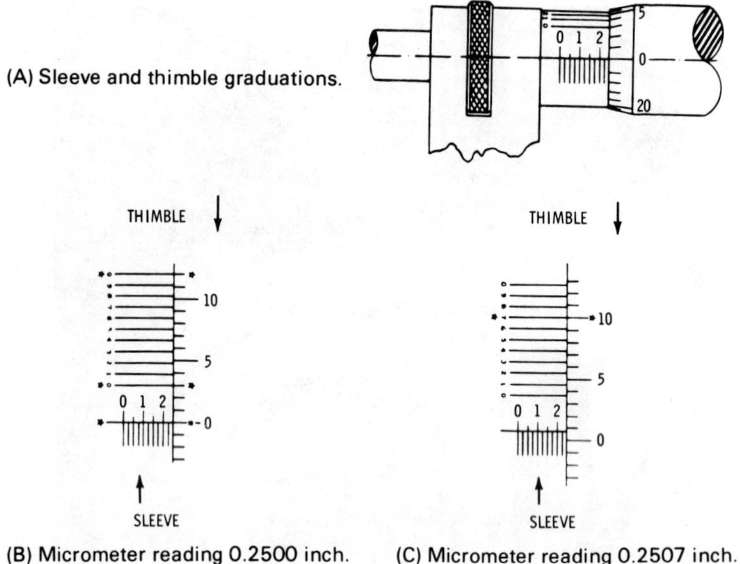

(A) Sleeve and thimble graduations.

(B) Micrometer reading 0.2500 inch.

(C) Micrometer reading 0.2507 inch.

Courtesy L. S. Starrett Co.

Fig. 4-6. A micrometer graduated in ten-thousandsths of an inch.

Example: In the above illustration, the vernier has been moved to the right 1.000 plus 0.400 plus 0.025, which is equal to 1.425 inches, as shown on the bar. The eleventh line on the vernier coincides with a line on the bar, as indicated by the stars. Therefore, 0.011 inch is added to the reading on the bar, giving a total reading of 1.436 inches.

BEVEL PROTRACTORS

This instrument is equivalent to a bevel with a protractor in addition, which adapts it to all kinds of work in which angles are to be laid out. In general, there are two kinds of bevel protractors, as follows:

1. Those without a vernier.
2. Those with a vernier.

The bevel protractor without a vernier is suitable for angles that

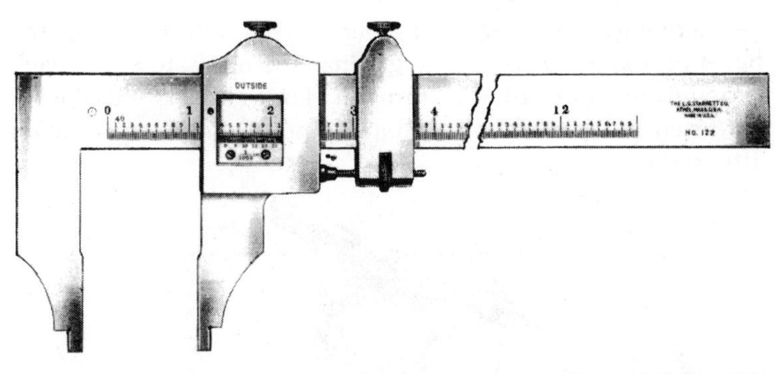

Courtesy L. S. Starrett Co.

Fig. 4-7. A vernier caliper. Fine adjustment of the points is made by clamping the thumbscrews at the right and turning the knurled nut on the horizontal screw.

Courtesy L. S. Starrett Co.

Fig. 4-8. Detail of a calipers with a vernier. This vernier permits readings to 0.001 inch.

do not require a high degree of accuracy (Fig. 4-9). The dial of the bevel protractor is accurately graduated from 0° to 90° to each extremity of an arc of 180°. It turns on a large central stud, which is hardened and ground, and can be clamped rigidly in any position after setting.

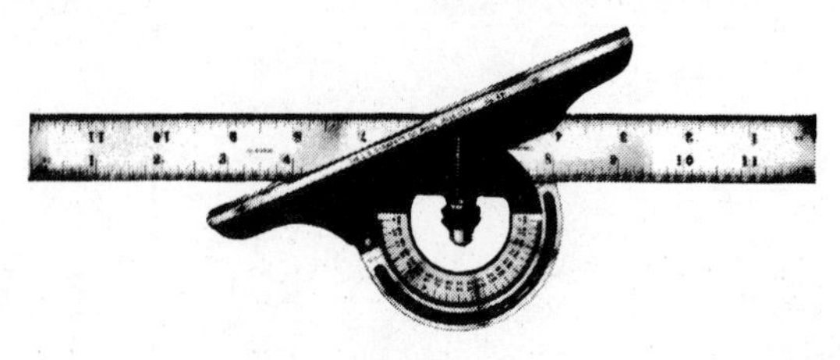

Courtesy L. S. Starrett Co.

Fig. 4-9. Bevel protractor. It turns on a large central stud, which is hardened and ground, and can be rigidly clamped in any position desired. The dial is accurately graduated in degrees over an arc of 180°, reading 0° to 90° from each extremity of the arc.

The *universal bevel protractor* with a vernier is graduated in degrees throughout the entire circle (Fig. 4-10). The swivel turns on a large central stud, which is hardened and ground, and can be rigidly clamped by a thumbnut. The vernier increases materially the adaptability of the protractor for obtaining finer measurements. Readings to five minutes (5′), or 1/12 of a degree, can be obtained.

The dial of the protractor is graduated both to the right- and left-hand sides of zero to 90 degrees. The vernier scale is also graduated to the right and left of zero to 60 minutes (60′); each of the 12 vernier scale have graduations representing 5 minutes (5′). Any size angle can be measured because both the protractor dial and the vernier scale have graduations in opposite directions from zero (Fig. 4-11).

Since 12 graduations on the vernier scale occupy the same space as 23 degrees on the protractor dial, each vernier graduation is 1/12 degree, or 5 minutes, shorter than 2 graduations on the protractor dial. Therefore, if the zero graduation on the vernier scale coincides with a graduation on the protractor dial, the reading is in exact degrees; but if any other graduation on the vernier scale coincides with a protractor graduation, the number of vernier graduations must be multiplied by 5 minutes, and added to the

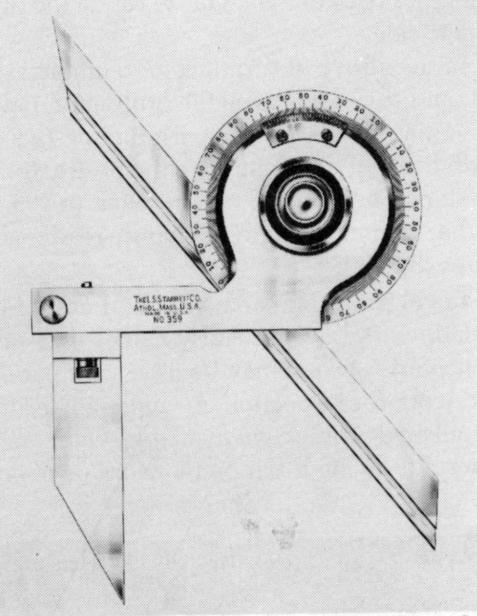

Courtesy L. S. Starrett Co.

Fig. 4-10. Universal bevel protractor with vernier. The vernier permits its use in obtaining fine measurements.

Courtesy L. S. Starrett Co.

Fig. 4-11. Detail of a protractor vernier. The figures are close to the graduations to facilitate reading the vernier.

number of degrees read between the zeros on the protractor dial and the vernier scale.

Example: In the above illustration, zero on the vernier scale is between "50" and "51" degrees on the protractor dial to the left of zero. Also, reading to the left, the 4th line on the vernier scale coincides with the "58" graduation on the protractor dial, as indicated by the stars. Therefore, 4 × 5 minutes, or 20 minutes, must be added to the number of degrees, giving a reading of 50 degrees and 20 minutes (50°20′).

Universal bevel protractors have several uses (Figs. 4-12 and 4-13). The blade with beveled ends enables measurement of an angle from the vertex. They may be used with parallels or knees for laying out work for inspection. An acute-angle attachment for laying out small angles quickly is available. One side of the tool is flat, which permits laying it flat on the work or paper.

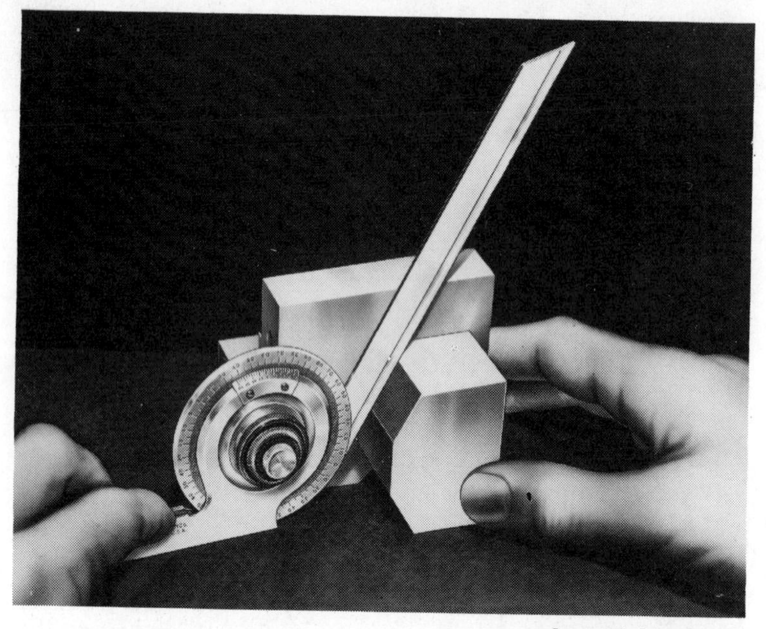

Courtesy L. S. Starrett Co.

Fig. 4-12. Uses of the universal bevel protractor. The blade with beveled ends permits measurements of angles from the vertex.

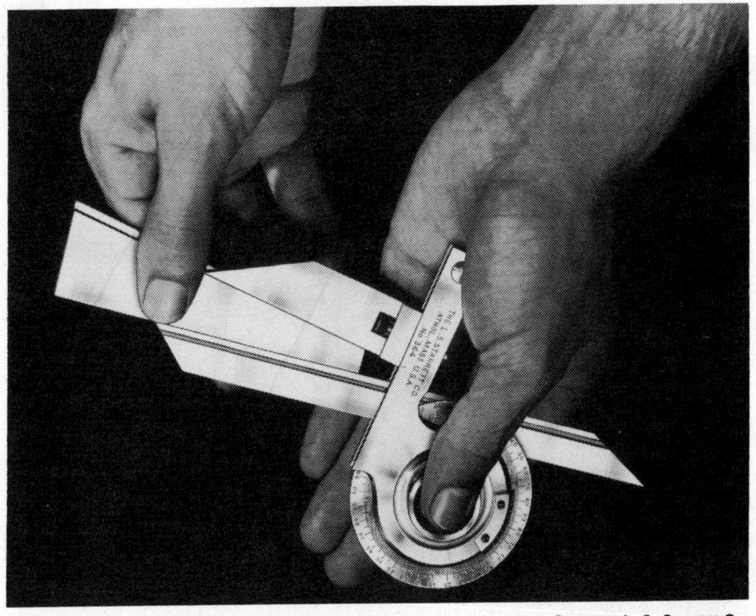

Courtesy L. S. Starrett Co.

Fig. 4-13. Universal bevel protractors may be used with parallels or knees for laying out work for inspection. The acute-angle attachment permits small angles.

DIAL INDICATORS

A dial indicator (incorrectly called a dial gage) is an instrument for indicating size differences, rather than making measurements, as the dial indicator ordinarily is not used to indicate distance. It can be used in combination with a micrometer to measure exact distances.

Variations in measurements are shown by the movement of a hand on the dial of the dial indicator. The dial is graduated in thousandths of an inch; that is, each division on the dial represents contact point movement of 0.001 of an inch.

The dial indicator (Fig. 4-14) is useful in testing shafts for alignment, for checking cylinder bores for roundness and taper, and for testing bearing bores. The dial indicates the alignment, or roundness, of the piece tested to within 0.001 of an inch. A skilled

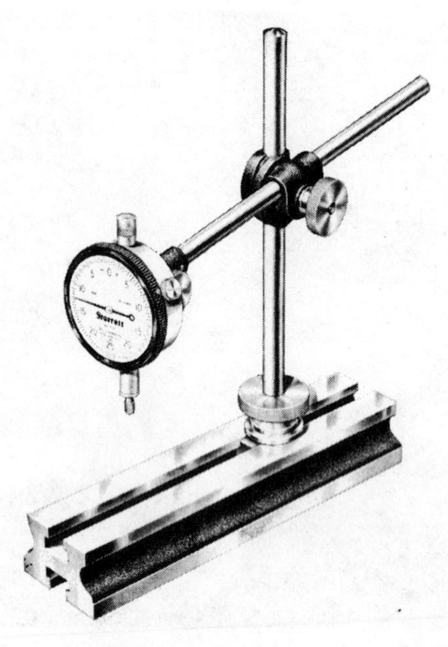

Courtesy L. S. Starrett Co.

Fig. 4-14. A dial test indicator with attachments.

workman can check alignment to within 0.00025 of an inch.

The dial indicator is extensively used in manufacturing and in service and repair work. Other uses are for straightening crankshafts, locating wrist-pin holes, determining the amount of shim to insert or remove, determining taper, checking play in bearings, reboring work, lining up magneto coil assembly, etc. (Fig. 4-15).

Another type of dial indicator is shown in Fig. 4-16. This is a precision gage utilized on the precision dial comparator. It can also be used on a number of measuring gages.

GAGES

A gage is often erroneously considered to be any measuring instrument. A gage is a *fixed* device that establishes a particular dimension, but it is not a measuring instrument.

However, some gages, a surface gage, for example, are adjusta-

Courtesy L. S. Starrett Co.

Fig. 4-15. A lathe operator using a dial indicator to center work.

ble and can be set to any desired dimension within their range. After it is set to a particular dimension, a gage becomes a fixed device and is properly called a gage.

Surface gage—A surface gage (Fig. 4-17) is a machinist's instrument for testing planed surfaces. It has a heavy base, grooved through the bottom and end, adapting it for use on circular work as well as flat surfaces. The spindle may be set upright or at an angle, or turned to work under the base. It can be sensitively adjusted to any position.

Layout work often includes lines scribed at a given height from a face of the work, or a continuation of lines around several surfaces. It can be used to scribe lines at a given height on any number of pieces when duplicate parts are being made. Thus, the height of a standard bearing may be transferred to the faces of any

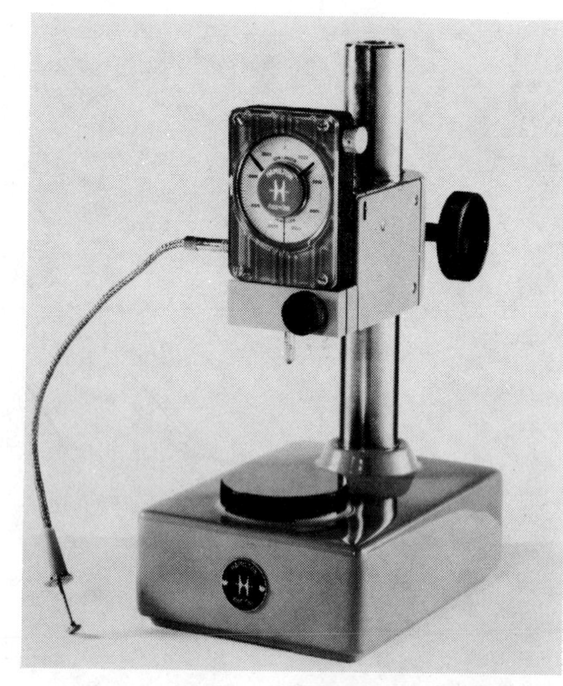

Courtesy Hamilton Watch Company

Fig. 4-16. A precision dial comparator. This instrument is direct reading to 0.00005 inch.

number of castings from which duplicate bearings are to be made (Fig. 4-18).

A clean surface plate and a combination square for obtaining the dimension are needed to set the surface gage properly (Fig. 4-19). All instruments should be absolutely clean. Three steps are necessary in setting a surface gage, as follows:

1. Adjust the standard to a convenient position.
2. Adjust the scriber to the approximate dimension.
3. Further adjust the scriber by turning the knurled adjusting screw on top of the base to the desired index line on the blade of the combination square.

Height gage—This gage is designed to measure or mark off vertical distances from a plane surface. The *vernier height gage* is indispensable for layout, jig, and fixture making because of its fine

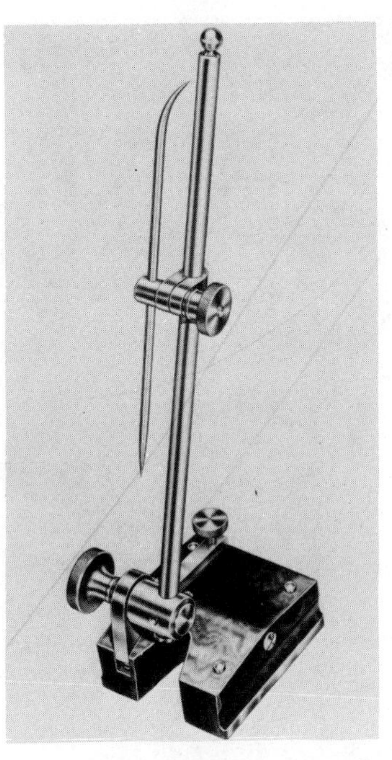

Courtesy L. S. Starrett Co.

Fig. 4-17. Universal surface gage.

adjustment, which permits extremely accurate measurements (Fig. 4-20). The location of center distances of jigs, dies, etc., can be accurately obtained by the use of toolmakers' buttons. A combination marker and extension may be used with the movable jaw for measuring and scribing lines on the work.

The end of the extension is beveled to a sharp edge for scribing lines (Fig. 4-21).

The vernier height gage is graduated to read in thousandths, by means of a vernier scale on the sliding jaw. Graduations on one side are for outside measurements, and graduations on the other side are for inside measurements.

The vernier depth gage is a similar precision instrument for measuring depths of slots, holes, etc. (Fig. 4-22).

Depth gage—The *micrometer depth gage* is another accurate instrument used for vertical measurements. It is also essential for

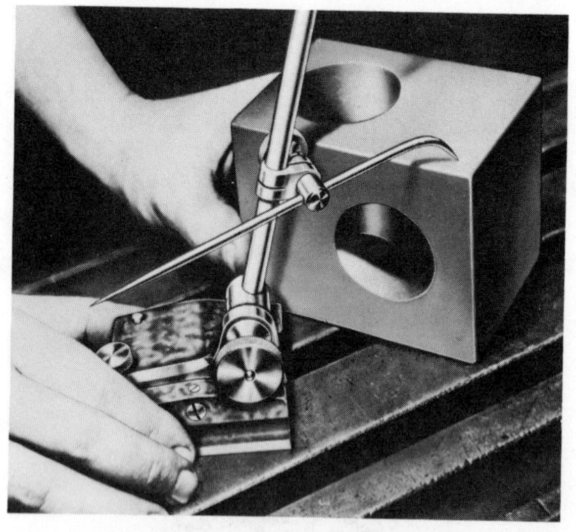

Courtesy L. S. Starrett Co.

Fig. 4-18. Machinist using a surface gage to level up a cast-iron block to determine the amount of work to be machined off.

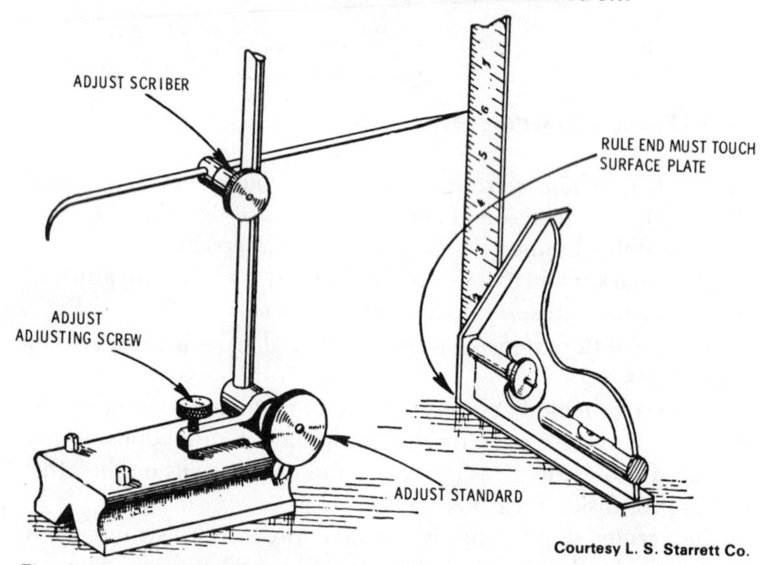

Courtesy L. S. Starrett Co.

Fig. 4-19. A method of setting a surface gage with the aid of a combination square. Make certain that the instruments and surface plates are absolutely clean.

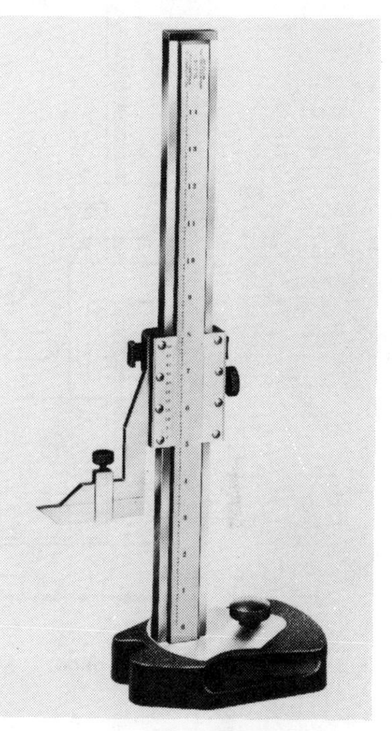

Fig. 4-20. Vernier height gage.

Courtesy L. S. Starrett Co.

jig and fixture making (Fig. 4-23). Depths of holes, slots, etc., can be measured with micrometer accuracy.

Snap gages—Snap limit gages are used for both *internal* and *external* dimensions (Fig. 4-24). The distance between measuring surfaces is fixed and represents the size stamped on the gage. The gages are available in different sizes and can be used for measuring duplicate parts in machine shop work. When these gages become worn, they can be closed in and reground, or lapped, to true size.

Plug gages—These gages are also called *go* and *not go* gages (Fig. 4-25). They have double ends, and have both a "go" end and a "not go" end; that is, when the work is at the correct size, one end of the gage will slip into it, but the other end will not. In some types of work, a given part may be required to be within certain size limits to be correct for a particular class of fit. A limit gage may be used for both the minimum and maximum sizes.

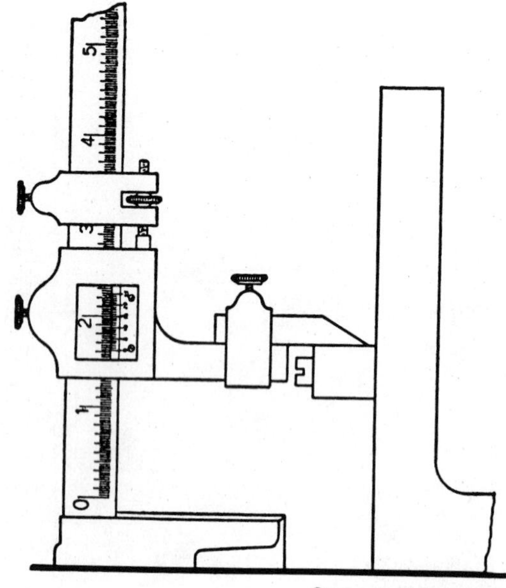

Courtesy L. S. Starrett Co.

Fig. 4-21. Vernier height gage.

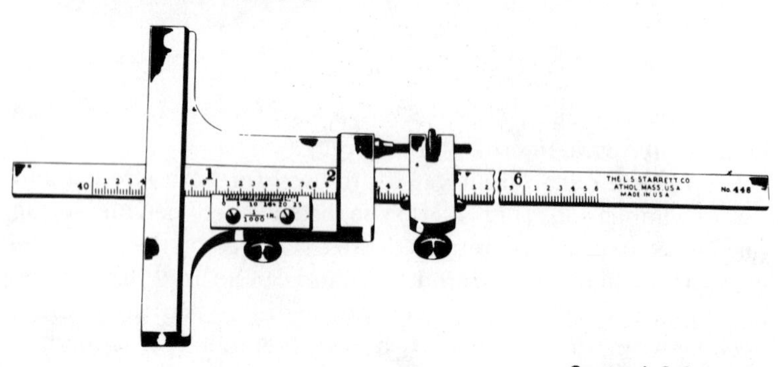

Courtesy L. S. Starrett Co.

Fig. 4-22. Vernier depth gage.

Ring gages—These gages are standard cylindrical gages (male and female). The ring, or external, gage is a bored ring (Fig. 4-26). It is also called a collar gage. These gages can be made to both "go" and "not go" dimensions.

Taper gages—This type of gage is made of metal and has a

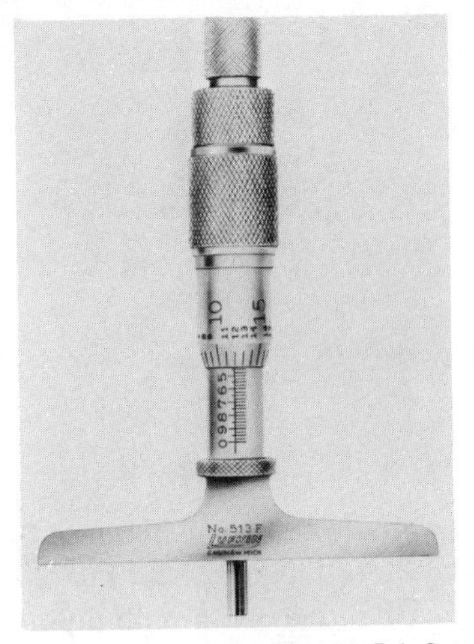

Courtesy Lufkin Rule Co.

Fig. 4-23. Micrometer depth gage.

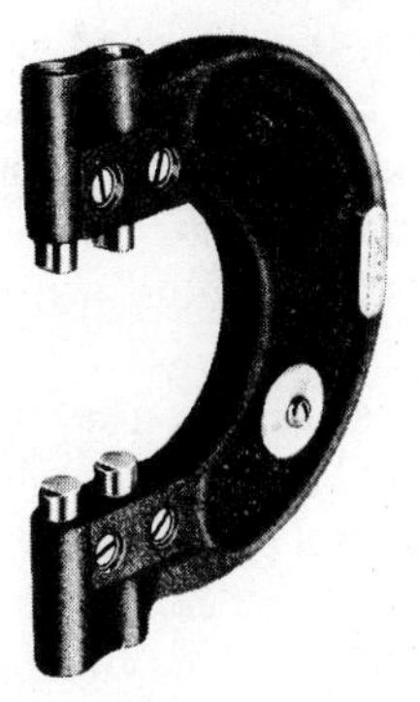

Courtesy Greenfield Tap & Die Co.

Fig. 4-24. Adjustable snap limit gage. The size is stamped on the gage.

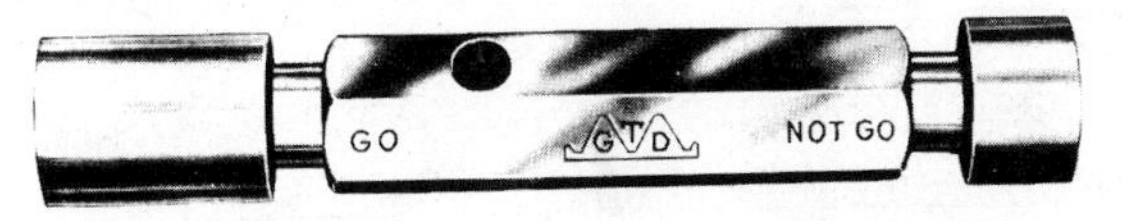

Courtesy Greenfield Tap & Die Co.

Fig. 4-25. A plug gage.

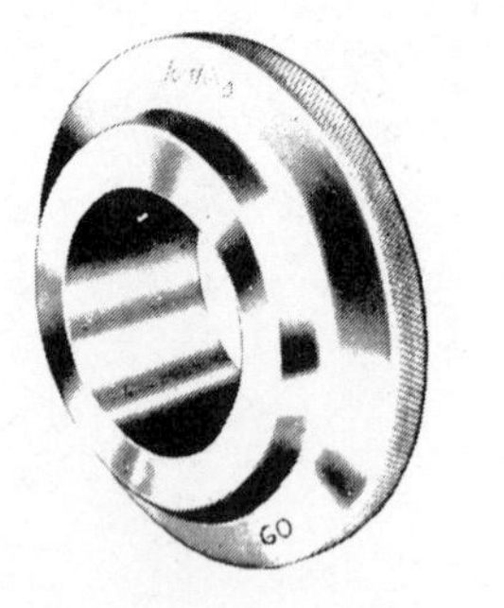

Courtesy Greenfield Tap & Die Co.

Fig. 4-26. A ring gage.

graduated taper. There are two types of taper gages. One has a tapered thickness, and the other type has a tapered width.

The taper gage with tapered thickness (Fig. 4-27) is used for bearing work and for gaging slots. The taper gage with the tapered width (Fig. 4-28) is used as a gage for tubing.

Center gage—This small tool features the standard angle (60°) for lathe centers and for threading tools for the American National Standard Screw Threads. The center gage has a 60° point at one end and a 60° tool (Fig. 4-29). It is used for testing angles of lathe centers, thread-cutting tools, and for setting tools at the correct angle relative to the work.

The large notch at the end of the center gage is used for testing lathe centers and threading-tool points. The small notches at the side of the gage are also used for testing tool points and for setting a threading tool at the correct angle to the work, by placing the opposite edge of the gage against the surface to be threaded. The tool is adjusted until the point fits into the notch in the gage.

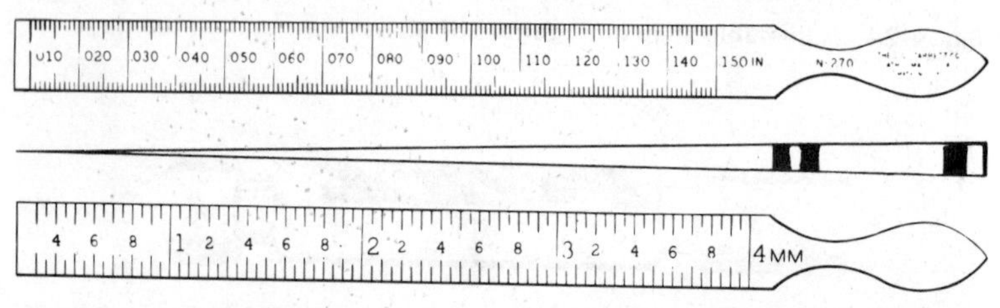

Courtesy L. S. Starrett Co.

Fig. 4-27. Taper gage with tapered thickness. Used for bearing work and for gaging slots.

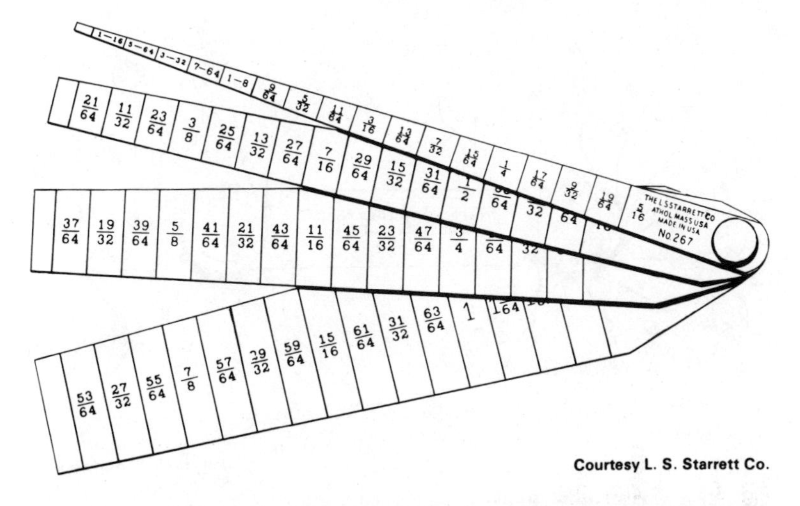

Courtesy L. S. Starrett Co.

Fig. 4-28. Taper gage with tapered width. Used as a tubing gage.

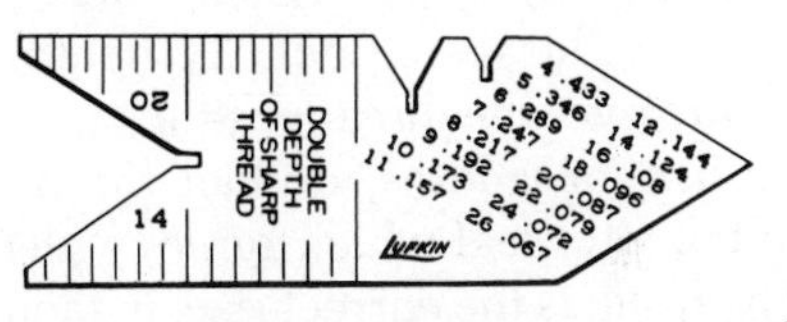

Courtesy Lufkin Rule Co.

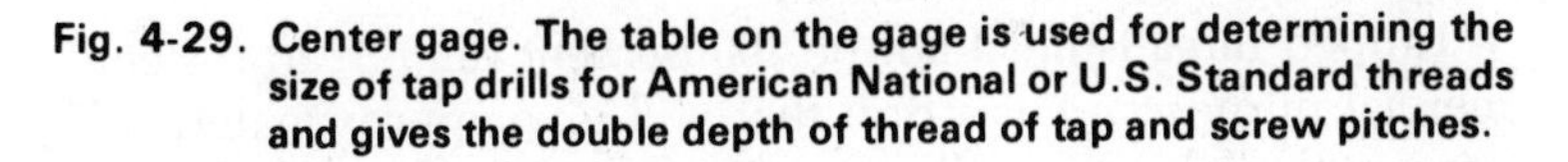

Fig. 4-29. Center gage. The table on the gage is used for determining the size of tap drills for American National or U.S. Standard threads and gives the double depth of thread of tap and screw pitches.

The center gage may also be used for setting internal threading tools. The end of the gage is placed against the face of the work; the side notches are at right angles to the hole to be threaded, provided the work has been faced true.

Screw-pitch gage—The number of threads per inch, or pitch, of a screw or nut can be determined by the use of the screw-pitch gage (Fig. 4-30). This device consists of a holder with a number of thin blades that have notches cut on them representing different numbers of threads per inch, the number being stamped on the blade. Some gages also have the double depth of thread (in decimals) stamped on the blade. This decimal number is equal to the depth of threads on the two sides of a tap. Thus, the workman

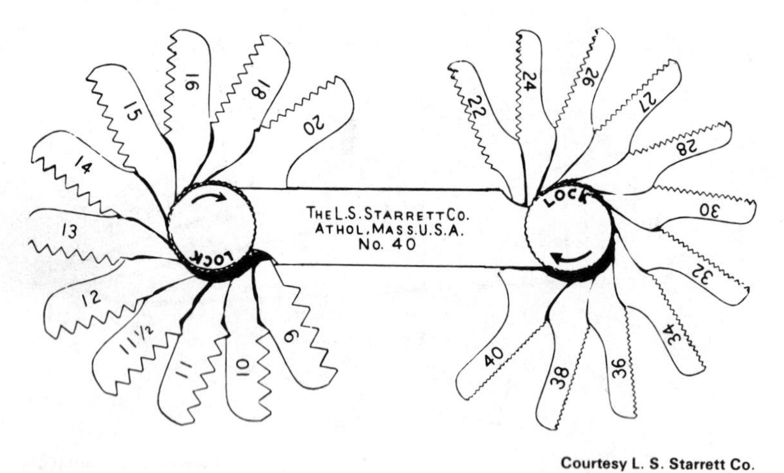

Courtesy L. S. Starrett Co.

Fig. 4-30. Screw-pitch gage.

can determine the size of tap drill to use, in order to leave a full vee thread for a tap having the same pitch. To determine the size of drill needed, measure over the threads of the tap with a micrometer; from its size in thousandths, deduct the decimal number stamped on the pitch-gage leaf which agrees with the pitch of the tap to be used. The result is the correct size, in thousandths, of the drill needed for a full vee thread.

Tapped threads need not be 100 percent full thread for commercial purposes. A tap drill that will give approximately a 75 percent thread is generally used. Sufficient stock in which to cut the threads must be left by the tap drill. A formula for finding approximate tap drill size is as follows:

$$\text{Tap drill size} = \text{Major diameter of thread} - \frac{(0.75 \times 1.299)}{\text{No. threads per inch}}$$

A more practical and simpler formula is:

$$\text{Tap drill size} = \text{Major diameter} - \frac{1}{\text{No. threads per inch}}$$

Tap drill sizes for full vee threads and for American National Standards threads can be calculated by the following:

1. Full vee threads. $d = D - \frac{1.733}{N}$

2. American National Standard Screw Threads.

$$d = D - \frac{1.2999}{N}$$

In the above formulas, D = major diameter of tap; d = minor diameter of tap; and N = number of threads per inch.

Tap and drill gage—The mechanic can use this gage to select quickly the tap drill size for the tap to be used. The correct tap drill leaves enough stock to cut a full thread without breaking the tap; thus, uncertainty of result and much time can be saved (Fig. 4-31).

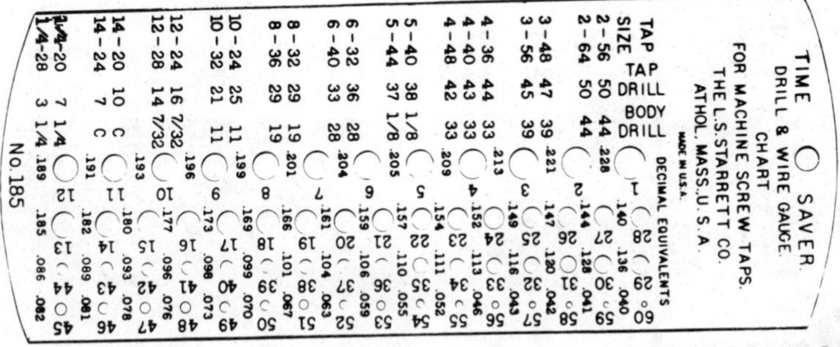

Courtesy L. S. Starrett Co.

Fig. 4-31. Tap and drill gage.

Thickness or feeler gage—Many varieties of these gages are on the market (Fig. 4-32). A Thickness gage consists of a number of thin steel leaves, which vary by thousandths of an inch. The leaves may be used singly or in groups; they enable the mechanic to form any desired thickness, within the limits of the tool. Standard thicknesses range from 0.0015 to 0.015, by thousandths.

Thickness, or feeler, gages are used extensively for setting ignition points, spark gaps, and valve tappets, and for checking ring clearances, piston clearances, etc. They are extremely valuable to the machinist and toolmaker for a variety of purposes.

Another form of thickness gage, or feeler stock (Fig. 4-33), is also available. Feeler stock is accurate, high grade, uniformly

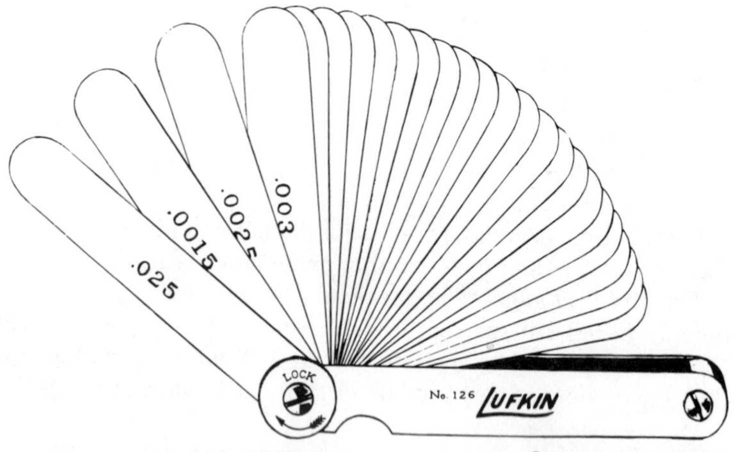

Courtesy L. S. Starrett Co.

Fig. 4-32. Thickness gage. The leaves of these gages include the most used sizes in automotive work.

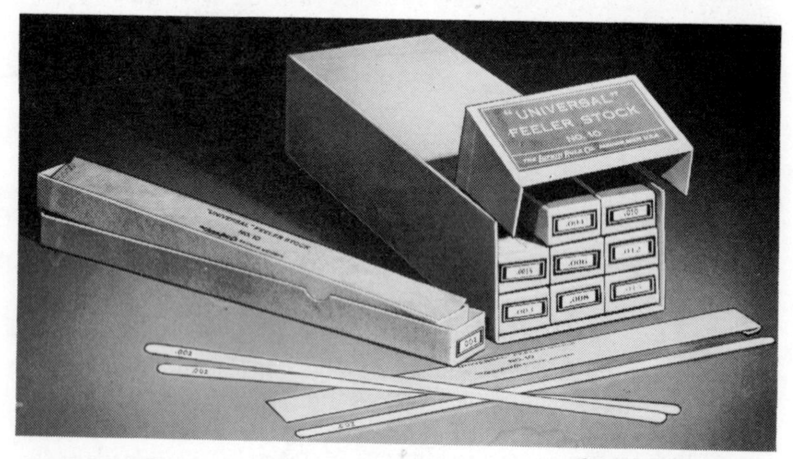

Courtesy Lufkin Rule Co.

Fig. 4-33. Feeler stock or thickness gages.

tempered, thickness gage stock. The size or thickness is marked on each piece in large, easily read figures. Feeler stock can be used in the same manner as the common thickness gage.

Wire gage—The American Standard Wire Gage for nonferrous metals is shown in Fig. 4-34. This gage is especially useful for

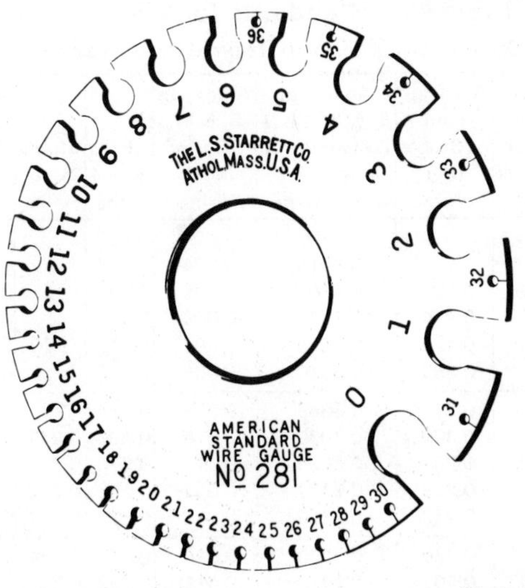

Courtesy L. S. Starrett Co.

Fig. 4-34. American Standard Wire Gage. It is generally standard for nonferrous metals such as copper, brass, and aluminum. These gages are useful for gaging sheets, plates, and wire.

electricians and others to gage sheets, plates, and wire made from the nonferrous metals, such as aluminum, brass, and copper. The decimal equivalents are stamped on the back of the gage.

U.S. Standard Gage for Sheet and Plate Iron and Steel—This gage is the recognized commercial standard in the United States for uncoated sheet, plate iron, and steel (Fig. 4-35). It is based on weight in ounces per square foot. The decimal equivalents are on the back of the gage.

Table 4-1 illustrates the various standards for wire gages. The Stub's Iron Wire Gage is commonly known as the Birmingham Gage and designates the Stub's soft-wire sizes. The Stub's Steel Wire Gage is used to measure drawn steel wire or drill rod.

Effects of Temperature and Weight on Precision Tools

The larger micrometers are more sensitive to temperature changes than are the smaller sizes of micrometers. Where accurate measurements are to be taken with a large micrometer, the

Table 4-1. Standards for Wire Gages

Dimensions of Sizes in Decimal Parts of an Inch.

No. of Wire	American or Brown & Sharpe for Non Ferrous Metals	Birmingham or Stub's Iron Wire	American S. & W. Co.'s (Washburn & Moen) Std. Steel Wire	American S. & W. Co.'s Music Wire	Imperial Wire	Stub's Steel Wire	U.S. Std. Gage for Sheet & Plate Iron & Steel	No. of Wire
7-0's	0.651354	...	0.4900	...	0.500	...	0.500	7-0's
6-0's	0.580049	...	0.4615	0.004	0.464	...	0.46875	6-0's
5-0's	0.516549	0.500	0.4305	0.005	0.432	...	0.4375	5-0's
4-0's	0.460	0.454	0.3938	0.006	0.400	...	0.40625	4-0's
000	0.40964	0.425	0.3625	0.007	0.372	...	0.375	000
00	0.3648	0.380	0.3310	0.008	0.348	...	0.34375	00
0	0.32486	0.340	0.3065	0.009	0.324	...	0.3125	0
1	0.2893	0.300	0.2830	0.010	0.300	0.227	0.28125	1
2	0.25763	0.284	0.2625	0.011	0.276	0.219	0.265625	2
3	0.22942	0.259	0.2437	0.012	0.252	0.212	0.250	3
4	0.20431	0.238	0.2253	0.013	0.232	0.207	0.234375	4
5	0.18194	0.220	0.2070	0.014	0.212	0.204	0.21875	5
6	0.16202	0.203	0.1920	0.016	0.192	0.201	0.203125	6
7	0.14428	0.180	0.1770	0.018	0.176	0.199	0.1875	7
8	0.12849	0.165	0.1620	0.020	0.160	0.197	0.171875	8
9	0.11443	0.148	0.1483	0.022	0.144	0.194	0.15625	9
10	0.10189	0.134	0.1350	0.024	0.128	0.191	0.140625	10
11	0.090742	0.120	0.1205	0.026	0.116	0.188	0.125	11
12	0.080808	0.109	0.1055	0.029	0.104	0.185	0.109375	12
13	0.07196	0.095	0.0915	0.031	0.092	0.182	0.09375	13
14	0.064084	0.083	0.0800	0.033	0.080	0.180	0.078125	14

instrument should be tested with a pin gage for correct adjustment. In very large micrometers (24 inches to 36 inches), the micrometer, while being tested, should be held in the same position in which it is to be held when taking the measurement.

The weight of the frame of a micrometer may cause a variation in readings of the instrument. It is also necessary in cold weather, when using the larger micrometers, to use a piece of waste or cloth between the hand and the frame. The heat transmitted from the hand may cause the frame to spring out of shape and cause a variation in readings.

Care should be taken not to press the two ends of the frame of large micrometers together. If this does occur, and the effect is tested with a dial indicator of large-scale amplification, the impor-

Table 4-1. Standards for Wire Gages (Cont'd)

No. of Wire	American or Brown & Sharpe for Non Ferrous Metals	Birmingham or Stub's Iron Wire	American S. & W. Co.'s (Washburn & Moen) Std. Steel Wire	American S. & W. Co.'s Music Wire	Imperial Wire	Stub's Steel Wire	U.S. Std. Gage for Sheet & Plate Iron & Steel	No. of Wire
15	0.057068	0.072	0.0720	0.035	0.072	0.178	0.0703125	15
16	0.05082	0.065	0.0625	0.037	0.064	0.175	0.0625	16
17	0.045257	0.058	0.0540	0.039	0.056	0.172	0.05625	17
18	0.040303	0.049	0.0475	0.041	0.048	0.168	0.050	18
19	0.03589	0.042	0.0410	0.043	0.040	0.164	0.04375	19
20	0.031961	0.035	0.0348	0.045	0.036	0.161	0.0375	20
21	0.028462	0.032	0.0317	0.047	0.032	0.157	0.034375	21
22	0.025347	0.028	0.0286	0.049	0.028	0.155	0.03125	22
23	0.022571	0.025	0.0258	0.051	0.024	0.153	0.028125	23
24	0.0201	0.022	0.0230	0.055	0.022	0.151	0.025	24
25	0.0179	0.020	0.0204	0.059	0.020	0.148	0.021875	25
26	0.01594	0.018	0.0181	0.063	0.018	0.146	0.01875	26
27	0.014195	0.016	0.0173	0.067	0.0164	0.143	0.0171875	27
28	0.012641	0.014	0.0162	0.071	0.0149	0.139	0.015625	28
29	0.011257	0.013	0.0150	0.075	0.0136	0.134	0.0140625	29
30	0.010025	0.012	0.0140	0.080	0.0124	0.127	0.0125	30
31	0.008928	0.010	0.0132	0.085	0.0116	0.120	0.0109375	31
32	0.00795	0.009	0.0128	0.090	0.0108	0.115	0.01015625	32
33	0.00708	0.008	0.0118	0.095	0.0100	0.112	0.009375	33
34	0.006304	0.007	0.0104	...	0.0092	0.110	0.00859375	34
35	0.005614	0.005	0.0095	...	0.0084	0.108	0.0078125	35
36	0.005	0.004	0.0090	...	0.0076	0.106	0.00703125	36
37	0.004453	...	0.0085	...	0.0068	0.103	0.00664063	37
38	0.003965	...	0.0080	...	0.0060	0.101	0.00625	38
39	0.003531	...	0.0075	...	0.0052	0.099	....	39
40	0.003144	...	0.0070	...	0.0048	0.097	....	40

tance of handling precision tools lightly may be demonstrated to a mechanic.

SUMMARY

Many accurate measuring tools are used in the shop. Machinist measurements must be done with a much greater degree of precision than is required for many other lines of work. The measuring tools most commonly used are micrometers, vernier, calipers, protractors, indicator, and gages.

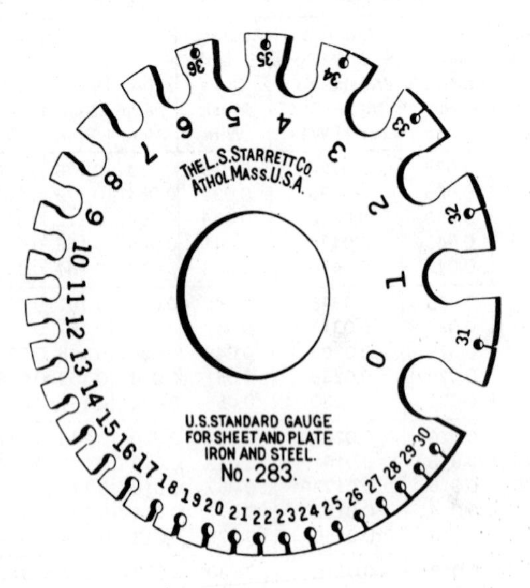

Courtesy L. S. Starrett Co.

Fig. 4-35. United States Standard Gage. This recognized commercial standard in the U.S. is used for uncoated sheet, plate iron, and steel. It is based on weights in ounces per square foot. Decimal equivalents of each gage number are stamped on the reverse side.

There are various types of micrometers designed to make accurate inside, outside, and depth measurements. Vernier micrometers are graduated in ten-thousandths. (0.0001) of an inch and are read in the same way as the micrometer calipers.

The vernier caliper is another precision measuring instrument used to take linear measurements as close as one-thousandth (0.001) of an inch. The vernier system of measurement is also used on other instruments such as height gages and bevel protractors.

Bevel protractors are used to make angular or circular measurements. Dial indicators are used for precision checking of alignment, roundness, and taper of manufactured parts.

A gage is a fixed device that establishes a particular dimension. Some gages are adjustable and therefore are more flexible for checking dimensions.

REVIEW QUESTIONS

1. Explain how to read a micrometer.
2. How is a vernier caliper used?
3. Describe the difference between bevel protractor and the universal bevel protractor.
4. How is the dial indicator used?
5. Name a few gages that are used by the machinist.

CHAPTER 5

Sheet-Metal Hand Tools and Machines

Sheet metal is made from all types of ferrous and nonferrous metals. It ranges in thickness from a few thousandths of an inch to over one inch. Generally, metal above 3/16 inch in thickness is referred to as plate rather than sheet metal. A worker in the sheet-metal trades primarily uses galvanized iron; tin plate; and aluminum, copper, and brass sheet. Sheet-metal workers in these trades are concerned largely with furnace and air-conditioning installations, metal trim, and roof work. Sheet-metal shops that specialize in installation and repair are established in towns of every size.

SHEARS

After the layout is made, one of the first operations is cutting the sheet metal with snips (Fig. 5-1). Several types of hand-operated and power-driven machines are also available for cutting sheet metal.

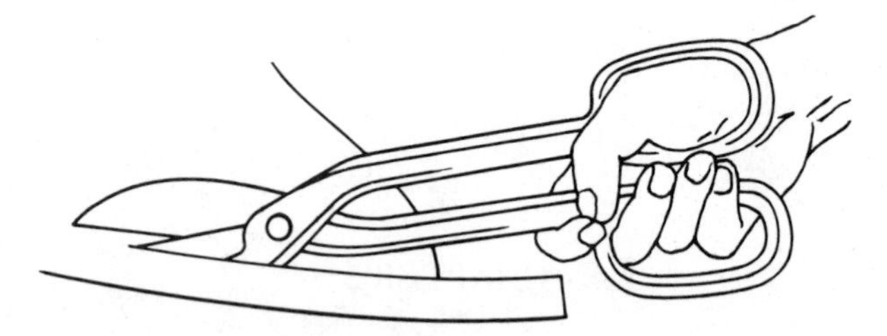

Fig. 5-1. Cutting sheet metal with straight hand snips.

Hand Snips

Hand snips are much like scissors, except that they are made in various sizes, usually 2 to 4 inches, depending on the length of the cutting blade. *Straight snips* are used to cut straight lines and outside curves; *hawksbill snips* (Fig. 5-2) are used to cut inside curves and intricate work. Never attempt to cut wire with snips.

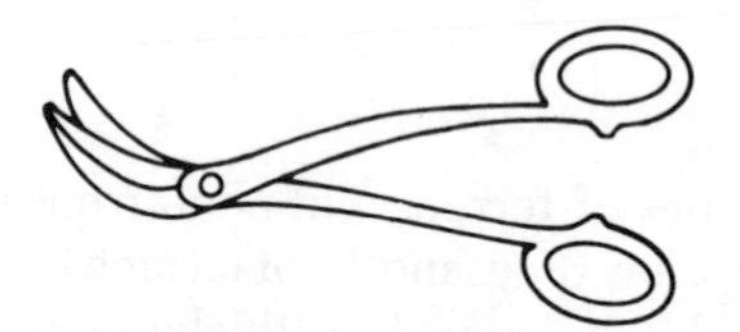

Fig. 5-2. Hawksbill snips.

Aviation snips are used to cut both straight and curved cuts in thin-gage sheet metal. They are made in three styles (Fig. 5-3). the serrated blade edge and compound-levered cutting action provides for an easy one-handed operation of the snips. NOTE: Right-hand snips cut to the left, and lefthand snips cut to the right.

Hand- or Foot-Operated Shears

A useful type of shears for almost any sheet-metal shop is the bench shear shown in Fig. 5-4. This type can be used to shear large sheets of metal that are too large to be handled conveniently with handsnips. One handle is held in a vise, while the other handle is moved to create the cutting action. A device for shearing rod stock is illustrated in Fig. 5-5. Both round and square stock may be sheared on the machine with a minimum of distortion of the stock.

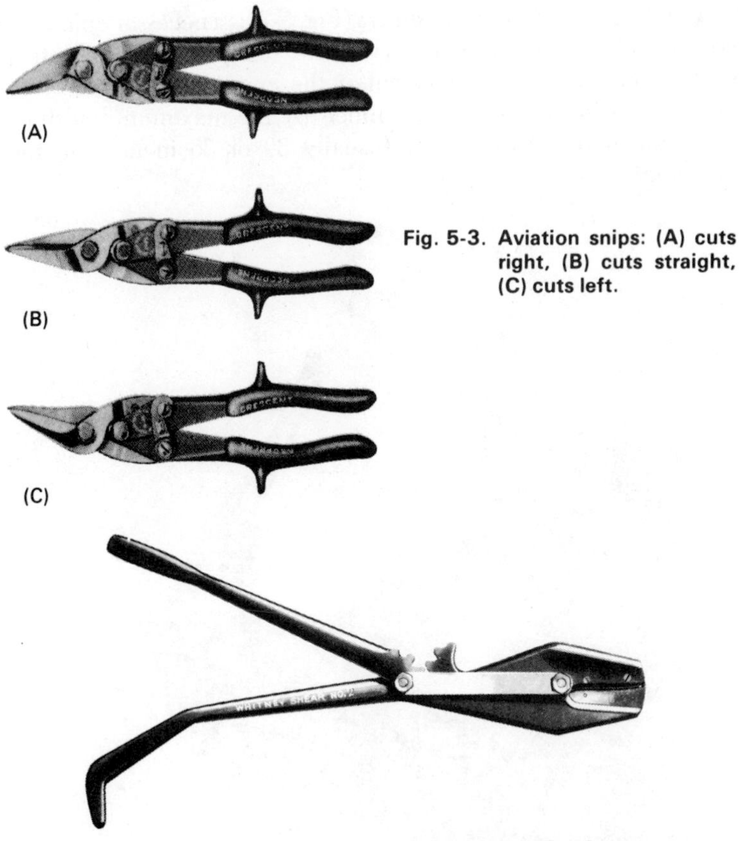

(A)

(B)

(C)

Fig. 5-3. Aviation snips: (A) cuts right, (B) cuts straight, (C) cuts left.

Courtesy Whitney Metal Tool Company

Fig. 5-4. Useful shears for shearing sheets of metal too thick to be handled by hand snips.

An entirely different type of shears, but a type that is extensively used in sheet-metal shops, is based on the throatless principle, as shown in Fig. 5-6. As the stock is passed through the shears, the sheared sections separate, going to the right and left and under and over the spiral shaped head. Because of the throatless principle, there is no limit on the width of stock to be sheared; and straight, irregular, or circular patterns can be followed. The upper cutter is serrated to feed the stock and is driven, while the lower cutter revolves freely in a sleeve.

A foot-operated *squaring shears* (Fig. 5-7) is a necessary piece of equipment in many sheet-metal shops. Usually, 16-gage mild steel sheet metal, or lighter, can be cut on the squaring shears. The size of the squaring shears is determined by the maximum width of sheet metal that can be cut (usually 30 or 36 inches) on the machine.

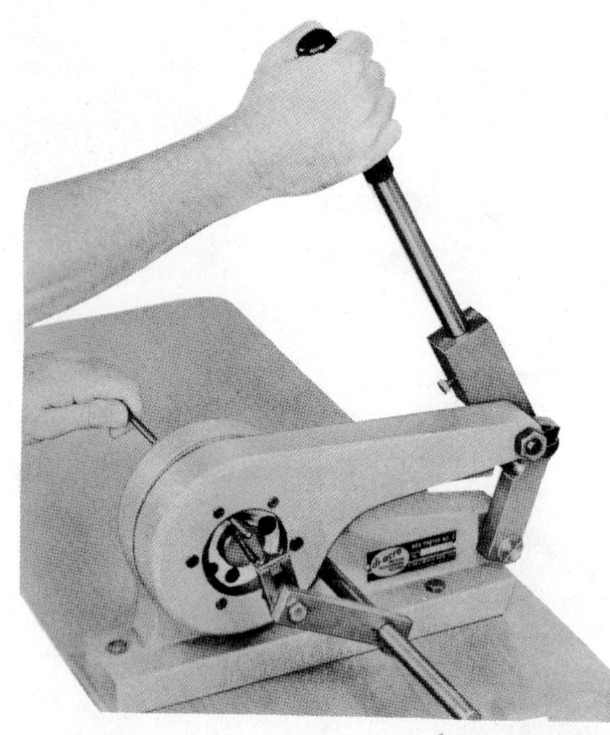

Courtesy Di-Acro Div. of Houdaille

Fig. 5-5. A rod parter. Note the holes for handling various sizes of rods.

Long sheets of metal are usually inserted from the back of the squaring shears. Most machines have a graduated scale for setting the front gage to the desired length. The front gage can also be used when several pieces of identical length are to be cut. After the long sheet of metal has been inserted from the rear, the side of the sheet should be pressed firmly in place, using both hands, and pressure applied to the foot pedal. *Always keep the fingers away from the cutting blade.*

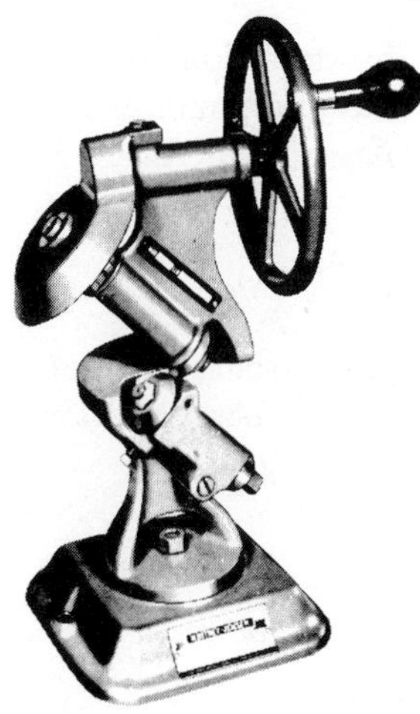

Courtesy Whitney Metal Tool Company

Fig. 5-6. Throatless shears (left) and the throatless principle used to shear sheet metal on the right.

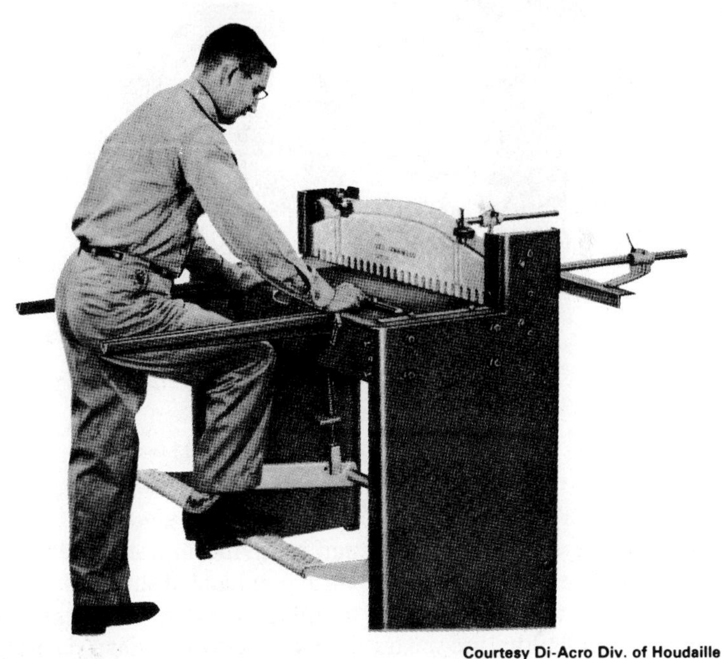

Courtesy Di-Acro Div. of Houdaille

Fig. 5-7. Foot-operated squaring shears.

Smaller lengths of metal may be cut by insertion from the front, against the back gage. Band iron, wire, or heavy gage metal should not be cut on the squaring shears. These may nick the blade and cause irregularities in the cut edge.

Motor-Driven Shears

Several types of motor-driven shears are used to cut sheet metal. A motor-driven throatless shears is shown in Fig. 5-8. A hydraulic operated angle-iron shears is illustrated in Fig. 5-9.

Courtesy Whitney Metal Tool Company

Fig. 5-8. Motor-driven throatless shears.

Large special-purpose shears can be found in the sheet-metal industries. Fig. 5-10 illustrates a large mechanical shears that can handle sheets or plates of metal up to ½ inch in thickness and 12 feet in width. Hydraulic shears are available in standard shearing capacities for 1½-inch plate.

Courtesy Whitney Metal Tool Company

Fig. 5-9. Hydraulic angle-iron shears.

PUNCHES

A punch is a blunt tool used to pierce a solid material. The cutting angle of the tool is 90°, as compared to the acute angle of the cutting edge of the shears. Shears actually cut the metal, but punches push or tear the metal (Fig. 5-11). Several types of punches are used in sheet-metal work, ranging from small hand punches to large motor-driven presses.

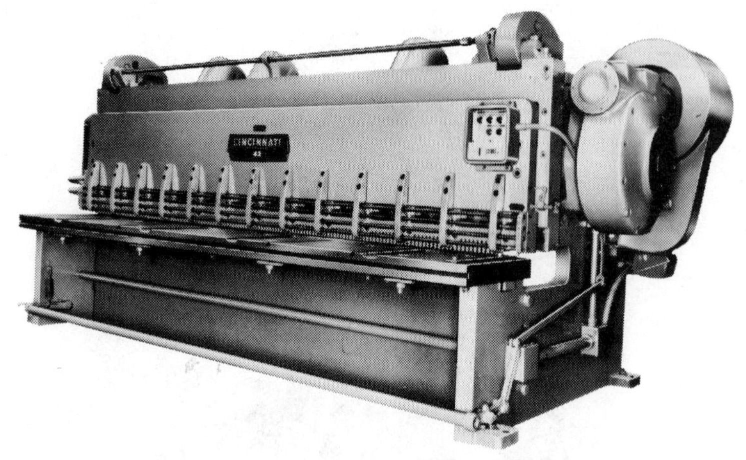

Courtesy Cincinnati Milacron Company

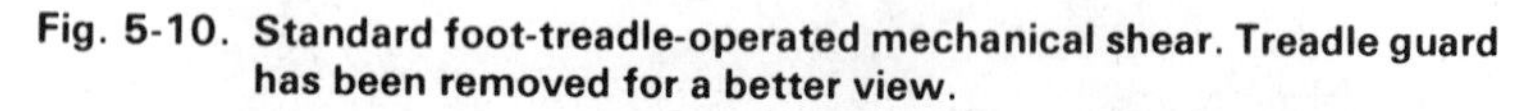

Fig. 5-10. Standard foot-treadle-operated mechanical shear. Treadle guard has been removed for a better view.

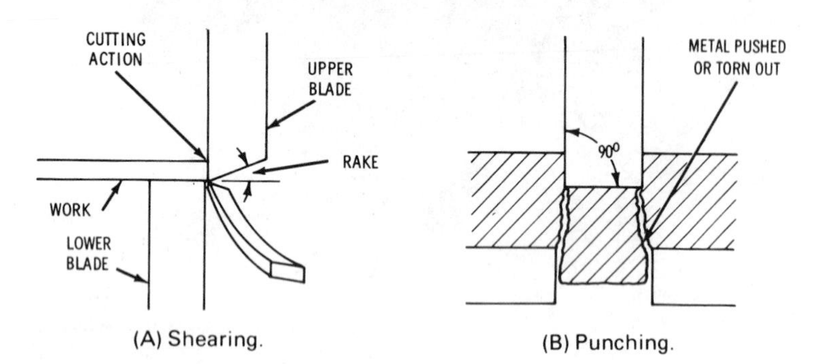

(A) Shearing. (B) Punching.

Fig. 5-11. Comparison of shearing and punching operations.

Hand Punches

The round-handled solid punch (Fig. 5-12) and the one-piece hollow punch (Fig. 5-13) are both commonly used on sheet metal. Some small hand-operated punches or machines can be adapted to the various hole sizes by a selection of interchangeable punches and dies (Fig. 5-14). Several models of bench-type punches are manufactured (Fig. 5-15). Various sizes of punches and dies can be used to produce the desired sizes and shapes of holes (Fig. 5-16).

Courtesy Whitney Metal Tool Company

Fig. 5-12. Round-handled solid punch.

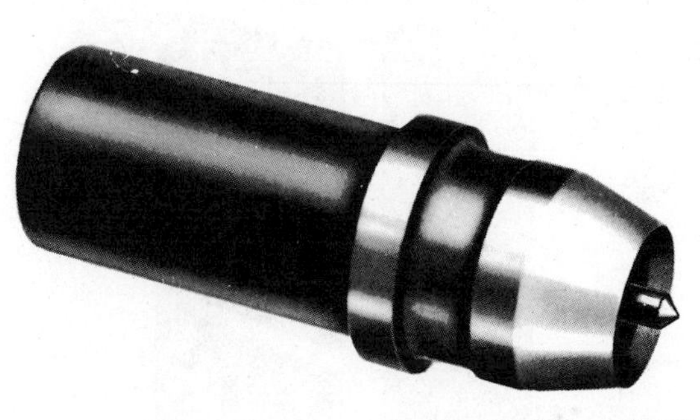

Courtesy Whitney Metal Tool Company

Fig. 5-13. One-piece hollow punch.

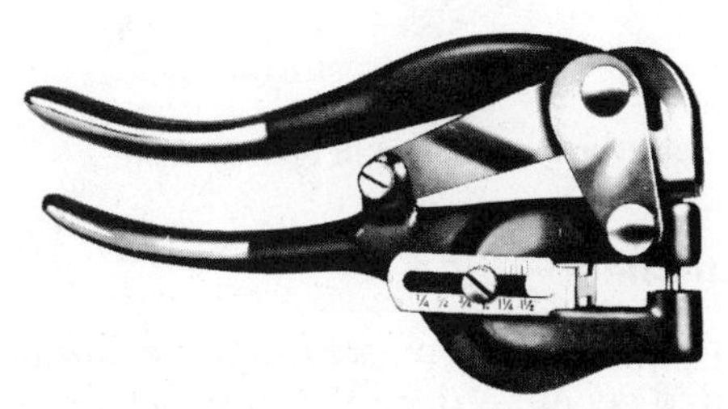

Courtesy Whitney Metal Tool Company

Fig. 5-14. A small hand-operated metal punch.

Courtesy Whitney Metal Tool Company

Fig. 5-15. A bench-type hand-operated metal punch.

Machine Punches

The power for machine presses may be supplied by lever, screw, or geared action. A deep-throat punch operated by lever action is shown in Fig. 5-17, and a deep-throat power punch press is shown in Fig. 5-18.

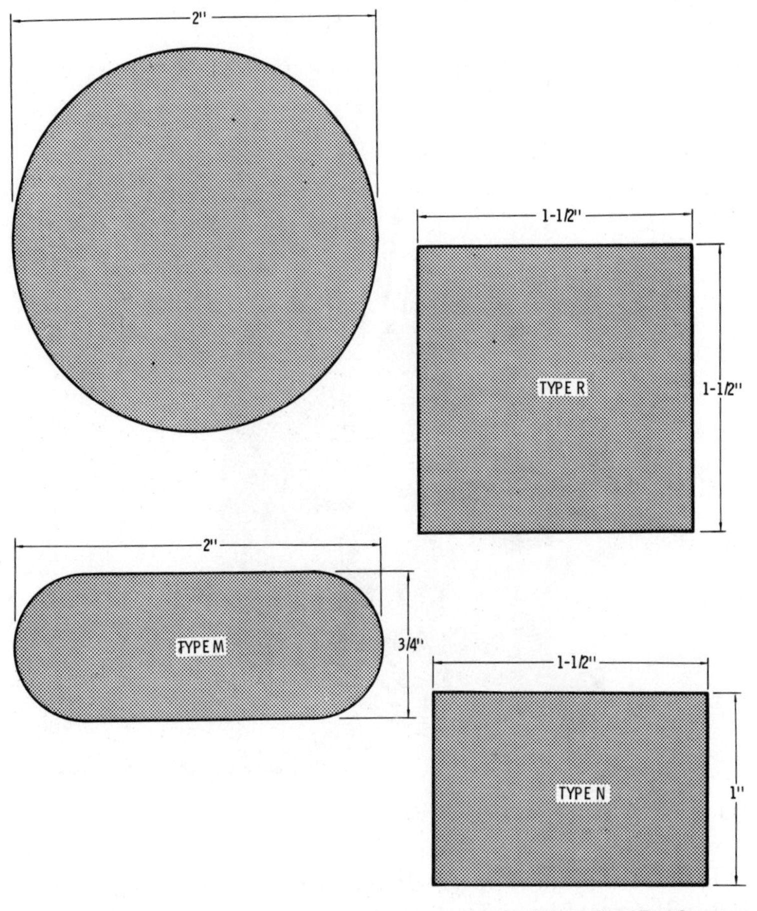

Courtesy Whitney Metal Tool Company

Fig. 5-16. Type of punches and dies than can be used on the bench-type, hand-operated metal punch.

Various types of punches are used on the machine punches. Some of these are shown in Fig. 5-19. A machine assembly for mounting the punch in the machine is shown in Fig. 5-20. The parts of the assembly are: plunger, coupling, stripper, punch, die, and socket. The punch is fastened to the plunger by the coupling nut, which bears against a shoulder on the punch. The die fits into a socket and is held in position by setscrews.

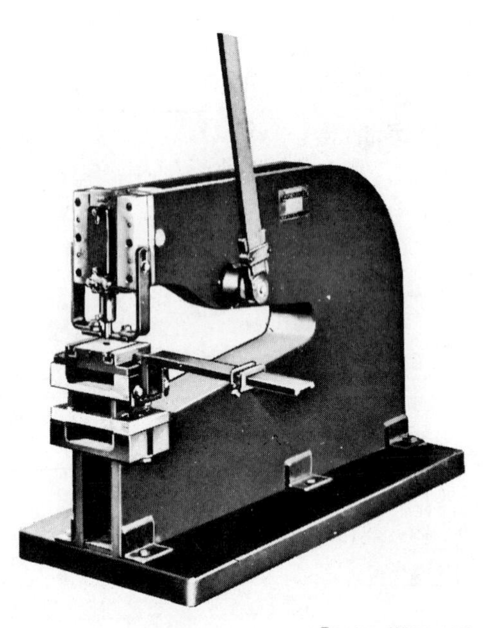

Courtesy Whitney Metal Tool Company

Fig. 5-17. A deep-throat lever press.

The basic design of power driven machine punches is shown in Fig. 5-21. Great force is necessary to drive a punch through a piece of metal. In machines that perform work of an intermittent nature—shears, punches, and presses—a heavy-rimmed flywheel is used to distribute the work of the machine over the entire period of revolution of the drive shaft. Thus, belt power is used to accelerate the speed of the flywheel during the greater part of the revolution of the drive shaft. The energy stored up in the flywheel is expended at the expense of its velocity during the portion of the revolution that work is done. This action is similar to that of a gasoline engine in which the flywheel keeps the engine turning, particularly during the compression stroke.

Courtesy Whitney Metal Tool Company

Fig. 5-18. A deep-throat power punch press.

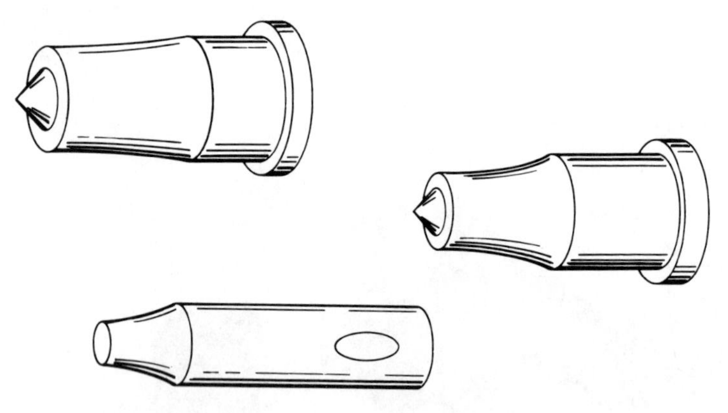

Fig. 5-19. Typical punches used on a punch press.

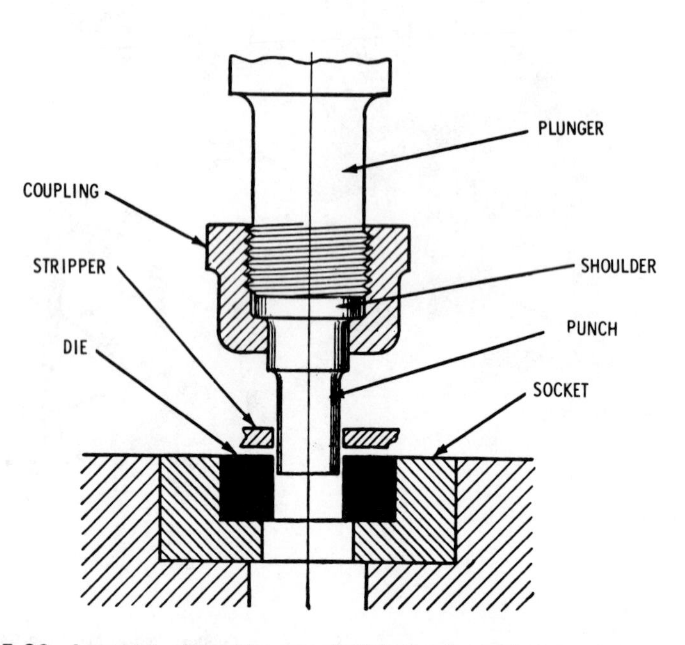

Fig. 5-20. Assembly of machine punch. This drawing shows the parts holding the punch, die mounting, and stripper.

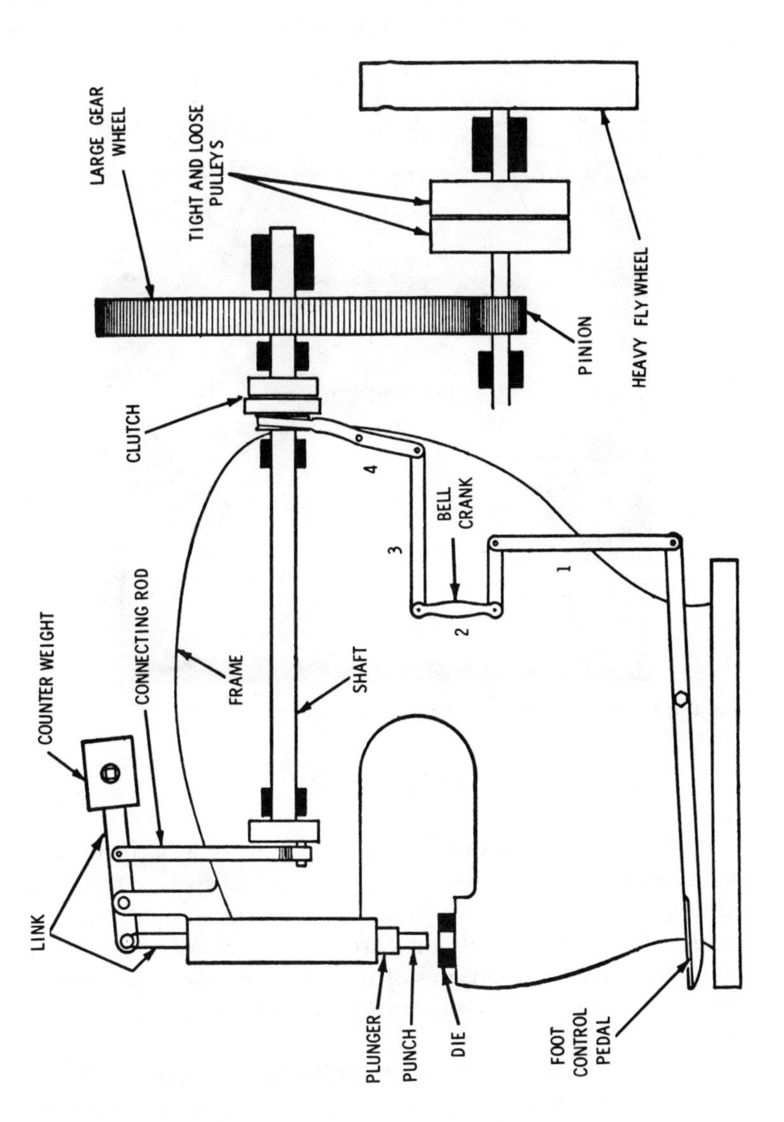

Fig. 5-21. Basic diagram of essential parts of a power-driven machine punch.

Machines may be equipped to punch a single hole, or many holes may be punched in a single operation, as in the panel shown in Fig. 5-22.

Fig. 5-22. Punching multiple holes in a panel.

SUMMARY

Snips are used to cut sheet metal. Hand shears or snips are much like scissors, except they are made in various sizes with shorter cutting blades. Straight snips are used to cut straight lines and outside curves. Hawksbill snips are used to cut inside curves and intricate work. Several types of motor-driven shears are used to cut sheet metal. Large special-purpose shears can be found in sheet-metal industries.

A punch is a blunt tool used to pierce a solid material. Shears actually cut the metal, but punches push or tear the metal. Several types of punches are used in sheet-metal shops, ranging from small hand punches to large motor-driven presses. Power for machine presses may be supplied by lever, screw, or geared action. Presses are equipped with dies and punches designed for specific forming, punching, and shearing operations.

REVIEW QUESTIONS

1. What are snips or shears?
2. What is the cutting action of a punch?
3. Describe the difference between plate and sheet metal.
4. What is the gage capacity of most hand and foot operated sheet metal machines?
5. What is the difference between a punch and a die?
6. Describe sheet metal.
7. What is a rod parter?
8. Why are throatless shears so named?
9. Describe a punch.
10. How are Hawksbill snips used?

CHAPTER 6

Sheet-Metal Forming Machines

In addition to the cutting and punching, operations performed with snips, shears, and punches, metal can also be shaped into arcs, spirals, circles, cylinders, and cones by means of forming equipment. These machines vary from small hand-operated forming rolls to the large motor-driven bending rolls and forming presses.

SLIP ROLLS

The standard forming rolls (slip-roll forming machine), shown in Fig. 6-1, is 18 inches in width (rolls 18 inches long). The machine is used to bend sheet metal to a curved form. Articles that are cylindrical in shape can be formed in this machine. The slip-roll forming machine consists of two geared rolls and one loose roll, which serves to bend the work after it has passed between the first two rolls. The distance between the geared rolls can be regulated

Courtesy Di-Acro Div. of Houdaille

Fig. 6-1. Hand-operated forming machine, or slip-roll former.

with adjusting screws by raising or lowering the lower front roll, depending on the thickness of the metal. The rear roll can be raised to bend the sheet metal to a smaller radius, or it can be lowered to form a cylinder with a larger radius. The lower front roll and the rear roll have grooves cut along the right-hand edges of the rolls that can be used either for forming wire or for forming cylinders with the wire edge already installed.

In principle, the forming of sheet metal is actually done by the three rolls. The two front rolls grip the sheet of metal and force it against the rear roll. This action bends the sheet of metal around the front upper roll, forming a cylinder. After the cylinder has been formed, the upper roll can be slipped out of place to remove the cylinder.

BAR FOLDER

Another sheet metal forming machine is the bar folder, which is used to turn a hem or lock on the edge of a piece of sheet metal (Fig. 6-2). The bar folder can also be used to prepare the edge of sheet metal to receive a wire. These machines have adjustments for regulating the width of the fold and the sharpness of the bend, and so they may also be used for preparing the edges of the metal sheets to receive a wire. The machine should always be properly adjusted before attempting to make a fold or to wire an edge.

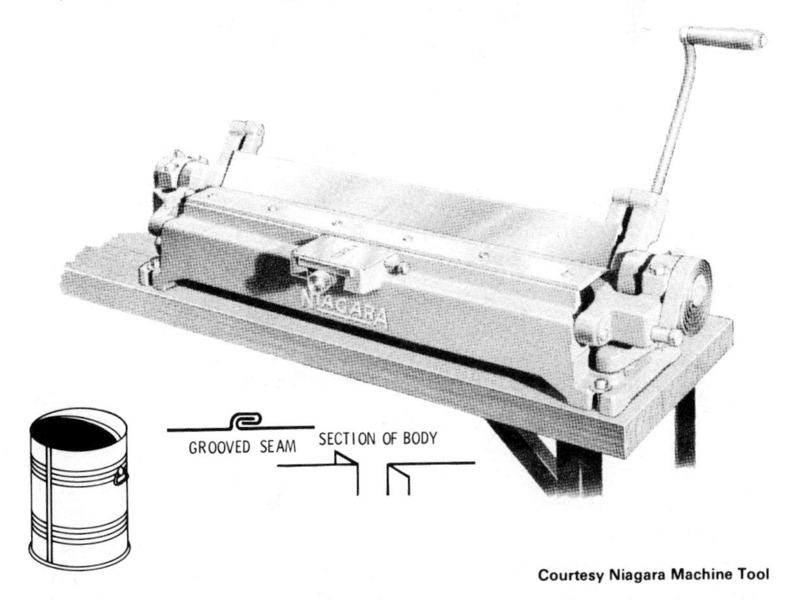

Courtesy Niagara Machine Tool

Fig. 6-2. Bar folder and work performed on the machine.

After a seam or hem has been folded on the bar folder, the seam should be closed down with a grooved wheel on the *grooving machine* (Fig. 6-3B), or a hand groover (Fig. 6-3A). The grooving rolls are made to fit various seam widths, and the proper roll should be used. A lock is first turned on the sheet of metal by

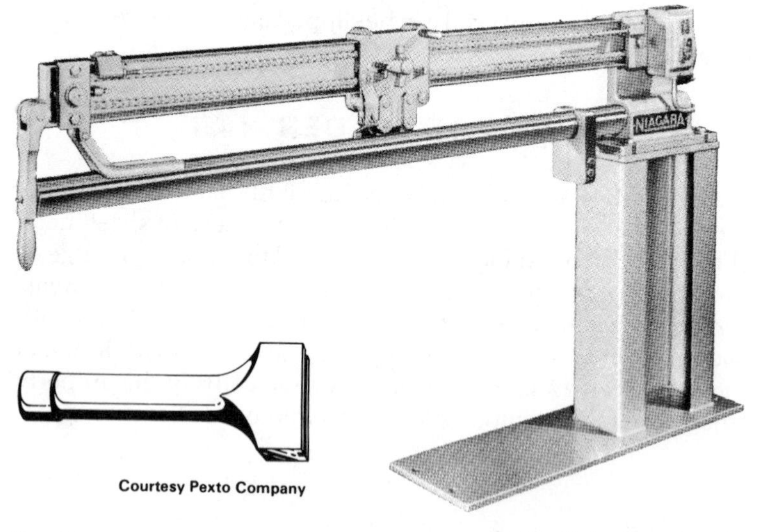

Courtesy Pexto Company

Fig. 6-3. (A) A hand groover

Courtesy Di-Acro Div. of Houdaille

(B) A grooving machine and the work it performs

means of the bar folder. The sneet of metal is then rolled into a cylinder by the forming rolls, or slip-roll former. The corresponding edges, as prepared in the cylinder, are snapped together and laid on the grooving horn in the grooving machine. The grooving rolls of the grooving machine are run over the seam lengthwise, effecting an operation called "grooving" or "seam closing" to complete the seam.

ROTARY MACHINE

A *burring machine* is used for turning an edge on cylinders of metal or on disks, such as can bottoms (Fig. 6-4). This is a difficult operation to master, but practice enables a worker to produce uniform flanges on sheet-metal bodies and prepares the burr for the bottoms prior to "setting down" and double seaming.

A *setting-down machine* is used to close the seams left by the burring machine, making the seams ready for double seaming (Fig. 6-5). This is a simple machine that has no adjustments except

Fig. 6-4. Using the burring machine to turn a burr.

Fig. 6-5. Work performed on a setting-down machine.

for thickness of the material. The machine turns down and compresses the flange of the bottom onto the flange at the end of the body, thereby forming a joint, which can be doubled over with a *double-seaming machine*. The successive operations performed on the setting-down machine and the double-seaming machine in forming a double seam are shown in Fig. 6-6.

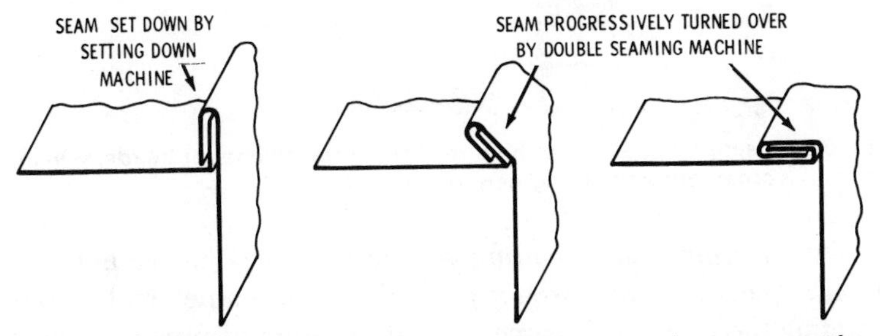

Fig. 6-6. Forming a double seam by progressive operations on a setting-down machine.

After the seams have been set down properly with the setting-down machine, the remaining flange must be turned against the body of the vessel (Fig. 6-7). The progressive turning over of the seam with the double-seaming machine makes a tight joint with a high degree of neatness and accuracy.

Fig. 6-7. Using the double-seaming machine to complete a double seam, which makes a tight joint.

A *beading machine* is simple to operate. It is provided with a series of rolls of various shapes. The impression in the body of a vessel corresponds with the shape of the bead in the rolls used. The operation of forming these impressions is called beading, which serves to ornament and strengthen tinware and other sheet-metal goods (Fig. 6-8).

Fig. 6-8. Using a beading machine to make impressions, or beads, which ornament and strengthen the work.

The *crimping and beading machine* is used in making and joining pieces of sheet-iron pipe of different diameters, by contracting the edge of the pipe, so that one joint of pipe will enter another. In joining the lengths of pipe, the ogee bead next to the crimp prevents the joints slipping beyond the impression made by the beading rolls. The crimp and bead on the edge of ordinary stove-pipe are made with this machine.

A *turning machine* is used to form a rounded edge for insertion of a wire (Fig. 6-9). The edge is then closed by a *wiring machine* (Fig. 6-10). The successive operations in making wired seam are shown in Fig. 6-11. Depending on the shape of the work, the seams for receiving the wire may be prepared on the bar folder or brake, instead of on the turning machine, and then closed on the wiring machine.

Fig. 6-9. Using the turning machine to form a rounded edge so that a wire can be inserted to make a wired seam.

Fig. 6-10. Using the wiring machine to close the rounded edge over the inserted wire and complete the wire seam.

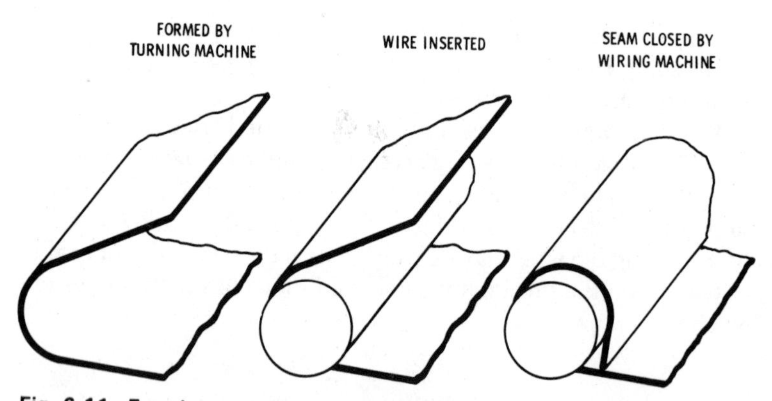

Fig. 6-11. Forming a wired seam by progressive operations performed on a turning machine and a wiring machine.

BRAKES

Formerly, sheet metal was bent entirely by hand. Hand-operated sheet-metal brakes are now widely used. Mechanical and hydraulic press brakes are also used where large pieces of sheet metal are to be formed.

Hand-Operated Brakes

Hand-operated sheet-metal brakes are useful for many different kinds of metal bending operations. Straight bending of sheet metal is easily performed on a brake. Box and pan bending can also be accomplished when the brake is equipped with a box and pan finger (Fig. 6-12). The radius finger (Fig. 6-13) permits the brake to be used for radius bending of sheet metal, in addition to straight bending, and for box and pan bending.

Press Brakes

Although tonnage is sometimes used to express the capacity of a brake, it is the flywheel energy (horsepower) that is required to exert the necessary pressure throughout the entire working stroke. The capacity of the brake required to form a single bend in a given material is dependent on (1) length of the work, (2) sharpness of

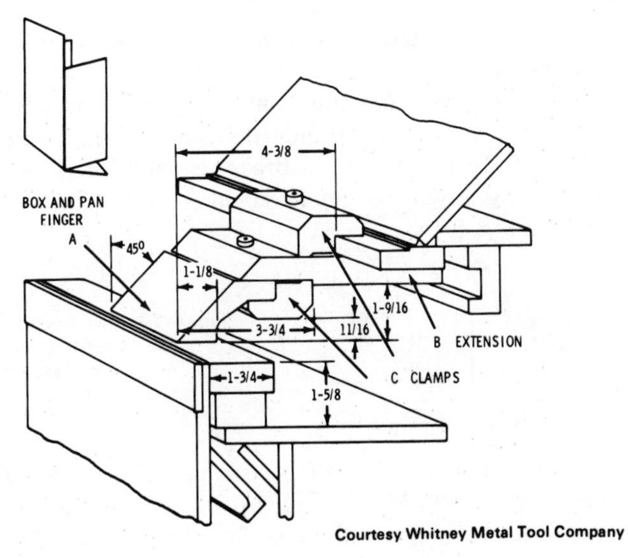

Courtesy Whitney Metal Tool Company

Fig. 6-12. Box and pan finger attachment for use on a sheet-metal brake.

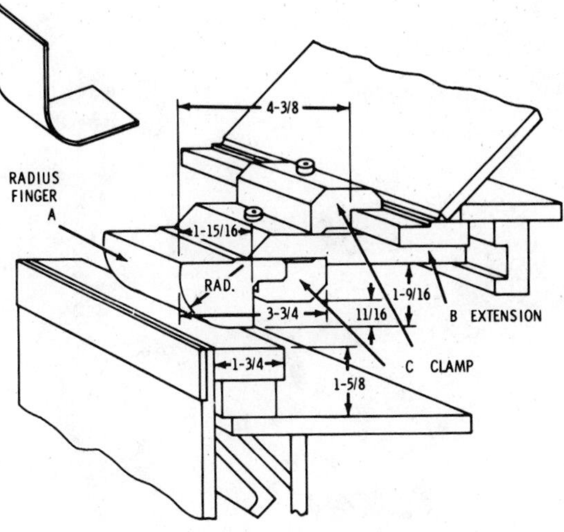

Courtesy Whitney Metal Tool Company

Fig. 6-13. Radius finger attachment for use on a sheet-metal brake.

the bend, and (3) thickness of the material. All these factors may vary widely.

The effective die opening controls the inside radius of a bend. The inside radius of a standard bend is equal roughly to the thickness of the material. Brake capacity requirements are changed by any variation from the standard inside radius, as shown in Fig. 6-14.

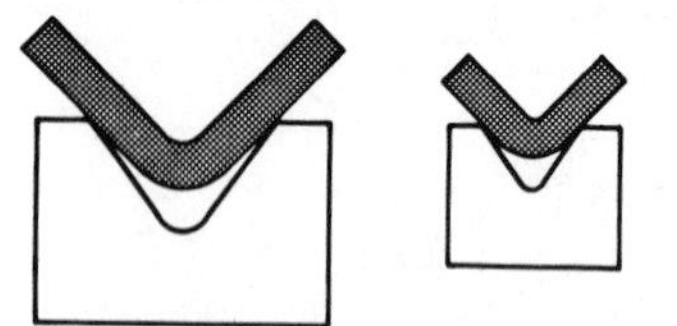

Courtesy Cincinnati Milacron Company

Fig. 6-14. The inside radius of a bend is controlled by the size of the die opening.

The distance between the two housings determines the rated length of a press brake (Fig. 6-15). Die surfaces may extend beyond the housings, so the actual working surface is greater than the rated length of the machine. The flange width is limited by the depth of the throat.

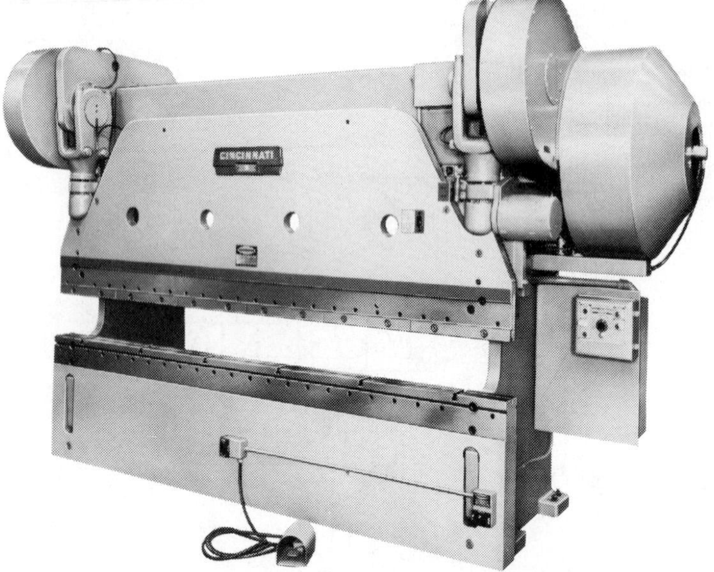

Courtesy Cincinnati Milacron Company

Fig. 6-15. Standard mechanical press brake. This is a 225-ton machine with 10 feet between housings.

A press brake can vary in thickness rating. Multiple bends in the thinner sheets of metal may require a brake of larger capacity than single bends in a thicker sheet of metal, as illustrated in Fig. 6-16.

Many forming operations can be performed on the press brake.

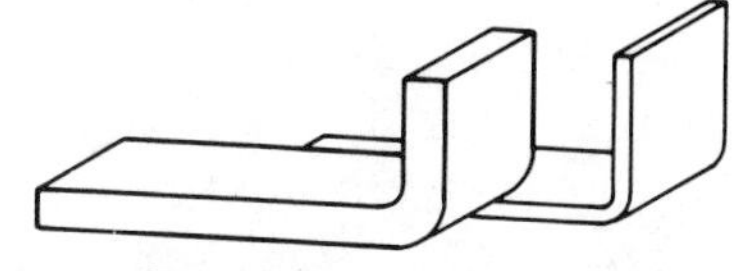

Courtesy Cincinnati Milacron Company

Fig. 6-16. Thickness rating of a press brake varies. Multiple bends on thin metal can require a larger capacity brake than a single bend on a thicker piece of metal.

These operations usually employ dies. Several successive operations may be required to complete the work (Fig. 6-17). Rims for large steel gears can be formed (Fig. 6-18), and tapers or cones can be formed accurately on the press brake (Fig. 6-19). Corrugating is still another operation that can be performed on the press brake (Fig. 6-20).

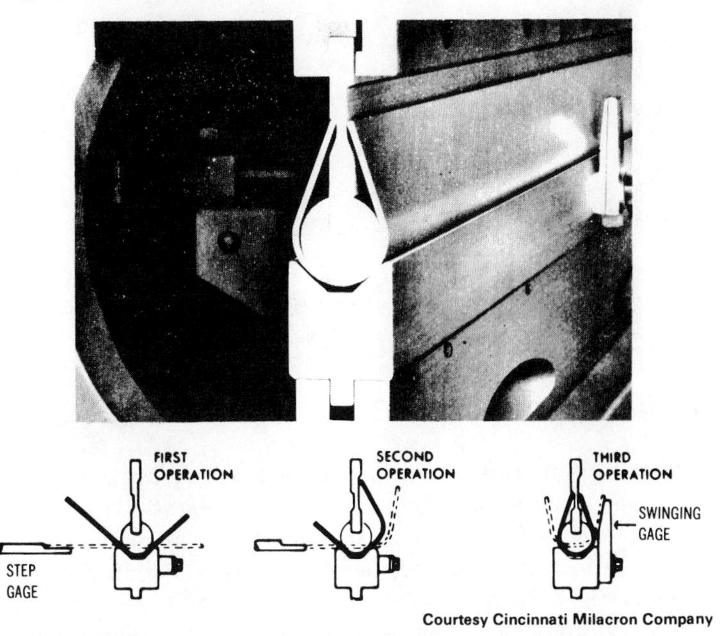

Courtesy Cincinnati Milacron Company

Fig. 6-17. Close-up view of forming a handrail and stiffener for a playground slide (top). Progressive steps in forming the handrail are shown below.

Courtesy Cincinnati Milacron Company

Fig. 6-18. A rim for a large steel gear is an example of work that can be formed on the press brake.

Fig. 6-19. A typical forming operation performed on the press brake.

Other forming operations that require a progressive series of operations to complete them are beading (Fig. 6-21) and tube forming (Fig. 6-22). The successive operations in radius bending are shown in Fig. 6-23, and the forming of a standing seam is shown in Fig. 6-24. All these operations require the use of dies.

Another type of forming operation is a deep-drawing process that can be used to form metal into simple and complex-shaped parts (Fig. 6-25). These parts can be produced on machines that

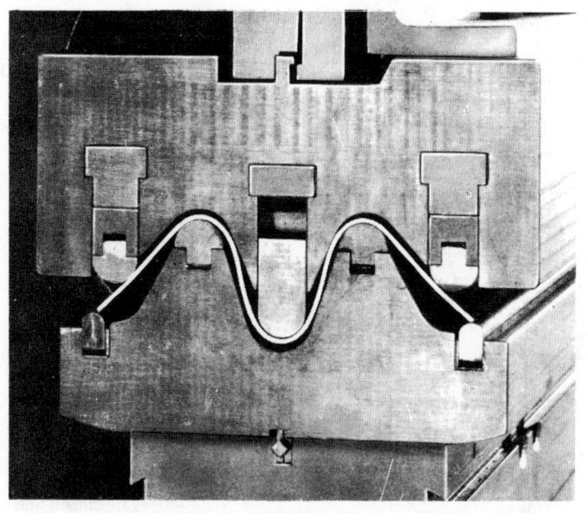

Courtesy Cincinnati Milacron Company

Fig. 6-20. A steel road guard being formed on a press brake.

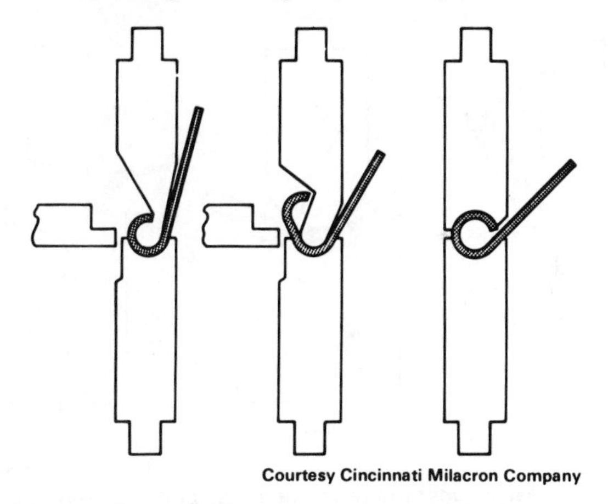

Courtesy Cincinnati Milacron Company

Fig. 6-21. Progressive operations on a press brake for forming a bead.

can be arranged for automatic handling of the work and for a push button automatic forming cycle; therefore, the deep-drawing process is adaptable to a high production rate for certain kinds of forming operations.

Both mechanical and hydraulic brakes are used extensively. Two hydraulic press brakes are shown in Fig. 6-26. Each machine

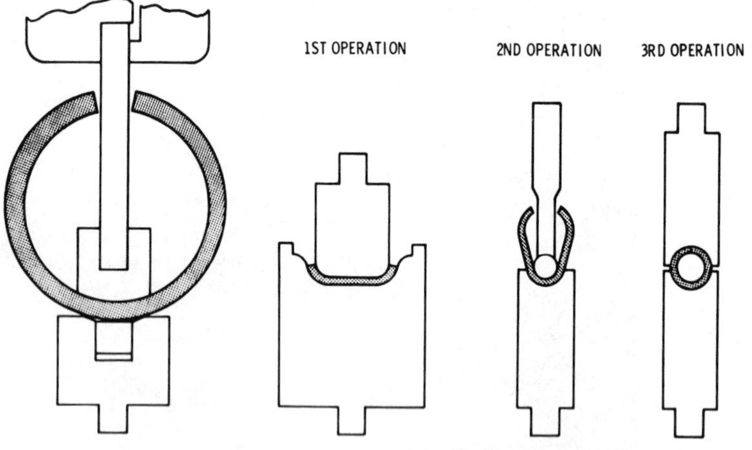

Courtesy Cincinnati Milacron Company

Fig. 6-22. Progressive operations on the press brake for forming a tube.

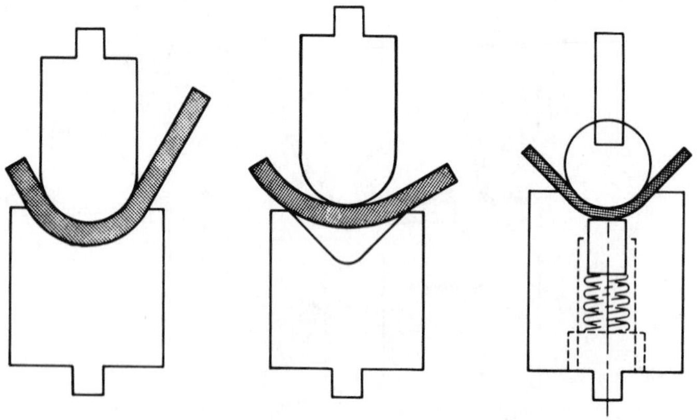

Courtesy Cincinnati Milacron Company

Fig. 6-23. Progressive operations on the press brake for forming a radius bend.

has a 2000-ton, 20-foot capacity. Connected as a single machine, it has a 4000-ton capacity and a 56-foot overall die length area.

SUMMARY

Metal-forming machines bend sheet metal, pipes, and tubes into various shapes such as arcs, spirals, cylinders, and cones. These

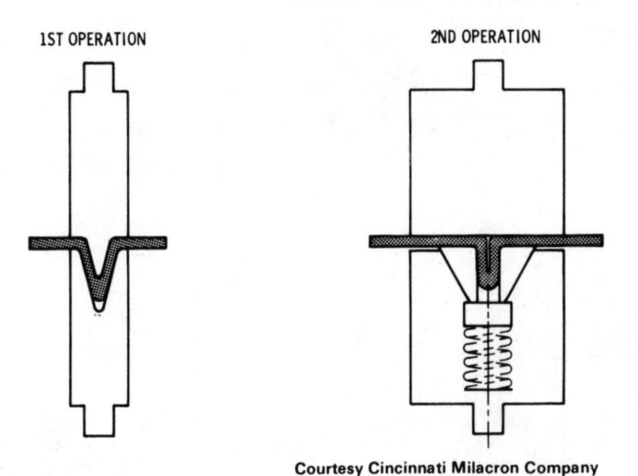

Courtesy Cincinnati Milacron Company

Fig. 6-24. Forming a standing seam in two progressive operations on the press brake.

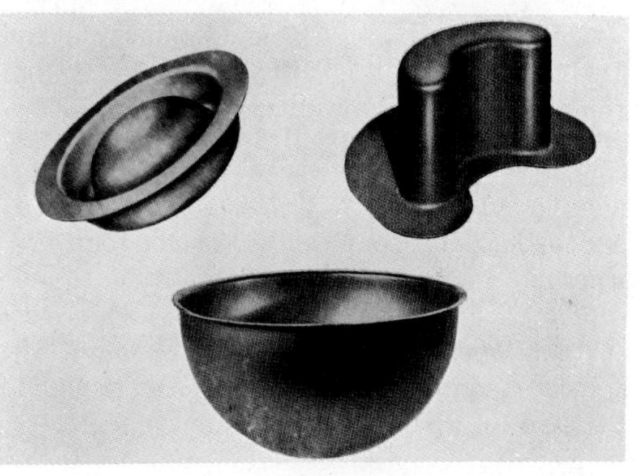

Courtesy Cincinnati Milacron Company

Fig. 6-25. Articles produced by a deep-drawing process on a Hydroform machine.

machines vary from small hand-operated forming rolls to the large motor-driven machines.

The standard hand-operated forming rolls are 30 inches in width and are used to bend sheet metal to a curved form. The distance between the geared rolls can be regulated with adjusted screws by raising or lowering the lower front roll, depending on the sheet

Courtesy Cincinnati Milacron Company

Fig. 6-26. A large hydraulic press. This one has a 4000-ton forming capacity.

metal, to a smaller radius; or it can be lowered to form a cylinder with a larger radius.

Hand-operated sheet-metal brakes are widely used in shops. Mechanical and hydraulic press brakes are in use where large pieces of sheet metal are to be handled. Even though tonnage is sometimes used to express the capacity of a brake, it is the flywheel energy that is required to exert the pressure in the working stroke. The capacity of the brake required to form a single bend in a given material depends on length of bend, sharpness of bend, and thickness of material.

REVIEW QUESTIONS

1. What size is the normal hand forming machine?
2. Explain the operations of a burring machine, a setting-down machine, and a grooving machine.

3. What is corrugating metal and why is it done?
4. Name a few shapes that can be made on a punch press.
5. How does the construction of a bar folder and a box and pan brake differ?

CHAPTER 7

Materials

In order that shop operations may be performed efficiently, every machinist and metal worker should have a general knowledge of the nature and properties of the materials with which he works.

Materials are generally classified as metallic or nonmetallic. Metallic materials are subdivided into a nonferrous category (e.g., copper, aluminum, and titanium) and a ferrous category (e.g., iron, steel, and various alloys). Nonmetallic materials include inorganic materials (such as ceramics, glass, and graphite); and organic materials (such as wood, rubber, and plastics).

PROPERTIES

A material is said to possess certain properties that define its character or behavior under various conditions.

Desirable Properties

Both static strength and dynamic strength are desirable properties in any material. Low cost is always desirable. Especially in the casting process, low cost may determine the material to be used, even though it may have some poor characteristics. For example, in cast metals the following characteristics are desirable:

1. Low melting temperature.
2. Good fluidity when melted.
3. A minimum of porosity.
4. Low reduction in volume during solidification (shrinkage).

Definition of Properties

Frequently used terms for expressing the properties of metals are:

1. *Brittle*—breaks easily and suddenly with a comparatively smooth fracture; not tough or tenacious. This property usually increases with hardness. The hardest steel is the most brittle; and white cast iron is more brittle than gray iron. The brittleness of castings and malleable work is reduced by annealing and/or tempering.
2. *Cold short*—the name given to a metal that cannot be worked under the hammer, or by rolling, or that cannot be bent when cold without cracking at the edges. Such a metal may be worked or bent at a high heat, but not at a temperature that is lower than dull red.
3. *Cold shut*—in foundry work when, through cooling, the metal passing round the two sides of a mold does not properly unite at the point of meeting.
4. *Ductile*—easily drawn out; flexible; pliable. Material, as iron, is "ductile" when it can be extended by pulling.
5. *Elastic limit*—the greatest strain that a substance can endure and still completely spring back to the original shape when the strain is released.
6. *Fusible*—capable of being melted or liquified by the action of heat.
7. *Hardness*—the ability to resist penetration or scratching.
8. *Homogeneous*—of the same kind or nature; hence, homo-

genous, as applied to boiler plates, means even grained. In steel plates there are no layers of fibers, and the metal is as strong one way as another.

9. *Hot short*—more or less brittle when heated; as hot-short iron.
10. *Melting point of a solid*—the temperature at which solids become liquid or gaseous. All metals are liquid, at temperatures more or less elevated, and they turn into gas, or vapor, at very high temperatures. Their melting points range from 39 degrees below zero Fahrenheit, the melting, or rather the freezing, point of mercury, up to more than 3000 degrees.
11. *Resilience*—the act or quality of elasticity; the property of springing back, or recoiling, upon removal of a pressure, as with a spring. Without special qualifications, the term is understood to mean the work given out by a spring, or piece, stressed similarly to a spring, after being stressed to the extreme limit within which it may be stressed repeatedly without rupture or receiving *permanent set.*
12. *Specific gravity*—the weight of a given substance relative to an equal bulk of some other substance which is taken as a standard of comparison. Water is the standard for liquids and solids, air or hydrogen for gases. If a certain mass is weighed first in air, then in water, and the weight in air divided by the loss of weight in water, the result is the specific gravity; thus, taking a 10-pound piece of cast iron, its weight suspended from the scale pan in a bucket of water is 8.6 pounds; dividing 10 by the difference 10 minus 8.6, or 1.4, the result is 7.14, which is the specific gravity of cast iron.
13. *Strength*—power to resist force; solidity or toughness; the quality of bodies by which they may endure the application of force without breaking or yielding.
14. *Tensile strength*—the greatest longitudinal stress a substance can bear without tearing apart.
15. *Toughness*—having the ability to absorb energy without failure; capable of resisting great strain; able to sustain hard usage. Material, such as iron, is said to be "tough" when it can be bent first in one direction, then in the other, without fracturing. The greater the angle it bends (coupled with the number of times it bends), the tougher it is.

METALS

A metal is any chemical element, such as iron, gold, or aluminum, which when dissolved in an acid solution and in a pure state carries a positive charge and seeks the negative pole in an electric cell. Metals are generally good conductors of heat and electricity and are generally hard, heavy, and tenacious.

Ferrous Metals

The ferrous materials are those materials that contain iron. The machinist has long been concerned with the useful properties of iron.

Iron—Pure iron (ferrite) is a relatively soft element of crystalline structure. Pure iron solidifies at 2782°F., the temperature remaining at that point for a very short period of time, depending on the rate of cooling and the mass of the metal. Then the temperature drops to 1648°F., where another pause occurs. On further cooling to 1416°F., the temperature again remains constant for a short time. No further pauses occur as it is cooled from 1416°F. to atmospheric temperature.

Certain changes take place in pure iron as it cools. Pure iron can exist in four solid phases having different physical characteristics. These forms of pure iron are known as:

1. *Alpha iron.* This iron is soft, magnetic, and incapable of dissolving carbon. Alpha iron occurs between atmospheric temperature and 1416°F.
2. *Beta iron.* This phase of iron is feebly magnetic at the higher temperatures, and nonmagnetic at the lower end of the range. It is intensely hard and brittle, and has almost no action on carbon. Beta iron occurs at 1416°F. to 1648°F.
3. *Gamma iron.* In this phase, the iron readily takes up carbon, especially as temperature increases. If gamma iron is cooled quickly past the critical point, the passage of the hard gamma iron to the soft alpha iron is retarded; the iron is then in an unstable hardened condition, ready to pass into the soft alpha form of iron.

   The presence of many foreign substances, such as carbon, nickel, and manganese, seems to help gamma iron resist passing into alpha iron; thus, the hard gamma iron is more stable and permanent at lower temperatures.

On the other hand, the presence of chromium, tungsten, aluminum, silicon, phosphorus, arsenic, and sulfur facilitates the passage of hard beta iron to the soft alpha form. As to hardness, gamma iron lies between that of alpha iron and beta iron. Gamma iron occurs between the temperatures of 1648°F. and 2554°F.

4. *Delta iron.* This form of iron has very little use. The liberation of heat at 2554°F. indicates that in changing from the delta to the gamma phase of iron, the internal structure of the metal changes. The appearance of a critical point at 2554°F. indicates that iron is in different states both above and below that point. Delta iron occurs between 2554°F. and 2782°F.

Pig iron is a compound of iron with carbon, silicon, sulfur, phosphorus, and manganese. The carbon content of pig iron is from 2 to 4.5 percent. This occurs in two forms, partly in solution or combined, and partly distributed throughout the mass in the form of graphite or uncombined carbon.

Cast iron generally cannot be formed and shaped by pressure, and rolled or drawn into shapes that are useful. It is remelted pig iron. The carbon content is over 2 percent, which indicates that it is not malleable at any temperature. Cast iron is widely used in industry for castings. There are four types of cast iron:

1. *Gray cast iron.* This is the kind most used in ordinary castings. This soft cast iron contains a high percentage of graphite, which renders it tough with low tensile strength; it breaks with a coarse-grained dark or grayish fracture. The color is due to the presence of flattened flakes of graphite scattered throughout the material. If the flakes of graphite are large and numerous, the tensile strength is low. The size and amount of graphite flakes are determined by their formation during solidification. If solidification occurs rapidly, less carbon will separate as graphite; therefore, there will be increased hardness because of increased amounts of combined carbon. Gray cast iron contains 2.5 to 3.5 percent carbon.

2. *White cast iron.* Iron of somewhat lower carbon content (2.0 to 2.5 percent) is called white iron. This iron completely retains its combined carbon throughout the casting. Therefore, graphite is not formed, resulting in a casting of extreme hardness and brittleness. Where hardness is desired, and brittleness can be tolerated, white cast iron can be used for machine parts.

   If some graphite formation has occurred, darker patches appear in the white iron. This indicates reduced hardness in those areas. Such an iron is called *mottled* iron.
3. *Malleable cast iron.* Where it is desirable that complicated parts of a machine be ductile, malleable cast iron may be used rather than forged parts. Malleable cast iron castings may be bent or distorted, within the limits of the material, without breaking.

   Castings are made of hard, brittle, white cast iron, and annealed, that is, converted into malleable iron. In the annealing process, the excess carbon is eliminated by heating in an extended heat treatment at about 1650°F. for several hours. The carbon in the form of graphite is absorbed, converting the cast iron into a form of steel.
4. *Wrought iron.* By definition, wrought iron is a low-carbon steel containing a considerable amount of slag. It differs from steel in the method of manufacture in that it is not entirely molten. It contains 1 to 2 percent of slag. White iron is used to produce wrought iron; the impurities are removed by the puddling process.

   The presence of sulfur causes wrought iron to be brittle or "red short" when hot. The presence of phosphorus causes the iron to be "cold short" at ordinary temperatures. Wrought iron softens and welds at 1600°F.; it can be forged at still lower temperatures.

Steel—This is a general term that describes a series of alloys in which iron is the base metal and carbon is the most important added element. The simple steels are alloys of iron and carbon, containing no other elements, and ranging from 0 to 2.0 percent carbon. The "pure" steels (iron and carbon alloy) have never been made in quantity.

The commercial *plain carbon steels* contain manganese as a third alloying element, and small quantities of silicon, phosphorus, sulfur, and traces of other elements. The term plain carbon steel is used for steels containing from a few hundredths of a percent carbon to 1.4 percent carbon. The properties depend on both the carbon content and the heat treatment.

Low-carbon steels are ordinarily used either in the "rolled" condition or in the annealed condition, while high-carbon steels are used where extreme hardness is desired. Increasing the content of steel up to a certain percentage increases its strength; beyond that point the strength decreases. For example, mild steel containing 0.1 percent carbon has a tensile strength of about 50,000 pounds per square inch. A carbon content of 1.2 percent increases the tenacity to nearly 140,000 pounds per square inch, which is probably the limit for carbon steel; a 2.0 percent carbon content gives it a tensile strength of about 90,000 pounds per square inch. A further gradual increase in carbon content causes the material to acquire rapidly the characteristics of cast iron.

Plain carbon steels seldom contain more than 1.4 percent carbon. A carbon content of 2.0 percent is the theoretical upper limit. Various elements other than carbon are added to give steel its desired properties. The effects of adding these elements are as follows:

1. Phosphorus enhances the hardness of steel and makes it better able to resist abrasion. Steel high in phosphorus is weak against shocks and vibratory stresses. Thus, phosphorus is considered a harmful impurity in steel boiler plate.
2. Sulfur interferes with the shaping and forging of steel because it increases the brittleness of steel while hot, making it "red short." Sulfur content of steel should not exceed 0.02 to 0.05 percent.
3. Manganese increases the strength, hardness, and soundness of steel. If a considerable proportion of manganese is present, steel acquires a peculiar brittleness and hardness that makes it difficult to cut. It has a neutralizing effect on sulfur.
4. Nickel increases both the strength and toughness of steel.
5. Aluminum improves the soundness of ingots and castings.
6. Vanadium renders steel nonfatigable. It gives great ductility,

high tensile strength, and high elastic limit, making the steel highly resistant to shocks.

Vanadium steels contain 0.16 to 0.25 percent vanadium. These steels are specially adapted for springs, car axles, gears subjected to severe service, and for all parts that must withstand constant vibration and varying stresses.

Vandium steels containing chromium are used for many automobile parts—springs, axles, drive shafts, and gears. Most chrome-vanadium steels contain 0.20 to 0.60 percent carbon. Many heat-treated forgings are made from these steels.

7. Molybdenum is sometimes specified for high-speed steel. Molybdenum steel is suitable for large crankshafts and propeller shafts, large guns, rifle barrels, and boiler plates.

High-speed steel is so named because it can remove metal faster when used for cutting tools, as in the lathe. The high-speed steel retains its hardness at higher temperatures. Such tools may operate satisfactorily at speeds that cause the edges to reach a red heat.

These steels contain 12 to 20 percent tungsten, 2 to 3 percent chromium, usually 1 to 2 percent vanadium, and sometimes cobalt. The carbon content is within rather narrow limits (usually 0.65 to 0.75 percent). The most used steel is referred to as 18-4-1 steel, which means that its content is 18 percent tungsten, 4 percent chromium, and 1 percent vanadium. Another favorite is 14-4-2 steel. Steels containing 18 percent tungsten are best for most purposes, but steels lower in tungsten content are somewhat cheaper.

Stainless steel resists oxidation and corrosion when correctly heat treated and finished. It is not absolutely corrosion resistant. Stainless cutlery and surgical and dental instruments contain 12 to 14 percent chromium.

Cast steel used in steel castings is stronger than cast iron. Steel castings made of stainless steel resist oxidation at temperatures up to 1800°F. or higher, depending on the chromium content.

Cast steel does not pour as sharply as iron. The shrinkage for steel castings is greater than that for iron castings because of the high temperature of pouring the steel. In machine design, abrupt changes in wall sections should be avoided.

Nonferrous Metals

Nonferrous metals are metals other than iron. In addition to the metals already mentioned, the important nonferrous metals are copper, zinc, tin, antimony, lead, and aluminum.

Copper—This is one of the most useful metals in itself and in its various alloys, such as brass or bronze. Copper has a brownish-red color and both ductile and malleable; it is very tenacious and one of the best conductors of heat and electricity.

Pure copper melts at 1980°F.; commercial copper melts at 1940°F. The strength of copper decreases rapidly with rise of temperature above 400°F.; its strength is reduced to about half, between 800° to 900°F. The heat conductivity of copper is greater than that of all other metals, except silver; it is also next to silver in electrical conductivity.

Zinc—In the form of ingots, zinc is called spelter. Zinc is brittle at ordinary temperatures; it is ductile and malleable between 212° and 300°F., and it again becomes brittle at 410°F.

When zinc is used for lining cisterns and for coating iron water pipes or sheet iron for making cisterns, and when the water with which it comes in contact is soft and contains a slight acid, the metal is gradually corroded or eaten away. Zinc tarnishes when exposed to moist air and is corroded when in contact with soot and moisture. Zinc is used for eaves, gutters, etc. When rolled into thin sheets, zinc is useful on roofs because of its lightness and ease of handling.

Tin—The melting point of tin is 450°F. It is ductile and easily drawn into wire at 212°F. (the boiling point of water). Tin has a low tenacity, but it is very malleable and can be rolled into very thin sheets. Tin is used as a protective coating for iron and copper, and for lining lead pipes used for conveying drinking water, because of its high resistance to tarnishing when exposed to air and moisture.

Antimony—A hard brittle metal that resembles tin. It combines readily with other metals, forming alloys that are extensively used commercially.

Lead—Other metals have largely replaced lead in the plumbing industry. Lead is the heaviest of the common metals; it melts at 621°F. Lead is soft enough to be cut with a knife; it is malleable and ductile, but compared with other metals, it is not a good conductor of heat or electricity. Lead has a low tensile strength. It is extensively used to alloy with other metals for bearings and solders.

Bismuth—This is a remarkable metal because of two properties: its specific gravity decreases under pressure, and it expands on cooling. The melting point is about 520°F. Bismuth is frequently used with antimony in type metals because it fills the molds completely on solidification.

Aluminum—This metal is the lightest of the common metals. Aluminum occurs in nature in the form of hydrates and silicates, but it is commercially prepared by the aid of electricity from cryolite and bauxite.

The metal is not corroded by atmospheric influences or fresh water, also resisting nitric acid; but it is decomposed by alkalies, in sea water, and by dilute sulfuric acid. Aluminum is malleable, ductile, and a good conductor of heat and electricity. Thermal expansion of aluminum is slightly more than twice that of steel and cast iron.

Refractory Metals

Tantalum, tungsten, and molybdenum are commonly called refractory metals because their melting points are above 2000°C. (3632°F.). Because of their high melting points and their reactivity at extremely high temperatures, these metals are produced by powder metallurgy rather than by smelting. Hydraulic presses compact the metal powders to form bars of suitable size and shape for further processing. The bars are fragile and like chalk when they are removed from the press, but they are made into strong metal by the sintering process. The sintering operation consists of heating the bars in furnaces from which the air is excluded. This causes the particles of powder to fuse together without actually melting, and starts a regular metallic crystal growth.

Improvement and development of forming, fabricating, and welding techniques have increased the use of these metals. Examples of these achievements are drawn seamless tubing and the adaptation of the inert gas arc welding process to make welding of tantalum and molybdenum relatively easy.

Tungsten and Molybdenum—The most extensively used metals of the ten refractory metals are tungsten and molybdenum.

Tungsten has the highest melting point of all the metals. Most of the uses of these two metals are due to their high melting points and their ability to retain strength and stiffness at high temperatures.

Electronic tubes use tungsten for filaments, heaters, anodes, and seals through glass; molybdenum is used for grids, anodes, and support members. Tungsten and molybdenum are used for electrical contacts in automotive ignition, vibrators, telegraph relays, and other devices in which the contact parts are in practically continuous service. The pure metals may be used in contacts, but they may be used in combination with silver or copper to form metals having high arc-resisting qualities.

Tungsten is the principal ingredient of Fansteel 77 Metal, a heavy material that approaches tungsten in density but is easily machinable. Fansteel 77 Metal is used for rotors, flywheels, balance weights, and other rotational control parts where maximum weight or inertia, accompanied by high strength, is required for installations with limited space.

Both tungsten and molybdenum are used as heating elements in electric furnaces where working temperatures of 1600° to 2000°C. are required. This is above the range of nickel-chrome alloys. Siliconized coating has been developed to permit the use of molybdenum heating elements up to 1650°C. (3000°F.).

Tungsten electrodes are used to maintain the arc in inert gas of atomic hydrogen welding. Consumption of tungsten electrodes is very low because of the high melting point and low vapor pressure of the metal when used in helium, argon, or hydrogen atmospheres.

Tungsten and molybdenum heating elements are used in vacuum equipment for deposition of thin metallic or nonmetallic coatings by vaporization. Products coated in this manner are mirrors, television tubes, headlight reflectors, photographic lenses, and many similar items.

Tungsten and molybdenum are available in square and rectangular bars, sheet and plate, rods, wire, and metal powder. Molybdenum seamless tubing is available in a wide range of diameters and wall thicknesses. Tungsten carbide powder is also available.

Tantalum—An element best known for its almost complete resistance to corrosion and chemical attack is tantalum. Very few acids have even the slightest effect on tantalum.

Tantalum has the ability to immobilize residual gases in electronic tubes at high temperatures. Other desirable properties for this purpose are its high melting point, low vapor pressure, thermal expansion, and ease of fabrication.

A third important property of tantalum is its ability to form highly stable anodic films. This action, combined with immunity to the corrosive action of electrolytes, is the basis of tantalum rectifiers, arresters, and electrolytic capacitors.

Columbium is a sister metal and occurs in the same ores with tantalum. It has properties generally similar to tantalum, although in a lesser degree.

Most of the metals with high melting points tend to be brittle and difficult to form or fabricate. Tantalum and columbium are exceptions. They are as malleable and ductile as mild steel, and all drawing, rolling, and forming operations are performed with the cold metal.

Because of its easy workability and its immunity to corrosion, tantalum has been widely used for surgical implants in the human body. Sutures are made of braided or monofilament wire. Severed nerves are repaired with fine wire and foil. Tantalum plate is used to repair skull injuries, and woven tantalum gauze is used in hernia operations.

Tantalum capacitors have long been used in telephone service, and the recent trend toward miniature components, along with the advent of television and other electronic applications, has greatly increased the use for tantalum capacitors. These capacitors are made from either sheet or foil.

Tantalum and columbium are available as square or round bars, rods, sheet and plate, foil, wire, powder, and as carbides. Tantalum tubing is available in a wide range of diameters as seamless, butt welded, or seam welded.

Tantalum equipment is recommended for operations involving chlorine or its compounds, including hydrochloric acid. It will not react with bromine, iodine, or other compounds. It has extensive use with sulfuric or nitric acids, hydrogen peroxide, and a large number of other acids, organic or inorganic compounds, and salts under conditions in which any other metal would be corroded quickly or would contaminate the purity of the processed material.

Nonferrous Alloys

A nonferrous alloy is a mixture of two or more metals, containing no iron. The result of such a mixture is generally a mechanical mixture, but it may be combined chemically. With respect to properties, the mixture may be regarded as forming a new metal. The number of possible alloys is unlimited. Some of the more important alloys are considered here.

Brass—There are many varieties of brass. It is a yellow alloy composed of copper and zinc in various proportions. Small percentages of tin, lead, and other metals are included in some varieties of brass. In general, the composition of brass is determined by its desired color. The percentage of zinc in the various varieties of brass is as follows:

Red	5%
Bronze color	10%
Light orange	15%
Greenish yellow	20%
Yellow	30%
Yellowish white	60%

Brass may be classified as (1) high brass and (2) low brass, meaning that the alloy has a high or low percentage of copper. The so-called brasses contain 30 to 40 percent zinc, being suitable for cold rolling. The low brasses contain 37 to 45 percent zinc and are suitable for hot rolling.

The commercial brasses are given various degrees of hardness by cold rolling, being designated as quarter-hard, half-hard, and full-hard. Tensile strength is variable, according to composition and treatment.

Bronze—This nonferrous alloy is an alloy of copper and tin. Many special bronzes have other ingredients. The greater the proportion of tin above 5 percent, the more brittle the alloy becomes.

Bronze is used instead of brass in some instances because of its better appearance and greater strength. A 1 to 6 percentage of tin is specified for bronze that is to be rolled cold and drawn into wire. If it is to be worked at red heat, 6 to 15 percent tin is specified. The percentages of tin specified for the following uses are: machine parts, 9 to 20; bell bronze, 20 to 30; and art bronze, 3 to 10. There are a number of special bronzes, such as phosphor, manganese, gun, and tobin bronze, the properties of which differ, adapting them to special uses.

Aluminum—Although there is no limit to the number of alloys of aluminum that could be produced, commercial manufacturing considerations require that the number of alloys be as small as possible to provide the necessary combinations of properties to meet the needs of industry.

The elements commonly used in the production of casting alloys of aluminum are copper, silicon, magnesium, nickel, iron, zinc, and manganese. The strength of aluminum may be increased by addition of proper amounts of some of these elements. For example, alloys containing magnesium in suitable proportions, as the hardener, are even more resistant to corrosion than the aluminum-silicon alloys.

Aluminum alloy castings are poured both in sand molds and in permanent metal molds. In addition, certain alloys are cast in pressure die-casting machines, which also use metal molds or dies. Permanent molds, or dies, are practical only where a large number of identical castings are required. The minimum number that will justify the production of a metal mold, or die, varies greatly with the nature of the casting.

Babbitt metal—This is an alloy of tin, antimony, and copper, discovered in 1839 by a goldsmith of Boston named Issac Babbitt. The United States granted Babbitt $20,000 for the right to use his formula in government work, and Massachusetts Charitable Mechanics Association awarded him a gold medal in 1841. Babbitt's formula is a good one. Unfortunately, competition and high-priced materials have encouraged adulteration, and the genuine formula is not always followed unless the alloy is subject to chemical analysis.

Other Nonferrous Alloys—Some of these alloys are used for special purposes in industry.

Monel metal is an alloy of copper and nickel and a small percentage of iron. Its melting point is 2480°F., and it may be forged at 165° to 1100°F. An important use for monel metal is in ship propellers.

Muntz metal is an alloys containing 60 percent copper and 40 percent tin. It can be used for purposes in which a hard sheet brass is desirable.

Tobin bronze is an alloy containing 58 to 60 percent copper, about 40 percent zinc, and a small percentage of iron, tin, and lead. Its tensile strength is about 60,000 pounds per square inch.

Delta metal is similar in composition and properties to Tobin bronze.

White metal is a term applied to various alloys containing mainly zinc and tin, or zinc, tin, and lead. It is used for bearings.

Tantung is a trade name for a series of alloys that have great hardness, strength, and toughness, and resistance to wear, heat, impact, corrosion, and erosion, even at extremely high temperatures. These alloys are composed chiefly of cobalt, chromium, and tungsten, with either tantalum or columbium carbide and other components added. A carbide of either tantalum or columbium imparts a low coefficient of friction, a self-lubricating action which minimizes wear.

One of these alloys, *Tantung G,* is widely used in tipped lathe tools and milling cutters and in solid bits. It is available in rods and bars that can be used as tool bits or converted into punches, rollers, drills, and other special tools or wear-resisting parts.

NONMETALS

A plastic is a nonmetallic material that can be readily molded into intricate shapes. There are hundreds of plastic products on the market, and new plastic products are being developed continually.

Bakelite was one of the early products, and is used for many purposes. It is a phenol resinoid having high mechanical resistance.

Formica is a laminated plastic product having many uses. It can be purchased in almost any size and shape desired—tubing, bars, rods, sheets, etc.

Many other plastic products are on the market, such as acetate, vinyl, nylon, polyethylene, *Teflon*, and many others. All these products have properties that adapt them either to a specific item or a variety of items.

TESTS OF MATERIALS

The purpose for which a material is to be used determines the kinds of properties that the material must possess. Steel used in construction of bridges, buildings, and certain types of machines should possess strength, toughness, and elasticity as desirable properties. In some kinds of tool manufacture, a high degree of hardness, in addition to strength and toughness, may be required. The mechanical properties of metals may be changed by using alloy elements and by heat treatment. Specific tests have been devised to evaluate the properties of metals so that materials can be selected wisely.

Elasticity and Plasticity

These two properties are similar in that they both indicate deformation of a material while under load. Elasticity is the property that permits a material to be deformed under a load but causes it to return to its original shape when the load is removed (Fig. 7-1).

Plasticity is similar to elasticity in that the material is deformed under load, but the material retains some of the deformation after the load has been removed (Fig. 7-2). This indicates that the elastic limit of the material has been exceeded.

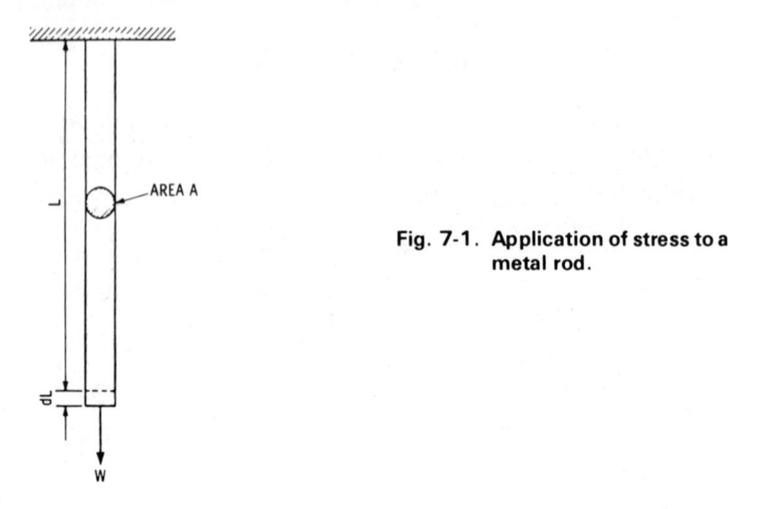

Fig. 7-1. Application of stress to a metal rod.

Stress and Strain

Stress is the load per unit area (pounds per square inch). If the weight *W* has caused the metal rod to stretch a small amount, indicated by *dl*, the rod will return to its original length *L* when the weight is removed if the metal has elasticity. However, if the bar is longer than the original length after the weight has been removed, the metal has exhibited plasticity (Fig. 7-1).

If the weight *W* is increased further and causes a reduced cross-sectional area in the rod, the rod may break at this point. Thus, strain is the deformation per unit of length (measured in inches per inch) that results from a given stress (Fig. 7-2).

This relationship between stress and deformation is known as the *elastic limit*. There is an elastic limit for all materials—a point beyond which complete recovery after stressing is not possible; thus, the material is permanently deformed.

Tensile Strength

Testing machines for tensile strength have been designed so that gradually increasing loads may be applied to the material, together with apparatus for measuring the corresponding deformation. Some of these machines automatically plot the relationship between stress and strain.

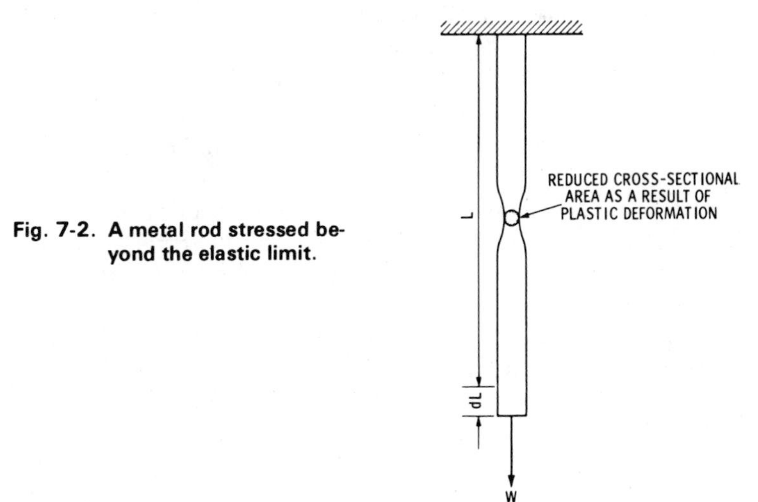

Fig. 7-2. A metal rod stressed beyond the elastic limit.

Ductility

The percent elongation is a measure of ductility. This property is the ability of a metal to permit plastic elongation.

Toughness

Toughness is the ability of a metal to assume deformation without rupture. Medium-carbon steel has a higher degree of toughness than high-carbon steel. However, the maximum strength of high-carbon steel is greater than the medium-carbon steel.

Hardness

Hardness is a property that predicts the behavior of the tested material. There are three main types of hardness: (1) penetration hardness, (2) wear hardness, and (3) rebound hardness.

Commercial hardness testers are widely used for testing metals. They are widely used for inspection; correlation with other properties, such as plasticity, toughness, and tensile strength; and in the specifications of materials for a definite purpose or use.

Brinell Hardness Test—The Brinell hardness tester (Fig. 7-3) has been widely used. A 10-mm hardened steel ball is pressed into the material under a hydraulic load. Depending on the material being tested, a measured load (3000-kg load for iron and steel) is applied for a short time interval (not less than 15 seconds). The diameter of the indentation of the ball is then measured in two directions. The *Brinell hardness number* is a ratio of the applied load and the surface area of the indentation of the ball in the material being tested.

The Brinell hardness test may be considered a destructive test for some conditions. The test indicates that the softer the metal, the larger the indentation of the ball, and the lower the Brinell hardness number.

Rockwell Hardness Test—The operations of the Rockwell hardness tester is similar to the Brinell tester in that an applied load presses a penetrator into the metal being tested (Fig. 7-4). In this tester, a hardened steel ball is used for some tests and a Brale, or diamond-tipped cone, is used on materials too hard to be tested with a ball. Also, a minor load is first applied and then a major load. The minor load produces an initial indentation; the dial is then set at zero, and the major load is applied for a time interval. Hardness numbers can be read directly from the indicating dial.

The Rockwell test is classed as nondestructive because the indentations are so small that it can be used for finished articles. It can be performed rapidly, and the accuracy is good. In the higher hardness range, it is considered more accurate than the Brinell test.

Shore Scleroscope—This instrument determines hardness by dropping a small diamond-pointed hammer on the surface of the material being tested and measuring the height of the rebound (Fig. 7-5). The hardness number is expressed in terms of the rebound distance. For a standard test, the specifications are given for the hammer. The height of rebound is given on a scale. The corresponding number on the scale is the *scleroscope hardness number*. The softer the material, the greater the deformation caused by the striking hammer, and the less energy will be available for rebound.

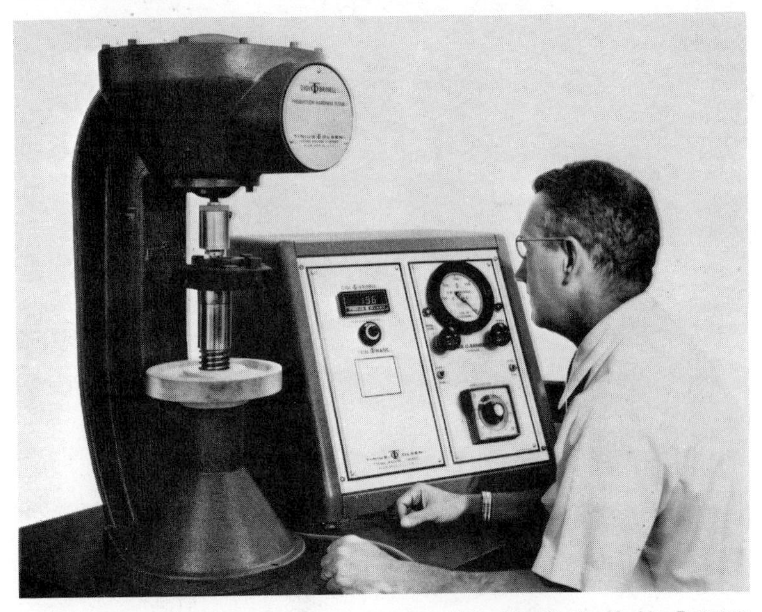

Courtesy Tinius Olsen Testing Machine Company

Fig. 7-3. Air-o-Brinell metal hardness tester with Digito-Brinell system for digital readout of Brinell values.

The machine is portable, and provides a nondestructive method of hardness testing. However, the precautions that must be observed may be a disadvantage for certain conditions.

File Hardness—This is the oldest of the hardness tests. When a sharp file is drawn slowly and firmly across the surface, the material is considered "file hard" if the file does not "bite" into the surface. If the file does bite into the surface, the material is considered softer than file hard. If the file cuts quickly and easily into the surface, the material is soft.

The disadvantages of the file test lie in the fact that there are

Courtesy Wilson Mechanical Instrument Div. of American Chain & Cable Company

Fig. 7-4. Rockwell hardness tester.

differences in the files used and in the operators, and in the fact that the hardness cannot be recorded as numerical data. The advantages, of course, are that the test is cheap, rapid, and nondestructive. A skilled inspector may be able to use the test to discard unsatisfactory pieces, without the use of more expensive and sensitive equipment.

Brittleness

This property is considered the opposite of toughness. A brittle material can undergo little or no plastic deformation. As the hardness of a material increases, the brittleness increases.

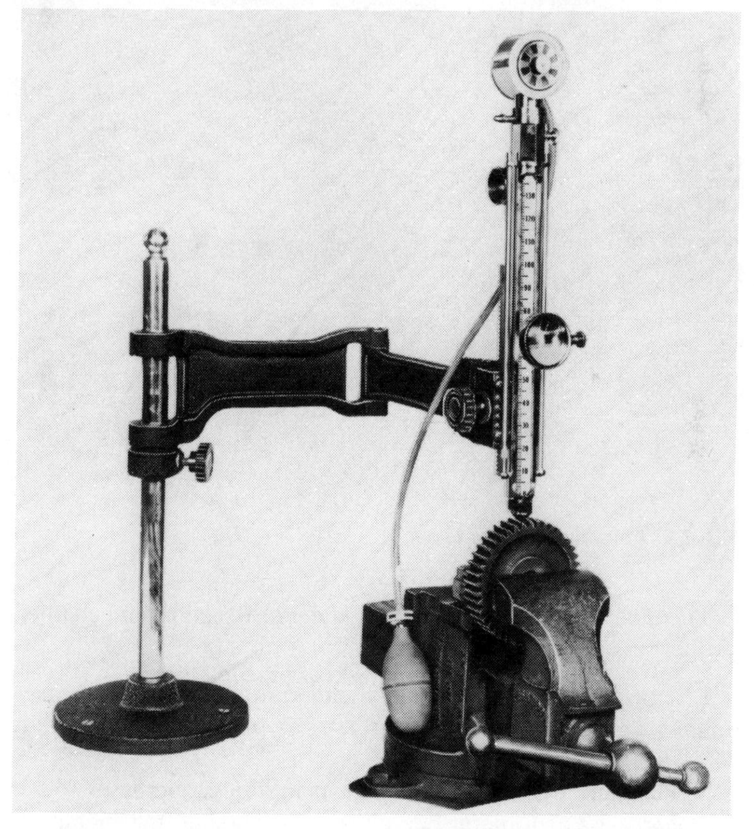

Courtesy The Shore Instrument & Mfg. Co.

Fig. 7-5. Store scleroscope.

Relationship Between Mechanical Properties and Hardness

The mechanical properties, such as elasticity, toughness, plasticity, ductility, and tensile strength, may be indicated by hardness tests of the metal. Fig. 7-6 shows one type of hardness tester. Thus, the prediction of these properties may be of as much value to the engineer as the actual hardness value of the metal.

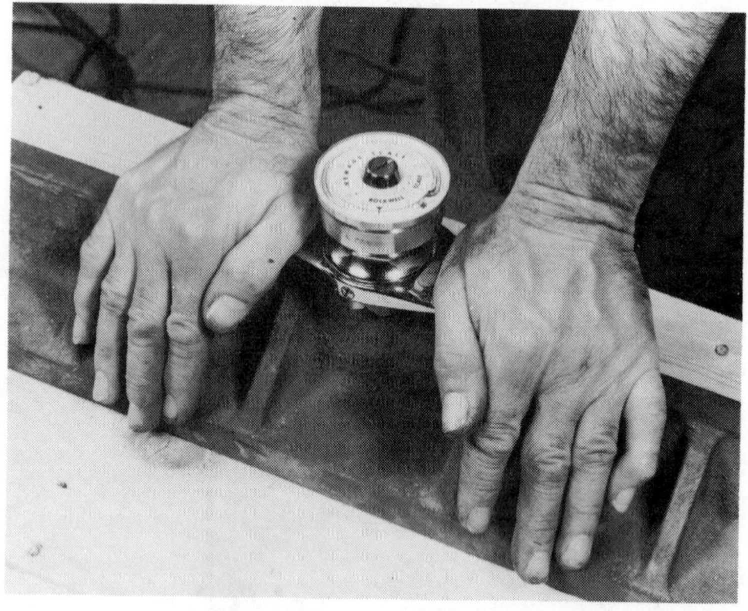

Courtesy Newage Industries, Inc.

Fig. 7-6. Portable hardness tester.

The degree of hardness of a given metal affects the mechanical properties as follows:

1. Toughness can be expected to increase as hardness decreases. In general, toughness decreases as plasticity and ductility decrease.
2. Plasticity of a metal increases as hardness decreases. When hardness of a metal becomes great enough, the metal will rupture before plastic deformation takes place.

3. Ductility of a metal decreases as hardness increases. Decrease in ductility may make fabrication of a metal more difficult.
4. Tensile strength increases as hardness increases.

Effects of Temperature

All the mechanical properties mentioned are affected by temperature. As temperature increases, tensile strength and hardness decrease. As temperature increases, plasticity and deformation also increase. The effects of temperature on the mechanical properties of metals provide the real basis for study of the entire subject of heat treatment of metals.

SUMMARY

Materials most commonly used in shop work are iron, steel, alloys, and plastic. Materials possess certain properties that define their character or behavior under various conditions. Both static strength and dynamic strength are desirable properties in any material. Cost is always a factor and in many cases may determine the material to be used.

A metal is a chemical element such as iron, gold, or aluminum. Metals are generally good conductors of heat and electricity, and are generally hard, heavy, and tenacious. Ferrous materials are those materials which contain iron. Pure iron is a relatively soft element of crystalline structure. Materials such as iron are said to be tough when they can be bent first in one direction, then in the other, without fracturing.

Steel is a general term which describes a series of alloys in which iron is the base metal and carbon the most important added element. Pure steel is iron and carbon alloy only, and has never been made in quantity.

Plastic is a nonmetallic material that can be readily molded into intricate shapes. *Bakelite* was one of the early products in plastic and is still used for many purposes. *Formica* is a laminated plastic product having many uses. Many other plastic products, such as acetate, vinyl, nylon, polyethylene, and *Teflon*, are being used in many ways.

REVIEW QUESTIONS

1. What materials are most commonly used in shop work?
2. Why is iron such a "tough" material?
3. What are the basic elements of steel?
4. Name a few nonmetallic materials used in various products.
5. What are nonferrous alloys, tensile strength, toughness, and hardness?
6. What is Babbit? Where is it used?
7. Do metals conduct heat easily? Why?
8. What is the general term used to describe a series of alloys in which iron is the base metal and carbon the most important added element?
9. What nonmetallic material is used to mold easily into intricate shapes?
10. How is *Formica* used in the workshop by machinists?

CHAPTER 8

Abrasives

Grinding, as we know it today, is not an old art. Most of the development in grinding has taken place since 1890. By definition, an abrasive is a substance such as sandpaper or emery that is used for grinding, polishing, etc.

STRUCTURE OF ABRASIVES

For centuries the only abrasives used were prepared from sandstone. With the development of harder metals and alloys, particularly steel, harder and more efficient abrasives were needed.

Natural Abrasives

Emery (Fig. 8-1) and corundum have been known for a long time as hard, natural minerals. They are naturally occurring forms of aluminum oxide, the corundum having the larger crystals and

(A) Turkish emery has a higher aluminum oxide content.

(B) American emery.

Courtesy American Abrasive Company

Fig. 8-1. Emery is a natural mixture of aluminum oxide and magnetite (Fe^3O^4). It is often preferred to synthetic abrasive for coated abrasive use. Emery is ideal for nonskid surfaces in conjunction with concrete or epoxy resins. It is also used for barrel finishing, pressure blasting, and general polishing operations.

containing fewer impurities. These readily available natural abrasives were used in manufactured grinding wheels until after the turn of the century when they were superseded by more efficient man-made abrasives. An almost negligible quantity of natural abrasives is in use in grinding wheels today (Fig. 8-2).

(A) Silicon carbide abrasive.

(C) Regular aluminum oxide abrasive.

(B) Bauxite, as mined.

(D) A modified form of aluminum oxide abrasive. Pure white in color, and contains about 98.6 percent pure aluminum oxide.

Courtesy Norton Company

Fig. 8-2. Examples of bauxite and fused abrasive as broken from the pig.

The principal sources of emery are Turkey, Greece, and Asia Minor. Corundum first came from India, and later deposits were found in South Africa, Canada, and the southern part of the United States. Shipments from the various localities, and even from the same locality, differed so widely in quality that it caused a variation in the quality of the wheels produced.

Manufactured Abrasives

As grinding became more refined, variation in quality presented a serious problem, which led to experiments in producing manufactured abrasives whose quality could be controlled.

Glue, shellac, and silicate of soda were first used as bonding materials to hold the abrasive grains together in the grinding wheel. Progress in the development of bonding materials was not made until the ceramic clays and firing kilns of the potter were adapted to grinding wheel manufacture.

COMPOSITION OF ABRASIVES

Formerly, all grinding wheels were made of emery. The cutting element in emery is crystalline aluminum oxide. New abrasives have replaced emery in grinding wheels, although it is still used for some forms of grinding.

Silicon Carbide

The commercial use of silicon carbide abrasives was developed by the Carborundum Company. In 1891, Edward G. Acheson, an electrical engineer who lived in Monongahela, Pennsylvania, produced a few ounces of small-sized bright crystals from a mixture of clay and powdered coke which he had heated in a small, crude electric furnace. He found that these crystals would scratch glass like a diamond. Chemical analysis showed the crystals to be silicon carbide (Fig. 8-2A). These crystals were sold for polishing precious gems, at a cost approaching that of the gems. The Carborundum Company developed the use of silicon carbide abrasive for grinding from this discovery.

Aluminum Oxide

Around the same time that silicon carbide abrasive was first produced, Charles B. Jacobs, chief engineer for Ampere Electro-Chemical Co. at Ampere, New Jersey, set out to make a synthetic corundum. *Bauxite* (Fig. 8-2B) was fused electrically into a hard material (crystalline aluminum oxide) similar to emery and corundum, but with the advantage that it could be produced in uniform grade and was of much higher purity (93 to 94 percent aluminum oxide). Its use in grinding wheels and as a polishing abrasive was developed by Norton Company. Today, approximately 75 percent of all grinding wheels are made with aluminum oxide abrasive of one type or another making it the most widely used abrasive for grinding wheels (Fig. 8-2C and 8-2D).

Diamonds

Actual synthesis of the diamond was achieved in 1955 by the General Electric Company, although the basic principles of the diamond's formation, heat and pressure, had been known for many years. Diamonds are the hardest materials found in nature, and attempts had been made for many years to reproduce them in laboratories.

Natural diamonds were developed for use in grinding wheels in the early 1930's to grind tungsten carbide, a material so hard that it resists the abrasive action of ordinary grinding wheels. The development of man-made diamonds for grinding wheels since 1959 has assured industry of a reliable supply and freedom from price fluctuations of natural diamonds. Diamond wheels are widely used today for grinding carbide, ceramics, glass, stone, and even some tool steels (Fig. 8-3).

USE OF ABRASIVES IN GRINDING WHEELS

Grinding is the process of disintegrating a material and reducing it into small particles of dust by crushing or attrition. Many of the newer abrasives have been developed for special grinding purposes.

Silicon Carbide Abrasives

Silicon carbide is made from pure silica sand and carbon, in the form of finely ground coke. These materials react when subjected to the high temperatures of the electric furnace to form the silicon carbide crystals, which are widely used today in a large variety of abrasive and refractory products (Fig. 8-4). Following are some of the brand names of silicon carbide abrasives:

1. Crystolon.
2. Carborundum.
3. Carbolon.
4. Carbonite.

Courtesy Norton Company

Fig. 8-3. Grinding a carbide cutter with a diamond grinding wheel.

Courtesy American Abrasive Company

Fig. 8-4. Black silicon carbide. Used for grinding and finishing nonferrous and nonmetallic materials, lapping and polishing, and in several coated abrasive products.

Aluminum Oxide Abrasives

Bauxite is the source of aluminum oxide, Al_2O_3, combined with water and associated with varying quantities of impurities (Fig. 8-5).

Arc-type electric furnaces are used for the manufacture of aluminous abrasives. The finished product is a large "pig" of crystalline aluminum oxide. Brand names of some of the aluminum oxide abrasives are:

1. Alundum.
2. Aloxite.
3. Lionite.
4. Borolon.
5. Exolon.

Diamond Abrasive

Crushed and sized diamonds as the abrasive in bonded grinding wheels came into use in the early 1930's for cutting tools, wear-resistant products, molded boron carbide, quartz, crystal, glass, porcelain, ceramics, marble, and granite. Diamond abrasive particles in grinding wheels enable the wheels to retain their shape, although the particles will dull eventually.

Both natural and man-made diamonds have fields of application in which they excel. However, the advent of man-made diamond abrasive has given users of diamond abrasives a wider selection of diamond grinding wheels adapted to specific uses.

Courtesy American Abrasive Company

Fig. 8-5. Regular aluminum oxide. This is tough abrasive, well suited for general-purpose grinding, nonskid surfaces, barrel finishing, lapping, pressure blasting, and polishing.

Diamond grinding wheels are made in three different bond types as follows:

1. Resinoid. The resinoid bonded diamond wheel is characterized by a very fast and cool cutting action.
2. Metal. The metal bonded diamond grinding wheel has unusual durability and high resistance to grooving.
3. Vitrified. The vitirified bonded diamond grinding wheel has a cutting action comparable to that of the resinoid diamond wheels and a durability approaching that of the metal-bonded diamond wheels.

The conventional manufactured abrasives (silicon carbide and aluminum oxide) are essential for the needs of many industries, not only for grinding and polishing operations, but also for use in other final abrasive products. They are the basic ingredients in grinding wheels, oilstones, pulpstones, abrasive-coated paper and cloth; also in nonslip floors and stair tile, porous plates, refractories, and refractory laboratory ware.

SUMMARY

Very few natural abrasives are used in industry because of the harder metals and alloys produced. Manufactured abrasives are used in grinding because uniformity and composition can be controlled. Aluminum oxide is the most widely used abrasive for the manufacture of grinding wheels.

Synthetic diamonds for use in grinding wheels were developed in 1959. Diamond wheels are widely used today for grinding carbide, ceramics, glass, stone, and even some steel tools.

Silicon carbide is another product used for grinding which is made from pure silica sand and carbon. These materials react when subjected to the high temperatures of the electric furnace to form the silicon carbide crystals widely used in various abrasives.

REVIEW QUESTIONS

1. Why are manufactured abrasives used in today's industry?
2. What is the most popular abrasive material used in industry?
3. Why are synthetic diamonds used in grinding or cutting tools?
4. What is an abrasive?
5. Where do you use abrasives in metalworking?
6. What is meant by bonding in a grinding wheel?
7. How is black silicon carbide used in the finishing of metal?
8. What is the source of aluminum oxide?
9. Why are diamonds used in the finishing of metals?
10. What does "vitrified" mean?

CHAPTER 9

Grinding

By definition, grinding is the process of disintegrating a material and reducing it into small particles of dust by crushing or attrition.

MANUFACTURE OF GRINDING WHEELS

Grinding wheels are manufactured either from sandstone, which occurs in nature, or from man-made abrasives. Sandstone was the chief abrasive material for many years.

Natural Grindstones

The natural grindstones are cut from sandstone, the most common of which is the Berea sandstone of Ohio. Natural grindstones are inexpensive and have been used largely in the glass, cutler, and woodworking edge-tool industries. Manufactured abrasive wheels, however, have largely replaced natural grindstones in these plants because these grinding wheels can be operated at higher speeds and the grain size, hardness, and structure can be controlled.

Manufactured Abrasive Grinding Wheels

Most of these grinding wheels are made from aluminum oxide and silicon carbide and are used in industries requiring modern high-speed grinding equipment.

Preparation of the Abrasive—The aluminum oxide and the silicon carbide abrasive wheels are prepared in a similar manner. The ore from the electric furnace is rough crushed to lumps about 6 inches in diameter. It is then shipped to the abrasive mill where the lumps are reduced to about ¾-inch diameter, or finer, by powerful crushers. Finally, the pieces are reduced in size to grains suitable for use in grinding wheels, coated abrasive products, etc., by passing them through a series of steel crushing rolls.

Any iron impurities that are present are removed by conveying the abrasive grains through magnetic separators. All fine abrasive dust and foreign particles are removed by washing thoroughly with steam and hot water. This operation is important because clean abrasive grains mix more uniformly with the bond. The abrasive is then dried in continuous rotary driers. The abrasive grains are accurately sized, or graded, by passing them over a series of vibrating screens. Standard grain size numerals range from 10 to 600, which refer to the "mesh" of the screen through which a particular grain size will pass. A grain that will pass through a screen having 10 mesh openings per linear inch is called a 10 grit size. This 10 grit screen therefore will have 100 openings per square inch (Fig. 9-1).

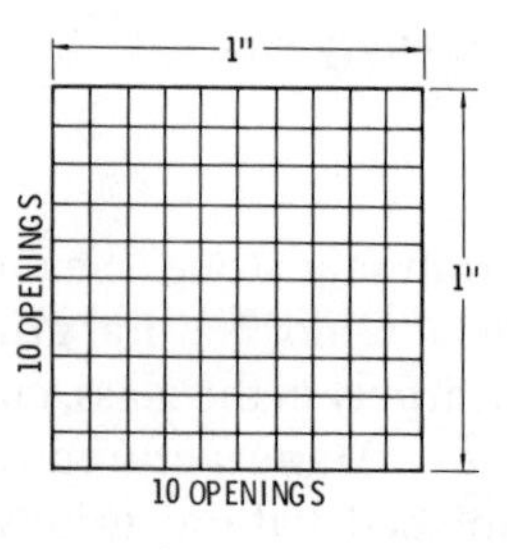

Fig. 9-1. Abrasive grain sizing.

The abrasive grain, as it comes from the mill, is inspected for capillarity, uniformity of size, strength, and weight per unit volume. This represents the finished abrasive product of the mill, and it is either transported directly to the various grinding wheel manufacturing departments or stored in huge storage tanks for later use.

Abrasive grains are uniformly distributed throughout the bond in the wheel. The structure of the grinding wheel refers to this relative spacing as dense, medium, or open, depending on the percentage of abrasive or pores (Fig. 9-2). Wheels of medium structure are best for hard, dense materials; open structures are best for heavy cuts and for soft, ductile materials that are easily penetrated and require good chip clearance.

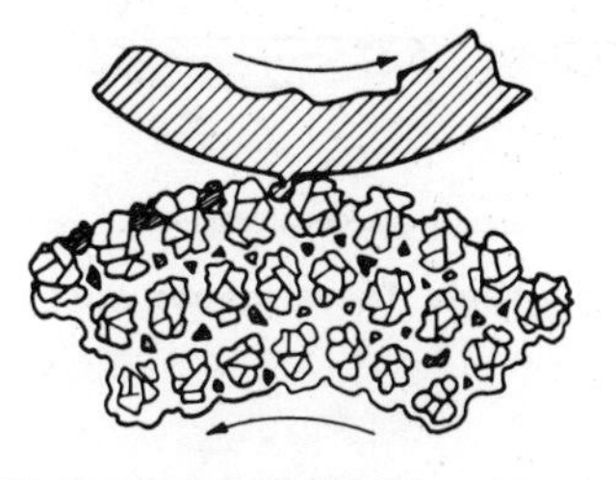

(A) Proper spacing provides effective chip clearance.

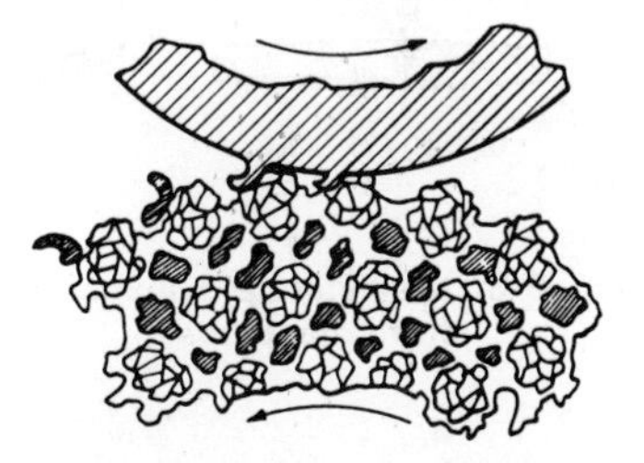

(B) Too-close spacing causes the wheel to load easily, and it will not cut effectively.

Courtesy Cincinnati Milacron Company

Fig. 9-2. Correct spacing of the abrasive grains within the grinding wheel is important for fast free-cutting action.

Shapes of Grinding Wheels—Grinding wheel manufacturers have standardized nine shapes and twelve faces for grinding wheels (Figs. 9-3 and 9-4). These wheels are made in a wide range of sizes to do most grinding jobs.

A great variety of "mounted pieces" and mounted wheels are made for precision grinding of small holes and for offhand grinding, as on dies (Fig. 9-5). These mounted wheels and "mounted points" vary in size from 1/16 to 1 1/2 inch in diameter.

Method of Mounting Grinding Wheels—Grinding wheels should be checked for balance before mounting. Out-of-balance wheels set up excessive vibration, causing chatter marks on the ground surface and excessive wear on the bearings and spindle.

Wheels with small holes are held by a single nut on the end of the spindle. This nut should be tightened firmly, but not excessively,

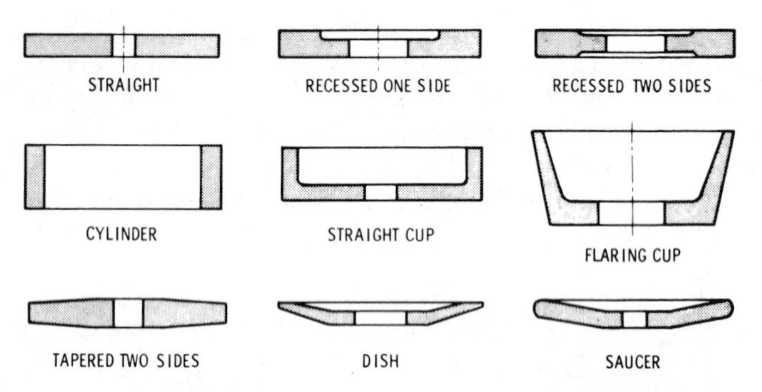

Courtesy Cincinnati Milacron Company

Fig. 9-3. Nine standard shapes for grinding wheels. These shapes will perform most jobs.

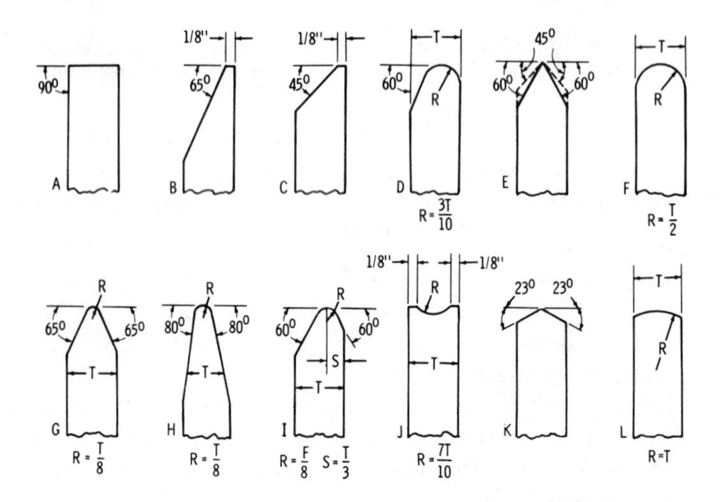

Courtesy Cincinnati Milacron Company

Fig. 9-4. Standardized grinding-wheel faces. These faces can be modified by dressing to suit the needs of the user.

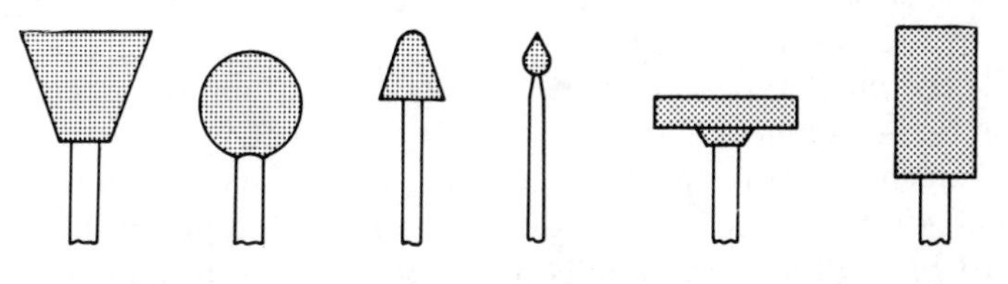

Courtesy Cincinnati Milacron Company

Fig. 9-5. Mounted points are tiny grinding wheels permanently mounted on small-diameter shanks. They may be as small as $\frac{1}{16}$ inch in diameter.

to avoid setting up excessive strains in the wheel. Large-hole wheels are held in place by flange screws around the sleeve. These should be drawn tight with the fingers. Then, the diametrically opposite screws should be tightened with a wrench (preferably a torque wrench) until all screws have been tightened uniformly, but not excessively. These screws should be checked occasionally for looseness. Grinding wheels should be mounted correctly for safe operation (Fig. 9-6).

Truing and Dressing the Grinding Wheel—A wheel should have enough of the cutting face removed, in preparation for grinding, to have it running true with its own spindle. When a wheel

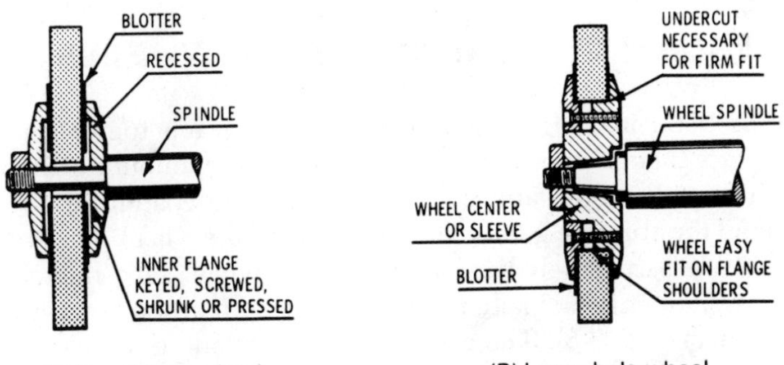

(A) Small-hole wheel.

(B) Large-hole wheel.

Courtesy Cincinnati Milacron Company

Fig. 9-6. Correct method of mounting a grinding wheel. Never omit the blotting-paper washers.

becomes loaded or glazed, it is then "dressed" to restore its original sharp and clean cutting face.

Three chief types of wheel dressers are in use on precision-grinding machines. These are: the diamond tool, abrasive wheel, and mechanical dresser. The diamond tool is most commonly used on all kinds of precision grinders (Fig. 9-7). All types of dressers must be held in firm mounts on the machine to true or dress a wheel accurately. This may be either an integral part of the machine or a special attachment for the machine. In many automatic machines, wheel dressing is part of the automatic cycle.

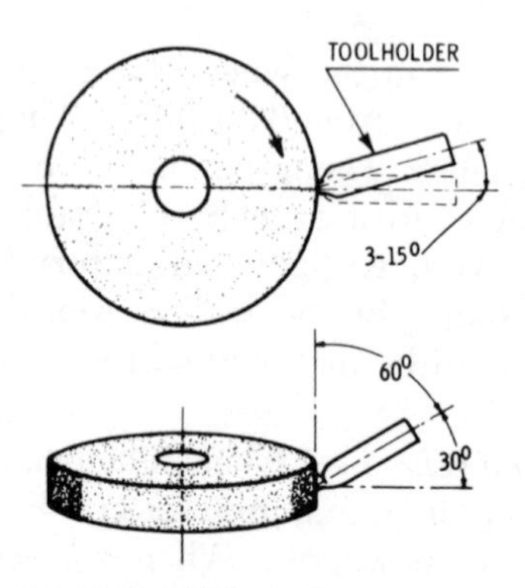

Courtesy Cincinnati Milacron Company

Fig. 9-7. Diamond truing tools must be canted, as shown, to prevent chatter and gouging of the wheel, and to maintain sharpness of the diamond abrasive.

BOND TYPES IN GRINDING WHEELS

The bond in a grinding wheel holds the grains together and supports them while they cut. The greater the amount of bond with respect to the abrasive, the heavier the coating of bond around the abrasive grains, and the "harder" the wheel. The abrasive itself is extremely hard in all grinding wheels, and the terms "hard" and "soft" actually refer to the strength of the bonding of the wheel (Fig. 9-8). The general types of grinding wheel bonds are:

1. Vitrified.
2. Organic.
   a. Resinoid.

b. Rubber.
c. Shellac.
3. Silicate.

Vitrified-Bond Grinding Wheels

About two-thirds of all the grinding wheels manufactured are made with a vitrified bond composed of clays and feldspars, and selected for their fusibiltiy (Fig. 9-9). During the "burning" process of manufacture, a temperature of 1270°C. is reached, which is sufficiently high to fuse the clays into a molten glass condition. Upon cooling, a span or post of this glass connects each abrasive grain to its neighboring grains to make a rigid, strong grinding wheel (Fig. 9-10).

Courtesy Norton Company

Fig. 9-8. Abrasive grain and bond are mixed in power mixing machines to ensure a homogeneous blending.

Organic-Bonded Grinding Wheels

These wheels are made with an organic bond such as resinoid, rubber, or shellac. Resinoid wheels are used primarily in high-speed rough grinding operations and make up by far the largest percentage of organic-bonding grinding wheels. Straight wheels are used widely on bench and pedestal grinders for offhand grinding of castings. Cup wheels and cones are used for cleaning castings and for weld grinding with portable grinders. Reinforced cutoff wheels are used on both cutting-off machines and portable grinders to remove gates and risers from casting, for cutting bar stock, and for cutting concrete blocks and other masonry materials.

Rubber-bonded grinding wheels are used as regulating wheels in centerless grinders to control the travel of the workpiece

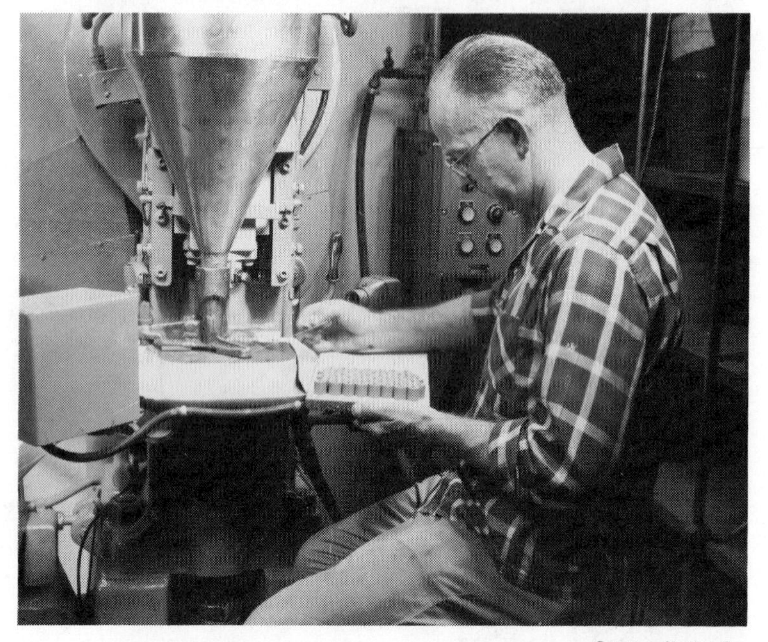

Courtesy Norton Co.

Fig. 9-9. Small grinding wheels are molded in automatic presses. The operator merely removes the pressed wheel between strokes. An electric-eye safety device stops the machine if his hand is under the plunger on the pressing stroke.

Courtesy Norton Company

Fig. 9-10. Grinding wheels emerging from an electrically heated tunnel kiln where the approximately 2500°F. temperature has fused the clay bond into a strong glass.

through the machine (Fig. 9-11). Rubber cutoff wheels are used where there must be a minimum burr left by the wheel. They should always be used with a good flow of coolant, while resinoid-bonded cutoff wheels can be used dry. Resilient rubber wheels are used in the hollow grinding of cutlery, as well as on work where finish is important.

Shellac-bonded wheels find their greatest used in grinding the rolls used for making steel, paper, plastics, etc. (Fig. 9-12). Shellac-bonded cutoff wheels are used to some extent in toolrooms for cutting off the ends of broken taps and drills. Burning of the heated-treated tool steel must be avoided and wheel wear is of minor importance.

Silicate-Bonded Grinding Wheels

These wheels derive their name from the bond, which is principally silicate of soda. Silicate-bonded wheels are considered rela-

Courtesy Norton Company

Fig. 9-11. Rubber-bonded wheels are used for the regulating wheels on centerless grinding machines; also for thin cutoff wheels.

tively "mild acting" and are still used to some extent in the form of large wheels for grinding edge tools in place of the old-fashioned sandstone wheels.

GRINDING WHEEL MARKINGS

A standard system for marking grinding wheels was adopted, in 1944, by the various grinding wheel manufacturers throughout the country. Each marking consists of six parts, arranged in the following sequence:

1. Abrasive type.
2. Grain size.
3. Grade.
4. Structure.

5. Bond type.
6. Bond modification symbol.

Abrasive Type

Manufactured abrasives fall into two distinct groups. Letter symbols are used to identify these two groups as follows:

1. A—Aluminum oxide.
2. C—Silicon carbide.

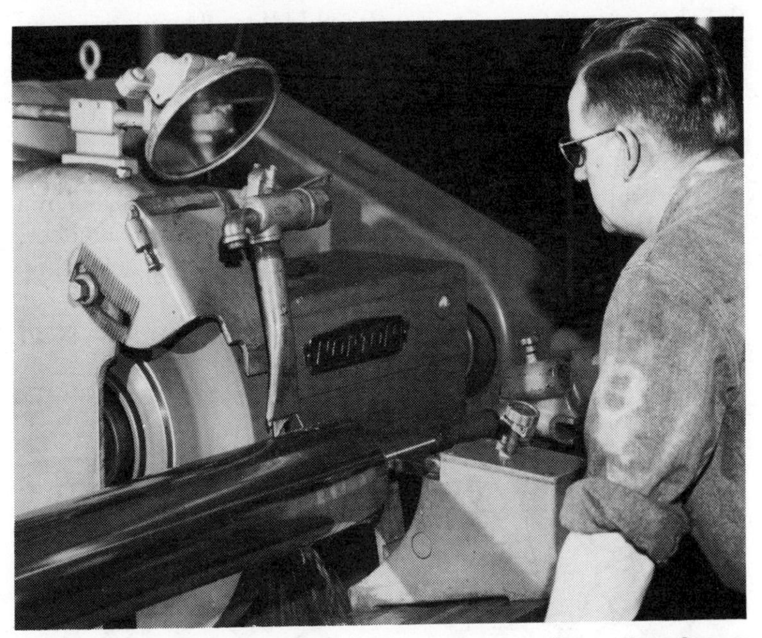

Courtesy Norton Company

Fig. 9-12. Shellac-bonded wheels are widely used for cutlery grinding and high-finish grinding.

In order to designate a particular type of either silicon carbide or aluminum oxide abrasive, the manufacturer may use his own symbol or brand designation as a prefix. Examples: Norton Company's 32A, 57A, 37C, or 39C.

Grain Size

Grain size in wheels is indicated by numbers ranging from 10 (coarse) to 600 (fine). If necessary, the manufacturer may use an additional symbol to the regular grain size. This is illustrated in the examples: 461 (46 grit, No. 1 combination) and 364 (36 grit, No. 4 combination).

Grade

The grade of a grinding wheel is indicated by a letter of the alphabet ranging from A to Z (soft to hard) in all bonds or manufacturing processes. (Grades A to H are soft; grades I to P are medium; and grades Q to Z are hard.)

Structure

Structure, or grain, spacing, in a grinding wheel is indicated by a number, generally from 1 to 12. The progressively higher numbers indicate a progressively "more open" or wider grain spacing (Fig. 9-13).

For example, a 3 structure indicates a dense or close grain structure; an 8 structure indicates a wide grain spacing. The letter

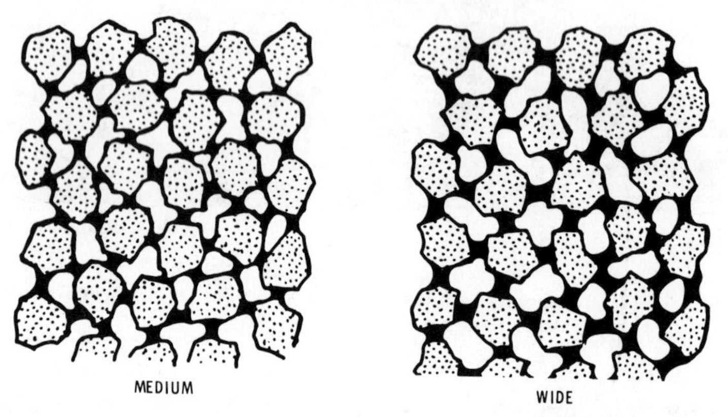

Fig. 9-13. Grinding-wheel structures showing medium (left) and wide (right) grain spacings.

"P" at the end indicates a "P" type product having extra large pores or voids. Examples: 32A30 - E12VBEP and 39C601 - H8VKP.

Bond or Process

Bond is designated by the following letters:

1. V—Vitrified.
2. B—Resinoid (synthetic resins such as Bakelite).
3. R—Rubber.
4. E—Shellac.
5. S—Silicate.

Bond Modification Symbols

This may describe a particular type of bond that distinguishes it as a variation from a basic bond. The position may be either omitted or shown according to the characteristics of the specified wheel. Examples: VG (Norton "G" type of vitrified bond) and B11 (Norton "11" type of resinoid bond).

FACTORS AFFECTING GRINDING WHEEL SELECTION

Selection of a grinding wheel for a given operation depends on a number of important factors. Some grinding wheels are designed for special purposes (Fig. 9-14).

Hardness of Material to be Ground

The kind of abrasive to be selected is determined by the nature and properties of the material to be ground.

Abrasive—Aluminum oxide is best suited for grinding steel and steel alloys. Silicon carbide grinding wheels are more efficient for grinding cast iron, nonferrous metals, and nonmetallic materials.

Grit size—Fine grit is best for hard, brittle, and difficult to penetrate materials. Coarse grit is best for soft, ductile, easily penetrated materials.

Grade—A relatively soft-grade grinding wheel is required for very hard, dense materials. Hard materials resist the penetration of the abrasive grains and cause them to dull quickly; a soft-grade wheel enables worn, dull grains to break away and expose newer, sharper cutting grains. Harder-grade grinding wheels should be used for soft, easily cut materials.

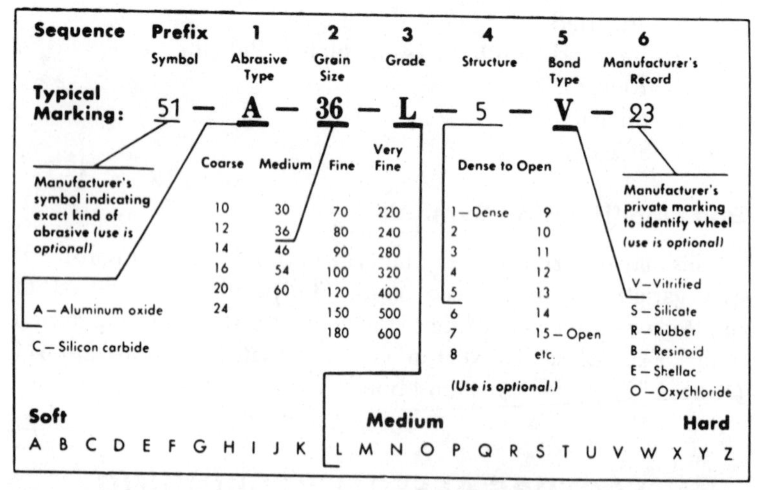

Courtesy Cincinnati Milacron Company

Fig. 9-14. Standard grinding-wheel marking system.

Amount of Stock to be Removed and Finish Required

Grit size and bond are important in selection of a wheel, depending on the amount of stock to be removed and the finish required.

Grit Size—A coarse grit is best for rough grinding or rapid stock removal. A fine grit size is best where close tolerances and a high finish are desired.

Bond—The vitrified bond is best suited for fast cutting and commercial finish. A resinoid, rubber, or shellac bond is usually best for obtaining a high finish.

Operation (Wet or Dry)

As a rule, wet grinding permits the use of grinding wheels at least one grade higher than for dry grinding without danger of burning the work from heat of friction. Water speeds up the work and reduces dust.

Wheel Speed

The grinding wheel speed (Fig. 9-15) is very important in the selection of a wheel and bond as follows:

1. For speeds less than 6500 sfpm, use vitrified-bonded wheels.
2. For speeds above 6500 sfpm, use organic-bonded wheels (resinoid, rubber, or shellac).

Area of Grinding Contact

The grit size and grade are influenced by the area of contact

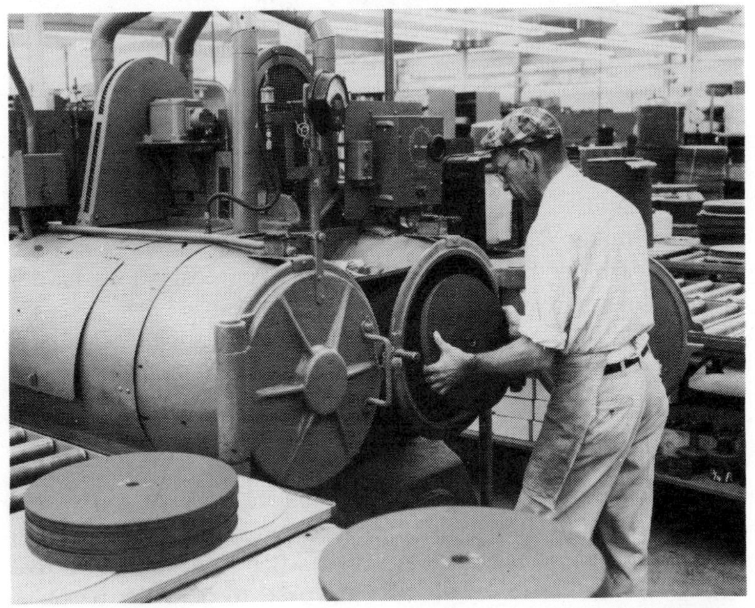

Courtesy Norton Company

Fig. 9-15. As a safety precaution, almost all grinding wheels, 6 inches and larger, are speed tested at 50 percent higher than their recommended operating speeds.

between the wheel and the work. As a general rule, the smaller the area of contact, the harder the grinding wheel should be. Therefore, a coarse grit wheel can be used for a large area of contact, while a fine grit wheel should be used where the area of contact is small.

Severity of Grinding Operation

This factor affects the choice of the abrasive as follows:

1. Use a tough abrasive, "regular" aluminum oxide for grinding steel and steel alloys under severe conditions. The use of a tougher abrasive is limited generally to grinding stainless steel billets and slabs with heavy-duty grinders.
2. Use a relatively mild abrasive for light grinding on hard steels.
3. Use an intermediate abrasive for a grinding job of average severity.

SUMMARY

Grinding refers to the shaping or smoothing of the surfaces of various materials and the shaping of cutting tools. The two basic abrasive materials used in manufacturing grinding wheels are silicone carbide and aluminum oxide. Silicon carbide is harder than aluminum oxide but is more brittle.

Abrasive grains are uniformly distributed throughout the bond in a grinding wheel. The structure of the grinding wheel refers to this relative spacing as dense, medium, or open, depending on the percentage of abrasive or pores. Wheels of medium structure are best for heavy cuts and for soft, ductile materials that are easily penetrated and require good chip clearance.

The size of the abrasive grain is determined by size of the mesh screen through which the abrasive particles will fall. The five types of bonding materials that are used to hold the particles together are: vitrified, resinoid, rubber, shellac, and silicate.

Grinding wheel manufacturers have standardized nine shapes and twelve faces for grinding wheels. These wheels are made in a wide range of sizes to do most grinding jobs. Each wheel should be checked for balance before mounting. Out-of-balance wheels set up excessive vibrations, causing chatter marks on ground surfaces.

The bond in a grinding wheel holds the grains together and supports them while they cut. The greater the amount of bond with respect to the abrasive, the heavier the coating around the abrasive grains, and the harder the wheel. The abrasive itself is extremely hard in all grinding wheels, and the terms hard and soft refer to the strength of the bond in the wheel.

A standard system for marking grinding wheels is used to designate the kind of abrasive, grain size, grade structure, bonding process, and the manufacturer's markings.

REVIEW QUESTIONS

1. Name the two abrasives most commonly used for making grinding wheels.
2. How is the grain size of an abrasive measured?
3. What is meant by the bond of a grinding wheel?
4. Name several common shapes of grinding wheels.
5. What type of abrasive is used to grind nonferrous and nonmetallic materials?
6. What is the most common natural grindstone?
7. How many shapes and faces are there for grinding wheels?
8. List at least six grinding wheel standard shapes.
9. Sketch at least four grinding wheel faces.
10. What do out-of-round grinding wheels do?
11. What does hard and soft refer to in terms of grinding wheels?
12. What are the three subdivisions of bonds used in organic grinding wheels?
13. List the six parts of the code used to mark grinding wheels.
14. What determines the kind of abrasive selected for a job?
15. Why is grinding wheel speed important?

CHAPTER 10

Cutting Fluids

Cutting fluids are primarily used in machining operations to reduce temperature and adhesion between the chip and tool. They also serve to keep the work piece cool to avoid thermal expansion and easier handling. When a lubricant is part of the cutting fluid, it provides a rust-proof layer to the finished work surface. Cutting fluids are useful in clearing chips away from the machining area.

COOLANT

A coolant is an agent (usually in liquid form), the sole function of which is to absorb heat from the work and the cutting tool. Water has the highest cooling effect of any cutting fluid. It may be used on materials which are tough or abrasive, but which have a great frictional effect and generate much heat in cutting. Typical materials of this nature are rubber tires and celluloid.

LUBRICANTS

A cutting fluid with the additional property that enables it to act as a lubricant is called a cooling lubricant. These materials usually consist of both cooling and lubricating agents, such as soluble oil or glycerin mixed in proper proportions. Cooling lubricants are used where cutting materials generate excessive heat and are, to a limited degree, tough and abrasive.

APPLICATION OF CUTTING FLUIDS

A stream of soluble oil is frequently used to increase the cutting capacity of an abrasive wheel by preventing it from glazing over, and by carrying off the heat generated by the friction of the wheel on the work. It is important that the coolant be properly applied to the work. A large stream of fluid at slow velocity is preferable to a small stream at high velocity. The cutting fluid should make contact with the work at the exact spot where the cutting action takes place—not above or to one side of the cutting tool (Fig. 10-1).

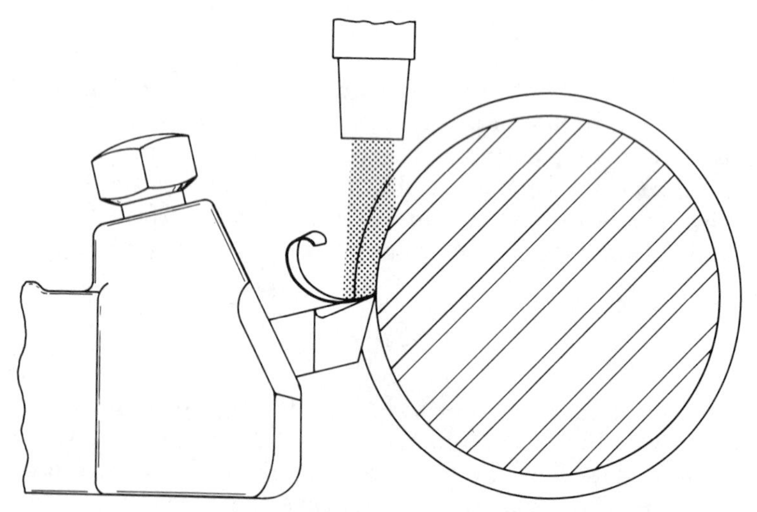

Courtesy South Bend Lathe, Inc.

Fig. 10-1. Cutting fluid should make contact with the work where the cutting action takes place.

Equipment

Limited amounts of cutting fluid may be applied with an oil can. A small paint brush can also be used to apply limited amounts of coolant in many instances.

On machine tools, cutting fluids keep heat from softening and ruining the cutting edge of the cutting tool. A cutting fluid cools the cutting tool and makes it cut more easily and smoothly. The fluid also tends to wash away chips, prevents undue friction, and permits faster cutting speeds.

Production machine tools and many general-purpose machines are equipped with oil pans, pumps, and reservoirs, which circulate the cutting fluid to the point of cutting. Spray-mist coolant systems use a water base fluid which is supplied to the cutting tool by compressed air. The compressed air atomizes the cutting fluid providing considerable cooling but little lubrication.

TYPES OF CUTTING FLUIDS

The characteristics of the most commonly used cutting fluids are as follows:

1. *Lard oil.* This is one of the oldest and best cutting oils. It provides excellent lubrication, increases tool life, produces a smooth finish on the work, and prevents rust. This coolant is especially good for cutting screw threads, drilling deep holes, and reaming. Its industrial application is limited because it is very expensive, tends to become rancid, and can cause skin irritation.
2. *Mineral-lard oil mixtures.* These mixtures are used in various proportions of lard oil and various petroleum base mineral oils in place of lard oil because they are more fluid, less expensive, and almost as effective.
3. *Mineral oils.* Petroleum base oils are compounded with chemicals to improve their lubricating and antiwelding qualities. They are also less expensive than lard oil and mineral oil mixtures.
4. *Water-soluble oils.* Their use is limited to rough turning operations. Although they carry away heat better than lard oil or

mineral oil, their lubricating qualities are poor. They are mineral oils treated so that they may be mixed with water to form an emulsion and provide an excellent low-cost coolant.

5. *Chemical cutting fluids.* These compounds are mixed with water and generally do not contain any petroleum products. The chemicals used provide excellent lubricating and anti-weld properties. These solutions are gradually replacing other cutting fluids in a variety of applications including several machining operations and surface grinding.
6. *Low-viscosity oils.* Oils such as kerosene are used for cutting tough nonferrous metals and alloys—materials such as the bronzes, certain aluminum alloys, and alloys containing a very small percentage of iron, for example, Monel metal. The heavier, more sulfonated oils are used for sawing tough steels and ferrous alloys because these oils are better able to withstand extreme pressure and abrading actions.

SOLID LUBRICANTS

Grease-type lubricants are used to lubricate the sides and back edge of a band saw and the guide bearings, particularly where high velocity is involved. These lubricants are often compounded with graphite and supplied in tubes. On some tubes, a threaded nozzle permits the tube to be screwed into a grease fitting associated with a saw-guide mounting. Contents of the tube can either trickle through or be squeezed through apertures in the mounting. The tube can remain fastened to the mounting to lubricate the blade until the tube is empty.

Wax or Soap Lubricants

The wax or soap lubricants are usually compounded of petroleum substances; many are impregnated with high film-strength lubricants. This lubricant is normally supplied in stick form, such as a cardboard tube, from which it is pushed out for use. A good solid lubricant is recommended for sawing metals, hardwoods, and other material where the use of surface lubricant is preferable to a circulating type.

Cutting oils are usually dripped onto the cutting edge of the tool at a rate of 6 to 40 drops of oil per minute, depending on the material being cut and the speed at which it is cut. Table 10-1 gives the cutting speeds and lubricants for both drilling and turning the various materials.

It is usually more convenient to use the fluid lubricants because they can be circulated automatically with a resultant saving in time. The fluid coolants and lubricants can be applied either in a stream or by dripping or spraying.

The solid lubricants are usually applied directly to the surface. For example, in band filing of nonferrous metals, the file segment may become loaded and difficult to clean. The solid wax lubricants may be applied directly to the filing surface to prevent loading.

SUMMARY

Water is practically the only agent whose sole function is cooling. Coolants are used as cooling agents to carry off heat generated by the friction of the grinding wheel on the work.

Table 10-1. Cutting Speeds and Fluids for Turning and Drilling With High-Speed Steel Cutting Tools

Material	Turning Speed in Ft. per Minute	Cutting Fluid	Drilling Speed in Ft. per Minute	Cutting Fluid
Aluminum	300-400	kerosene	200-330	kerosene
Brass	300-700	dry	200-500	dry
Bronze	300-700	compound	200-500	compound
Cast iron	50-110	dry	100-165	dry
Copper	300-700	compound	200-500	compound
Duralumin	275-400	compound	250-375	compound
Fiber	200-300	dry	175-275	dry
Machine steel	115-225	compound	80-120	compound
Malleable iron	80-130	dry or compound	80-100	dry or compound
Monel metal	100-125	compound	40-55	sulfur base
Plastics, hot-set molded	200-600	dry	75-300	dry
Rubber, hard	200-300	dry	175-275	dry
Stainless steel	100-150	sulfur base	30-45	sulfur base
Tool steel	70-130	compound	45-65	sulfur base

Courtesy South Bend Lathe, Inc.

Table 10-2. Cutting Fluid Recommendations for Sawing Various Materials

Material	Lubricant or Coolant
Aluminum; sheet, rod, bar, press forgings (2S, 3S, 4S, 11S, 17S)	coolant-lubricant
Autobestos and Raybestos (Brake Lining)	coolant-lubricant
Commercial Brass Sheet (SAE No. 70) quarter hard to extra spring	none
Atlas 90 and Auromet 55 Aluminum Bronze	light cutting oil
Felt (woolen or cotton)	none
Granite (Igneous)	coolant-lubricant
Optical Glass	coolant-lubricant
Magnesium Die Castings SAE 501	mineral oil
Bakelite (no filler)	none
Polystyrene	coolant-lubricant
Buna Rubber	glycerine and water
SAE (Carbon Steel) 1006, 1010, 1015, 1020, 1025, 1030	heavy cutting oil
SAE (Chromium-vanadium steel) 6130, 6135, 6140, 6150	heavy cutting oil
SAE (Stainless) 30-303	coolant-lubricant
Kestos Steel	heavy cutting oil
Meehanite	none

Other materials used primarily for cooling or lubricating properties include cutting oils, semisolid greases, and solid lubricants. Not only are cooling lubricants important in carrying away heat, etc., but it is also important that these materials are properly distributed to the cutting edge of the tool and the grinding stone.

REVIEW QUESTIONS

1. What is the purpose of coolants?
2. Name a few materials used primarily for cooling.
3. Name a few solid coolants used.

CHAPTER 11

Cutting Tools

All operators of machine tools should have a basic knowledge of the cutting action of the cutting tools. This understanding is necessary for the cutting tool to be ground properly and applied to the work correctly.

ACTION OF CUTTING TOOLS

Cutting tools employ a wedging action. All the power used in cutting metal is ultimately expended in heat. A tool that has been used on heavy cuts has a small ridge of metal directly over the cutting edge. This bit of metal is much harder than the metal being cut and is almost welded to the edge of the tool, indicating that an immense amount of heat and pressure was developed.

The fineness or sharpness of the edge (the angle of the two sides of the tool which make the edge) depends on the class of work (roughing or finishing) and on the metal being cut. It is not necessary to hone the edge of the cutting tool for heavy roughing cuts in steel. A fine edge lasts for only a few feet of cutting, then rounds off to a more solid edge and remains in approximately the same condition until the tool breaks down.

In high-speed production work, coolants help absorb the heat from the cutting edge of the tool. A steady stream of cutting compound should be directed at the point of the cutting tool so that it spreads and covers both the cutting tool and the work. A large pan collects the cutting compound, carries it to a settling tank, and then to a pump.

MATERIALS

There are several different materials used to make cutting tools or cutter bits. In order to machine metal accurately and efficiently, it is necessary to have the proper lathe tool ground for the particular kind of metal being machined, with a keen, well-supported cutting edge. Some of the materials used to make cutting tools are:

1. *Carbon tool steel.* These cutting tools are less expensive and can be used on some types of metal successfully.
2. *High-speed steel.* Cutter bits made of high-speed steel are the most popular cutting tools. They will withstand higher cutting speeds than carbon steel cutter bits. High-speed tools contain tungsten, chromium, vanadium, and carbon.
3. *Stellite.* These cutter bits will withstand higher cutting speeds than high-speed steel cutter bits. Stellite is a nonmagnetic alloy that is harder than common high-speed steel. The tool will not lose its temper, even though heated red hot from the friction generated by taking the cut.

   Stellite is more brittle than high-speed steel and requires less clearance, or just enough clearance to permit the tool to cut freely, to prevent breaking or chipping. Stellite is also used for machining hard steel, cast iron, bronze, etc.
4. *Carbide.* Tips of cutting tools are made of carbide for manufacturing operations where maximum cutting speeds are desired.

   a. *Tungsten carbide.* These cutting tools are efficient for machining cast iron, alloyed cast iron, bronze, copper, brass, aluminum, Babbit, and the abrasive nonmetallic materials, such as hard rubber, fiber, and plastics. These cutter bits are so hard that they cannot be ground satisfactorily on an ordinary grinding wheel.

b. *Tantalum carbide*. The term tantalum is applied to a combination of tungsten carbide and tantalum carbide. These cutting tools are similar to tungsten carbide tools, but are used mostly for machining steel.

c. *Titanium carbide*. This is a term applied to a combination of tungsten carbide and titanium carbide. Titanium carbide is interchangeable with tantalum carbide in its uses.

SHAPES AND USES OF CUTTING TOOLS

Nine of the most popular shapes of lathe cutter bits and their applications are shown in Fig. 11-1 as follows:

1. *Left-hand turning tool.* The opposite of the right-hand turning tool, this tool is used to machine work from left to right (Fig. 11-1A).
2. *Round-nose turning tool.* The round-nose tool is a convenient tool for turning in either direction and for reducing the diameter of a shaft in the center (Fig. 11-1B).
3. *Right-hand turning tool.* This tool is the most common type of cutting tool for general lathe work. It is used for machining work in which the cutting tool travels from right to left (Fig. 11-1C).
4. *Left-hand facing tool.* This tool is the opposite of the right-hand-side tool, and is used for facing the left-hand side of the work (Fig. 11-1D).
5. *Threading tool.* The point of this tool is ground to an included angle of 60°. It is used to cut screw threads (Fig. 11-1E).
6. *Right-hand facing tool.* This cutting tool is used for facing the ends of shafts and for machining work on the right side of the shoulder (Fig. 11-1F).
7. *Cutoff tool.* This tool is used for various classes of work that cannot be accomplished with a regular turning tool (Fig. 11-1G).

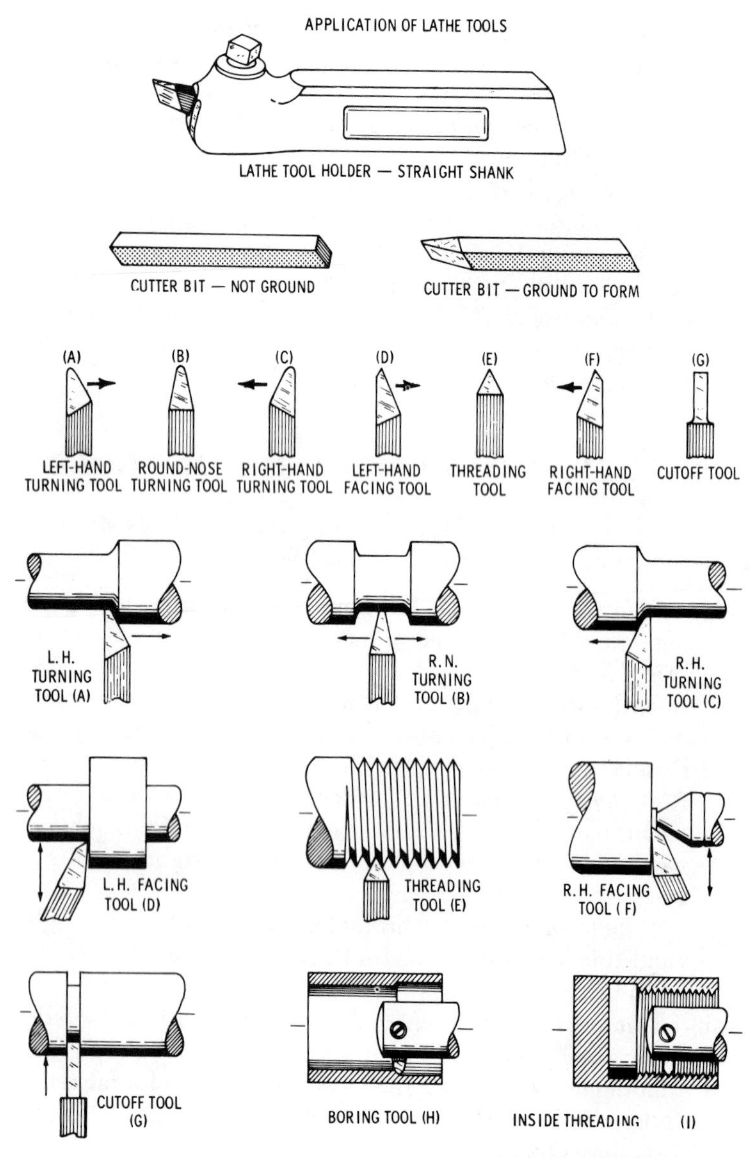

Courtesy South Bend Lathe, Inc.

Fig. 11-1. Nine of the most popular shapes of lathe tool cutter bits and their application.

8. *Boring tool.* This tool is ground the same as the left-hand turning tool, except that the front clearance angle is slightly greater, so that the heel of the tool will not rub in the hole of the work (Fig. 11-1H).
9. *Inside-threading tool.* This tool is ground the same as the screw-threading tool, except that the front clearance angle must be greater (Fig. 11-1I).

Heavy forged turning tools were used formerly, but in more recent years they have been replaced by toolholders and smaller tools made of high-speed steel and known as tool bits. These tool bits are generally used in a standard toolholder which presents them to the work at an angle of 15° to 20° with the horizontal axis of the work. This angle of inclination is important and must always be considered in the sharpening of every tool bit that is to be used in a toolholder.

TERMS RELATED TO CUTTING TOOLS

Base—Surface of the shank that bears against the support and takes the tangential pressure of the cut.

Chip Breaker—An irregularity in the face of a tool or a separate piece attached to the tool or holder to break the chips into short sections.

Cutting Edge—Portion of the face edge along which the chip is separated from the work.

Face—Surface on which the chip slides as it is cut from the work.

Flank—Surface or surfaces below and adjacent to the cutting edge.

Flat—Straight portion of the cutting edge intended to produce a smooth machined surface.

Heel—Edge between the base and the flank immediately below the face.

Neck—An extension of reduced sectional area of the shank.

Nose—A curved face edge.

Shank—Portion of the tool back of the face which is supported in the tool post.

CUTTING TOOL ANGLES

The cutting end of the cutting tool is adapted to its cutting requirements by grinding its sides and edges at various angles. Since the cutting tool is more or less tilted in the toolholder, the angles are classed as either tool angles or working angles.

Tool angles are those angles applied to the tool itself; whereas working angles are those angles between the tool itself and the work. Working angles depend, not only on the shape of the tool, but also on its angular position with respect to the work.

Tool Angles

The cutting tool itself must have both the rake angle and the relief angle ground at the proper angle, depending on the cutting requirements (Fig. 11-2).

Back Rake—All cutting tools employ a wedging action. The differences are in the angle of the two sides of the tool bit that form the cutting edge and the manner in which the cutting tool is applied to the work.

The edge of a pocketknife would be ruined in trying to cut a nail, even though the metal in the knife is much harder than that in the nail. A cold chisel, however, shows no sign of damage in cutting the same nail, although the chisel is usually a poorer grade of steel than the knife. Obviously, the difference is in the angle of the cutting edge.

Back rake is usually provided for in the toolholder by setting the tool at an angle. This angle may vary from 15° to 20°. Back rake may be varied for the material being turned by adjusting the toolholder in the post, through the rocker.

The inclination of the face of a tool is to or from the base. If it inclines toward the base, the rake angle is *positive*. If it inclines away from the base, the rake angle is *negative* (Fig. 11-3).

The cutting angle should be as large as possible for maximum strength at the edge and to carry heat away from the cutting edge. On the other hand, the larger the cutting angle, the more power is required to force it into the work. Thus, a compromise is necessary in arriving at the best rake angles (Fig. 11-4).

Top rake should be less for softer material, as there is a tendency

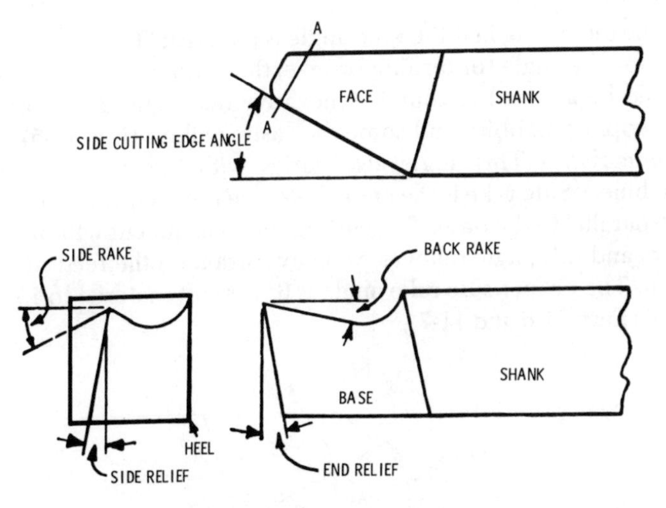

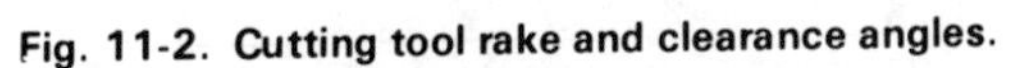
Fig. 11-2. Cutting tool rake and clearance angles.

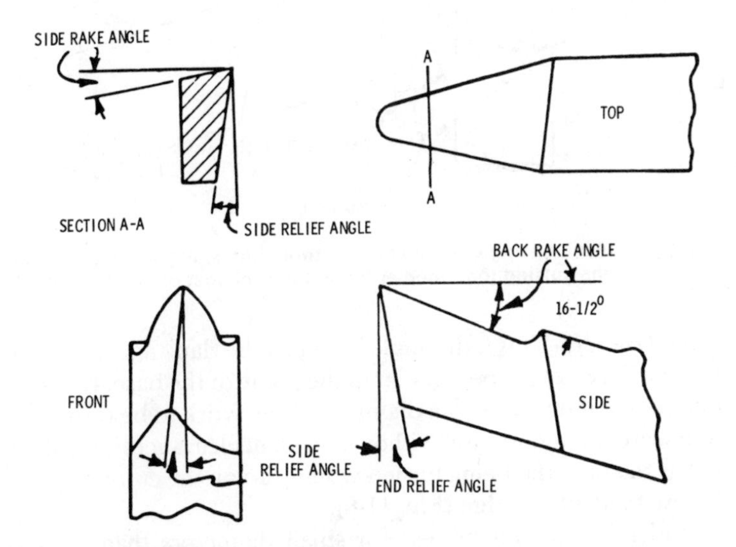

Fig. 11-3. A round-nose, right-hand cutting tool, suitable for roughing and general-purpose turning. By increasing the front rake and by using no side rake, it can be used for either right-hand or left-hand turning.

for the tool to dig in if the rake angle is too great. There should be no top rake angle for turning bronze; the cutting edge of the tool should be almost horizontal. A negative rake is used for turning soft copper, Babbitt, and some die casting alloys (Fig. 11-5).

Side Rake—This angle also varies with the material being machined. Side rake is the angle between the face of a tool and a line parallel to the base. The cutting tool will not cut without side rake, and this angle relieves excessive strain on the feed mechanism. The proper side rake angle is 6° for soft material to 15° for steel (Figs. 11-6 and 11-7).

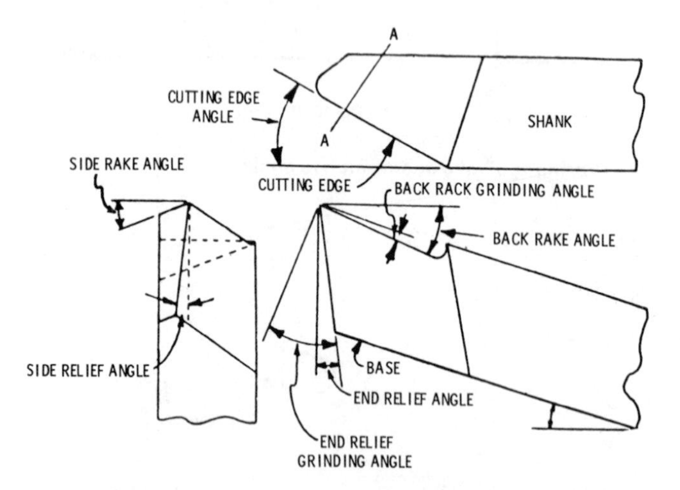

Fig. 11-4. Working angles of the cutting tool, that is, the actual angles with the cutting tool tilted either in the tool post or in the toolholder.

End Relief—This is the angle between the flank and a line from the cutting edge perpendicular to the plane of the base. End relief depends somewhat on the diameter of the work to be turned. If a tool were ground square, without any front clearance, it would not cut as the material being turned would rub on the cutting tool just below the cutting edge (Fig. 11-8).

End relief should be less for small diameters than for large diameters, and may range from 8° to 15°. Do not grind more front clearance than is necessary, as this takes away support from the cutting edge of the tool.

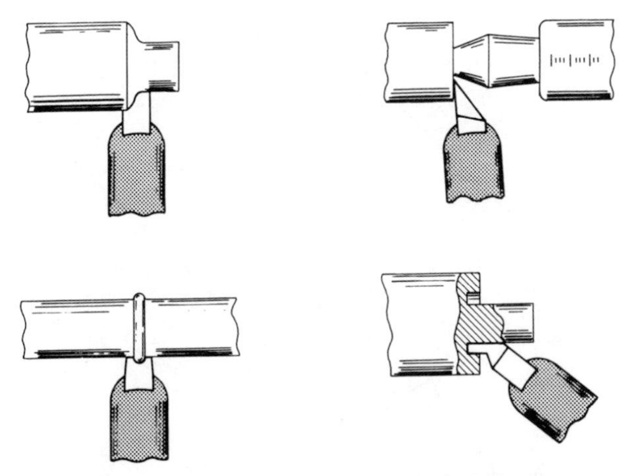

Fig. 11-5. Showing some of the special form cutting tools and their applications. These are called form tools to distinguish them from regular types. In form cutting tools such as A and D, side rake is not used. Front rake, however, should be used, except when turning brass. Form cutting tools wider than $1/8$ inch should not be used on steel. Form cutting tools as wide as $1/2$ inch can be used on brass, aluminum, and similar metals.

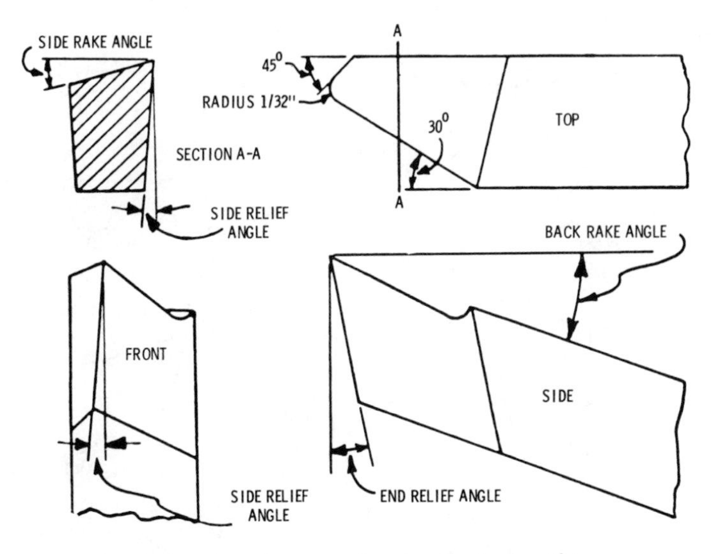

Fig. 11-6. A right-hand turning tool, showing the tool angles.

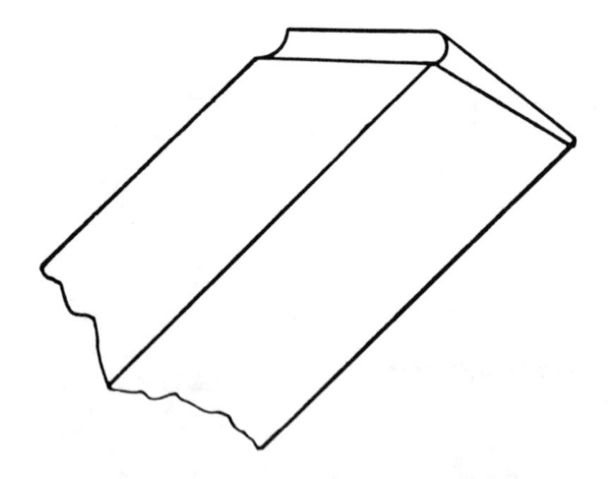

Fig. 11-7. A special roughing tool for heavy-duty work in steel. A large side-rake angle can be obtained without unduly weakening the tool, by grinding a groove along the edge of the tool, instead of grinding away the top of the tool at an angle. The radius of the groove largely determines the diameter of the chip curl. This entire groove should be honed smoothly because the entire face of the chip bears on the surface of this groove, and any roughness increases friction and heat.

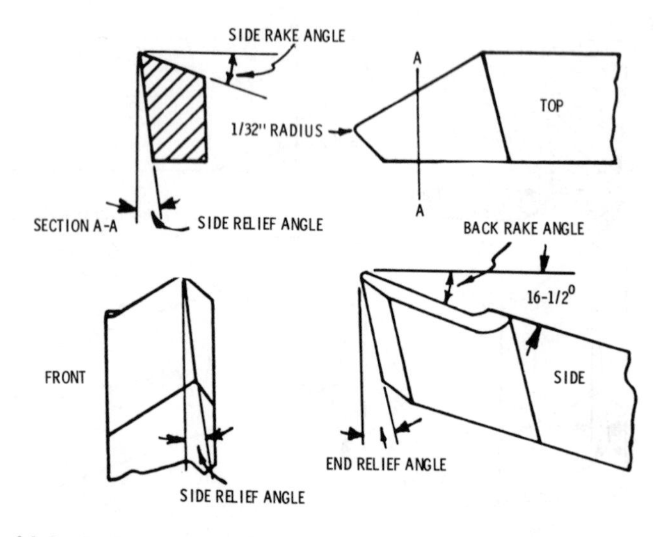

Fig. 11-8. A left-hand turning tool, showing the various tool angles.

Side Relief—This is the angle between the side of a tool and a line from the face edge perpendicular to the plane of the base (Figs. 11-9 and 11-10). In turning, the relief angle allows the part of the tool bit directly under the cutting edge to clear the work while taking a chip.

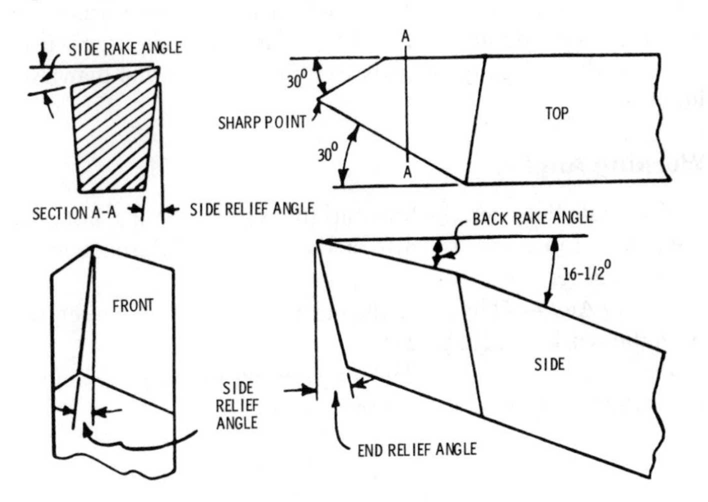

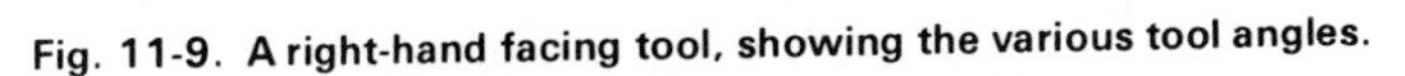
Fig. 11-9. A right-hand facing tool, showing the various tool angles.

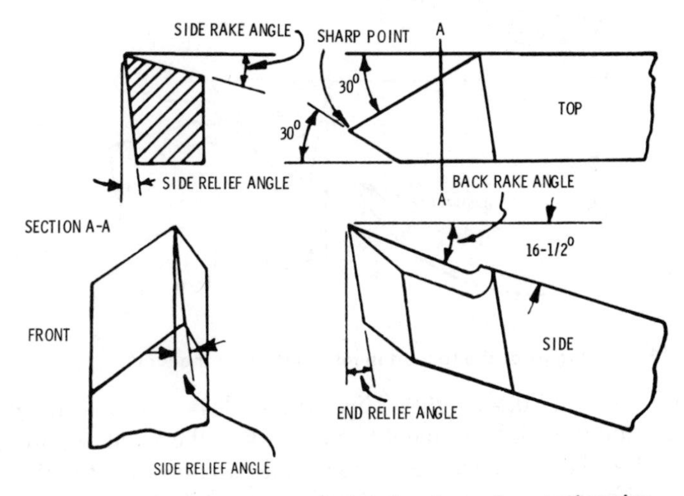

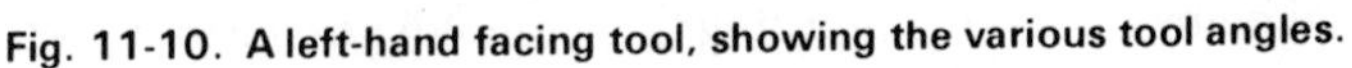
Fig. 11-10. A left-hand facing tool, showing the various tool angles.

Too much relief weakens the cutting edge for clearing the work while taking a chip; the high pressure exerted downward on the tool bit demands that the relief be as small as possible and still allow the tool bit to cut properly. Whenever the tool bit digs into the work, or refuses to cut unless forced, the relief of the tool bit should be checked. Digging in occurs most often during facing and threading operations. For light turning operations, it is usually better to allow slightly more than enough relief rather than risk too little relief.

Working Angles

The angle between the tool and the work depends, not only on the cutting tool angle, but also on the position of the cutting tool in the toolholder.

Setting Angle—The angle the cutting tool axis, when set, makes with the work axis (Fig. 11-4).

True Top Rate Angle—The top rate angle plus the setting angle gives the true top rake angle (Fig. 11-11).

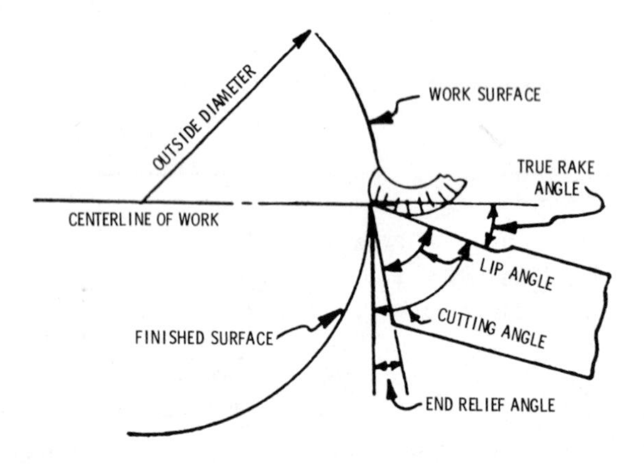

Fig. 11-11. Angles of the tool bit in relationship to the work.

Side Cutting Edge Angle—This is the angle between the face of the cutting tool and a tangent to the work at the point of cutting (Fig. 11-4): Here, again, the cutting angle can be varied to suit the material being turned by adjusting the toolholder in the tool post.

Angle of Keenness—This angle is the included angle of the tool between the face of the tool and the ground flank adjacent to the side cutting edge.

Table 11-1. Rake and Clearance Angles in Degrees

Material	Top Rake	Side Rake	Front Clearance	Side Clearance
Cast iron	5	12	8	10
Stainless steel	16½	10	10	12
Copper	16½	20	12	14
Brass and softer copper alloys	0	0	8	10
Harder copper alloys	10	0-2	12	10
Hard bronze	0	0-2	8	10
Aluminum	35	15	8	12
Monel metal & nickel	8	14	13	15
Phenol plastics	0	0	8	12
Various base plastics	0-5	0	10	14
Formica gear plastic	16½	10	10	15
Fiber	0	0	12	15
Hard rubber	0 or −5	0	15	20

HIGH-SPEED STEEL LATHE TOOLS

Depending on the job, the lip and clearance angles ground on high-speed tools vary in size. The tool is weakened by a clearance angle that is too great, while too little clearance may permit the tool to rub against the work. The lip, which is ground on top of the cutting tool parallel to the cutting edge, causes the chip to curl, and controls the chip. The lip should be ground ⅛ to ¼ inch wide and about ⅙ inch in depth. Coarse feeds and deep cuts require a large lip (Figs. 11-12 and 11-13).

The tool bit should be mounted in the toolholder at an angle of approximately 15½°. A 21° angle must be ground on the front of the tool bit (Figs. 11-14 and 11-15). A finish turning and facing tool may be used for finish turning an outside diameter and facing in one setting (Fig. 11-16).

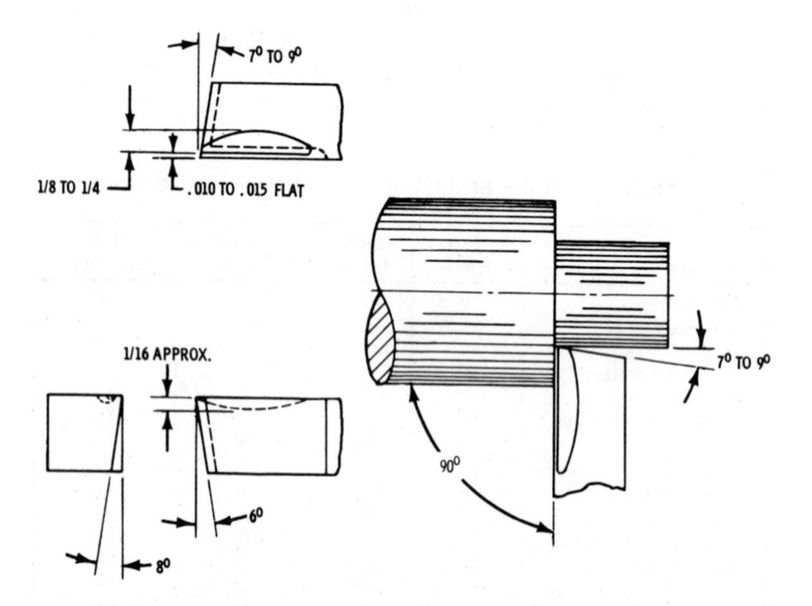

Fig. 11-12. Turning tool used in a four-way turret tool post, showing lip and clearance angles.

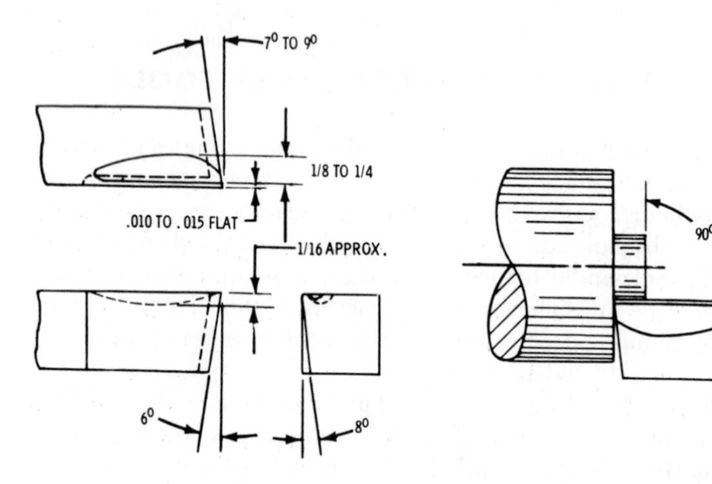

Fig. 11-13. Facing tool used in a four-way turret tool post, showing lip and clearance angles.

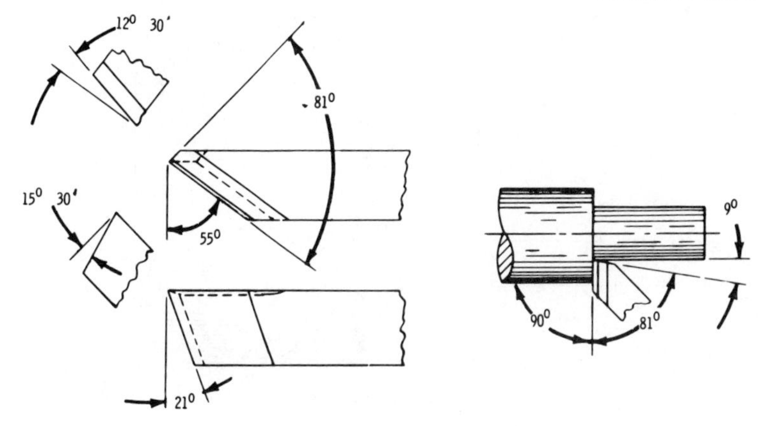

Fig. 11-14. Turning tool bit, as used in toolholder.

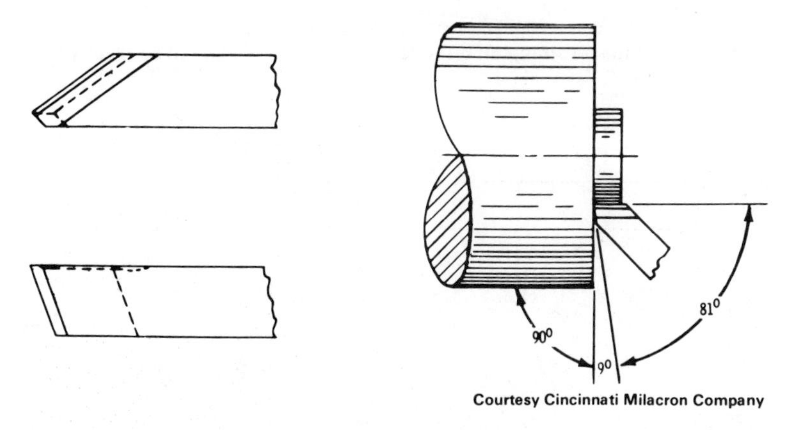

Courtesy Cincinnati Milacron Company

Fig. 11-15. Facing tool bit, as used in toolholder.

A *corner necking tool* can be used to neck an outside diameter and face in one setting (Fig. 11-17.) An *offset necking tool* is shown in Fig. 11-18.

The counterbore, or corner tool, can be used for boring steps in soft chuck jaws, as well as for counterboring. More clearance must be ground on the tool when smaller holes are to be counterbored (Fig. 11-19).

A *cutoff tool* is shown in Fig. 11-20. The angles on the cutoff tool will vary, depending on the width of the tool.

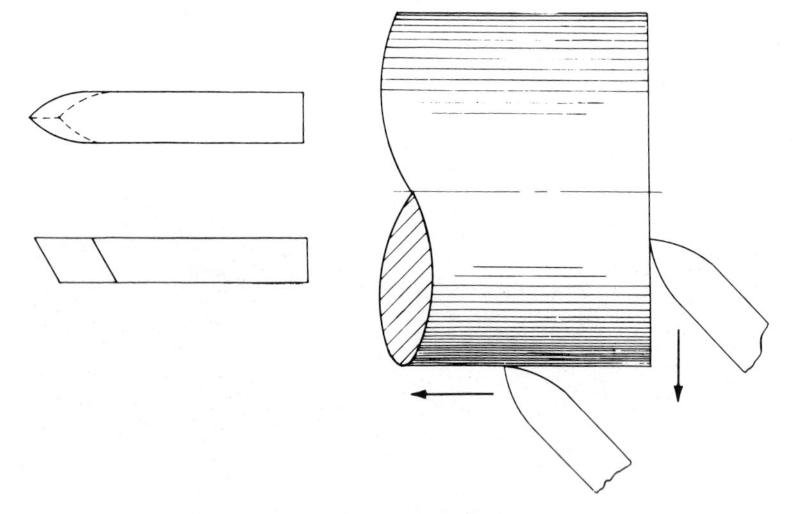

Fig. 11-16. Finish turning an outside diameter, and facing in one setting.

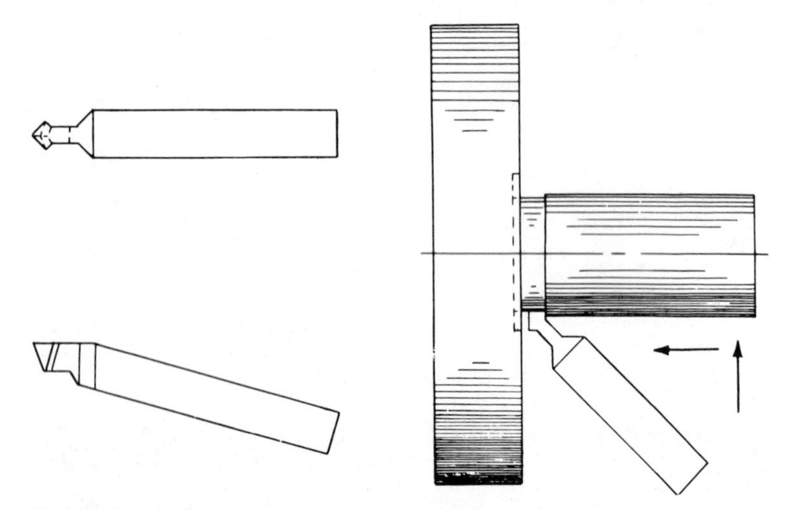

Fig. 11-17. Corner necking tool used to neck an outside diameter and to face in one setting.

Boring tools are held in special toolholders. The boring tool is designed primarily for small work. These tools are sharpened in much the same manner as other lathe tools (Fig. 21).

Boring bars are designed for larger work. The tool bit is inserted

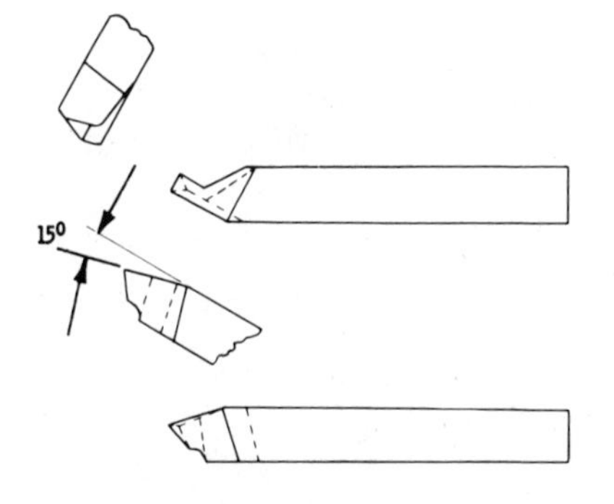

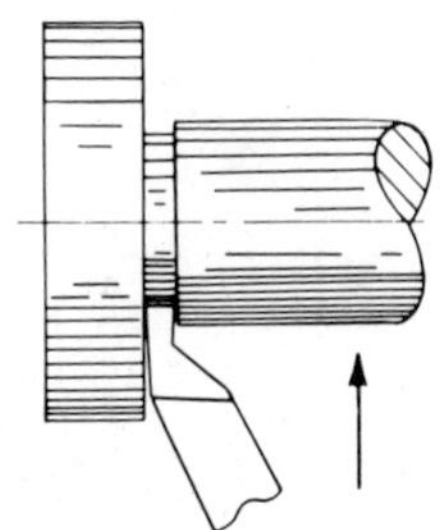

Courtesy Cincinnati Milacron Company

Fig. 11-18. Offset necking tool.

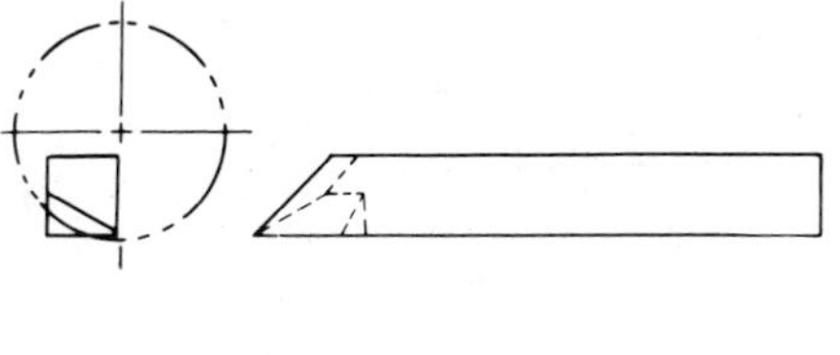

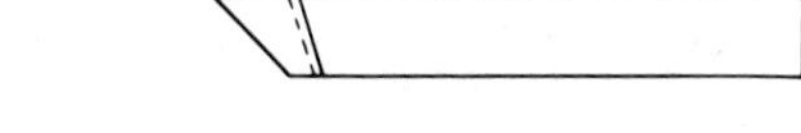

Courtesy Cincinnati Milacron Company

Fig. 11-19. Counterbore or corner tool.

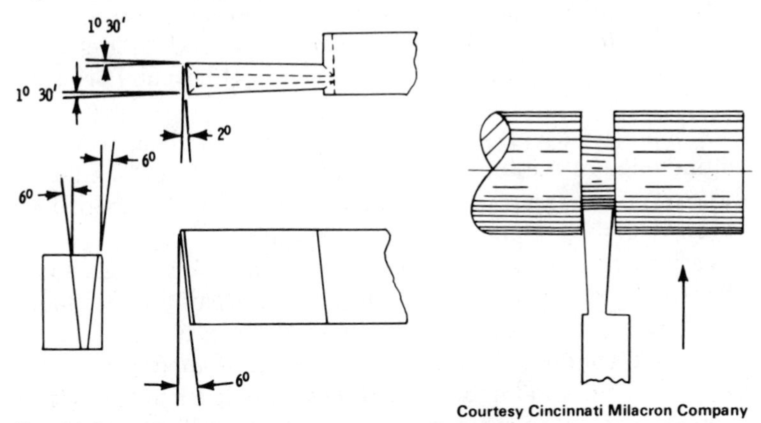

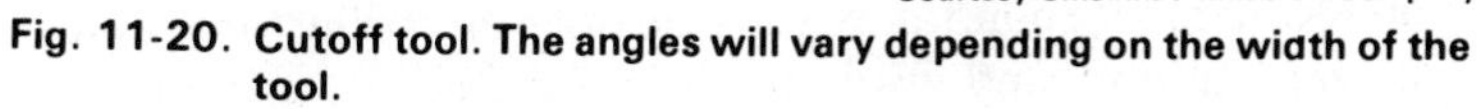

Courtesy Cincinnati Milacron Company

Fig. 11-20. Cutoff tool. The angles will vary depending on the width of the tool.

into a slot in the boring bar, being held by a set screw or other means. The bars are made with the slot at a different angle to the center line of the bar (45°, 60°, and 90°) for different types of work (Fig. 11-22).

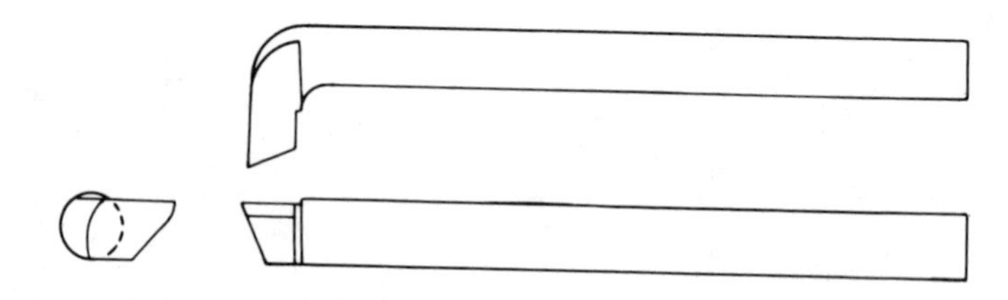

Fig. 11-21. Boring tool.

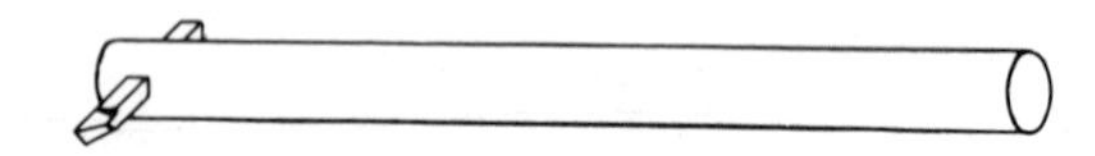

Fig. 11-22. Boring bar with tool bit.

SUMMARY

Cutting tools used in shop work are generally wedge shaped. A tool that has been used on heavy cuts develops a small ridge of metal directly over the cutting edge. This bit of metal is much harder than the metal being cut and is almost welded to the edge of the cutting tool, which indicates the amount of heat and pressure which has been developed.

In high-speed work, coolants help absorb the heat from the cutting edge of the tool. A stream of cutting coolant should be directed to the point of the cutting tool so that it spreads and covers both the cutting tool and the work.

Various materials are used in making cutting tools or bits. Some of the materials used to make cutting tools are carbon steel, high-speed steel, stellite, and carbide, which includes tungsten, tantalum, and titanium. Nine of the most popular shaped lathe cutter bits are left-hand and right-hand turning tools, left-hand and right-hand facing tools, round-nose turning tools, threading tools, cutoff tools, boring tools, and inside-threading tools.

REVIEW QUESTIONS

1. Why are coolants used with cutting tools?
2. Name at least four cutting tools used in maching shops.
3. What type of material is used to make cutting tools?
4. What is a tool angle or working angle?
5. What happens to the cutting tool edge after it has been used for a few minutes?
6. What is the purpose of a settling tank?
7. How is the round-nose turning tool used?
8. What is a tool bit?
9. What is a chip breaker?
10. Why is the rake angle of a cutting tool important?
11. What is meant by side rake?
12. Where do you use a corner necking tool?
13. Where is the counterbore, or corner tool, used?
14. What is a boring bar used for?
15. How is the cutoff tool used to cut off material?

CHAPTER 12

Cutter and Tool Grinders

A variety of metal cutting applications can be performed to close tolerances with modern precision tools. Unless the machine has been provided with suitably designed and prepared cutting tools, close tolerances are impossible. The machine, no matter how precise, can be no better than the tool with which it is equipped.

IMPORTANCE OF TOOL SHARPENING

If a cutter is used after it becomes dull, it deteriorates rapidly. Very little stock needs to be ground off if the cutter is sharpened at the proper time. Trained operators with suitable equipment, using properly specified grinding wheels, should be used to sharpen lathe and planer tools (Fig. 12-1). Proper sharpening also extends the life and increases the efficiency of twist drills.

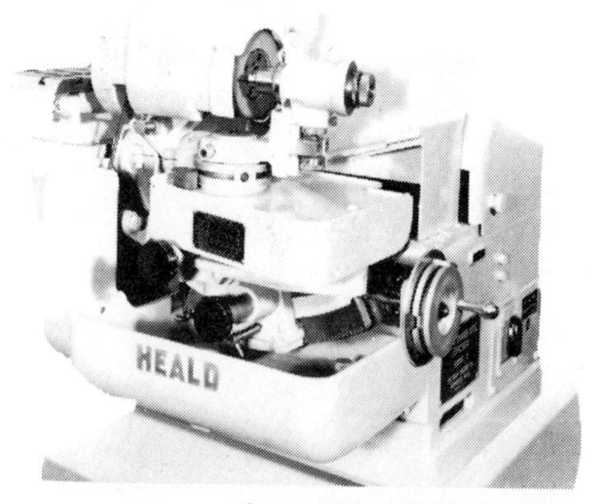

Courtesy Heald Machine Company

Fig. 12-1. Tool-sharpening machine. Direct-reading scales and dials permit grinding different angles, plus the tip radius, in one operation. An extra wheel and freehand grinding attachment is available on the left-hand side of the machine for sharpening odd-shaped tools, or for roughing shapes that will be precision finished on the right-hand side of the machine.

CUTTER AND TOOL SHARPENING

The universal cutter and tool grinding machines meet practically all toolroom requirements. Cutter grinding machines may vary from simple machines for sharpening the multitooth cutters to the universal grinding machines.

Lathe and Planer Tools

Either the offhand method or the machine method may be used to sharpen lathe and planer tools. The skill of the operator determines the accuracy of a tool ground by the offhand method (Fig. 12-2). Precise rake and relief angles can be reproduced by means of dial settings when tools are sharpened on a machine (Fig. 12-3). Localized overheating may crack the cutting tool unless it is kept constantly in motion across the grinding-wheel face.

Twist-Drill Sharpening

Machine grinding is a more accurate method than grinding by hand of sharpening twist drills. A twist drill usually cuts faster, lasts

Courtesy Heald Machine Company

Fig. 12-2. Freehand grinding on the left-hand side of the cutting tool sharpening machine.

longer, and produces more accurate holes when ground on a machine, as compared to grinding by hand.

The suggestion to an old-time mechanic that he use a machine to sharpen a twist drill would probably elicit the scornful retort as follows: "I can grind a drill as well by hand as with a machine." Undoubtedly, a skilled mechanic can produce a satisfactory drill point by exercising great care and taking plenty of time because this is not an impossible task. However, all drill operators are not skilled mechanics, and in this advanced age of mechanical knowledge, a modern up-to-date machine shop will not rely on a skilled mechanic for this precise operation; it will not tolerate the use of hand methods when there is an inexpensive machine available to do the job more accurately and quickly.

If we consider the essential points in connection with sharpening twist drills, the fallacy of any mechanic attempting to sharpen a drill by hand can be seen readily (Fig. 12-4). Drill sharpening requirements are as follows:

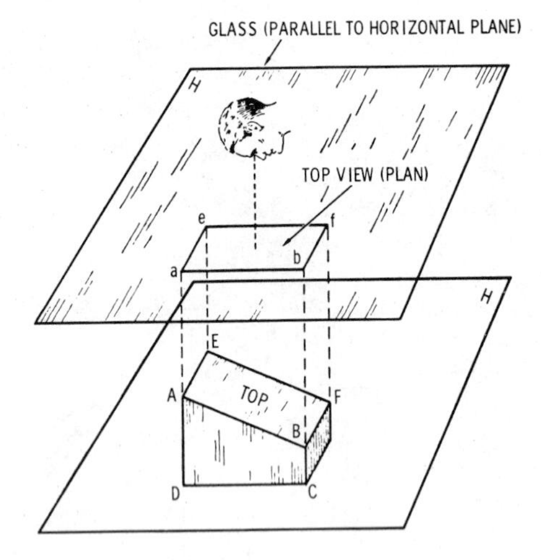

Courtesy Heald Machine Company

Fig. 12-3. Mechanical sharpening on the right-hand side of the cutting tool sharpening machine.

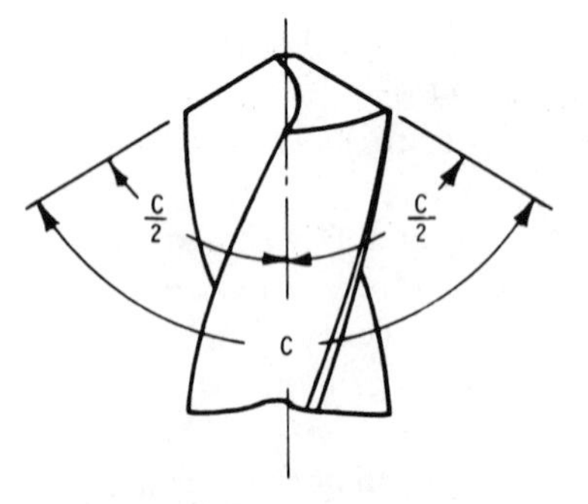

Courtesy National Twist Drill & Tool Company

Fig. 12-4. The two cutting edges (lips) should be equal in length, and should form equal angles with the axis of the drill. Angle C should be 135° for drilling hard or alloy steels. For drilling soft materials and for general purposes, angle C should be 118°.

1. The cutting lip of the twist drill should form equal angles with the axis of the drill. (The commercial standard angle is 59°.) A machinist must possess a pretty good eye, indeed, to gage an angle of 50° with the eye. Drill manufacturers have found 59° to be the best angle for all-around work.

If one lip of a twist drill is ground at 60° and the other at 59°, it is easy to see that the 60° lip will do all the cutting. Half-speed drilling, half the length of service between grindings, conical cone, double wear, and waste of the drill, etc., are all results of unequal lip angles (Fig. 12-5).

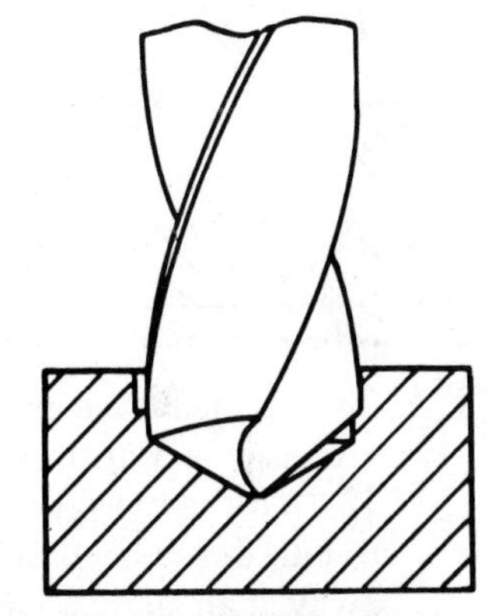

Fig. 12-5. A twist drill with the lips ground at unequal angles with the axis of the drill can be the cause of an oversized hole. Unequal angles also result in unnecessary breakage and cause the drill to dull quickly.

Courtesy National Twist Drill & Tool Company

2. The cutting lips should be of equal length. Even though the lip angles are equal, a twist drill with the cutting lip unequal in length makes oversize holes and causes strain on both the twist drill and the drill press (Fig. 12-6). When one lip is only slightly longer, that lip does all the cutting for that extra length. This means rapid wear, frequent regrinding, and slower drilling speed (Fig. 12-7).

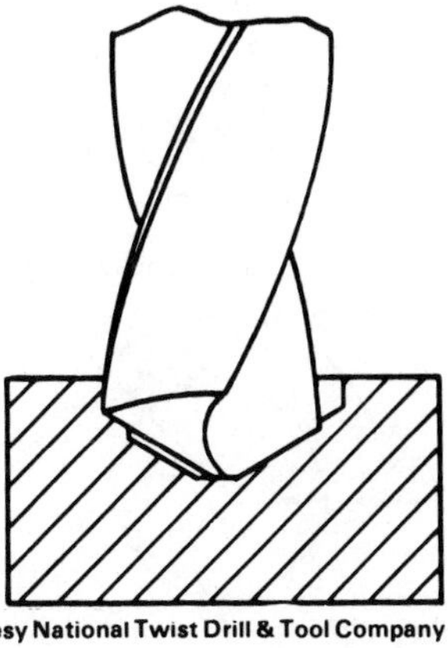

Fig. 12-6. This shows the result of grinding the drill with equal angles, but having the lips unequal.

Courtesy National Twist Drill & Tool Company

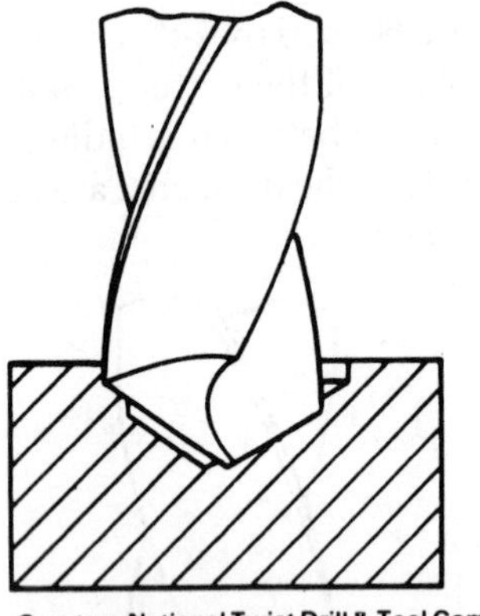

Fig. 12-7. This shows the result of grinding the drill with lips of unequal angles, and with lips of unequal lengths.

Courtesy National Twist Drill & Tool Company

3. The lip clearance angle, or lip relief angle, should gradually increase as it approaches the center of the twist drill (Fig. 12-8). If the twist drill is to penetrate the work, and its edge is to cut, the surface behind the cutting edge must be ground away at an angle, giving what is termed clearance. If there is no clearance angle, the tool will ride along the surface without entering the work. The clearance angle determines the effectiveness of the twist drill and its length of life (Fig. 12-9).

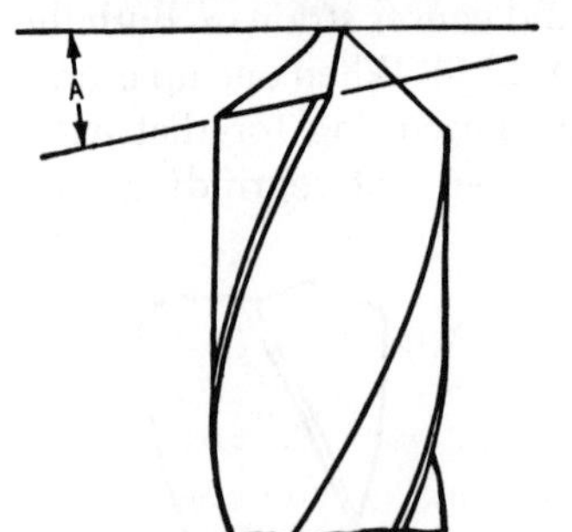

Fig. 12-8. The lip relief angle A should vary according to the material to be drilled and the diameter of the drill. Lesser relief angles are required for hard and tough materials than for soft, free-machining materials.

Courtesy National Twist Drill & Tool Company

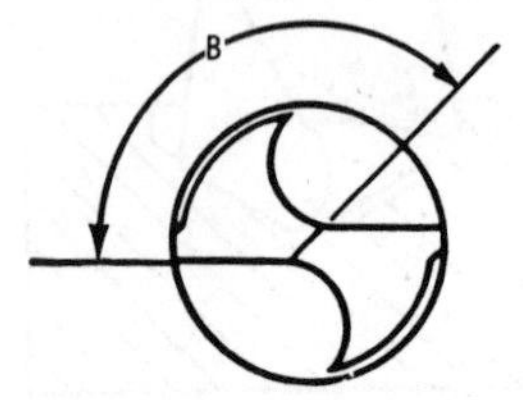

Fig. 12-9. The chisel point angle B increases or decreases with the relief angle, but it should range from 115° to 135°.

Courtesy National Twist Drill & Tool Company

Table 12-1. Suggested Lip Relief Angle at the Periphery

Drill Diameters	For General Purpose	Hard and Tough Materials	Soft and Free-Machining Materials
No. 80–No. 61	24°	20°	26°
No. 60–No. 41	21°	18°	24°
No. 40–No. 31	18°	16°	22°
No. 30–1/4"	16°	14°	20°
F to 11/32"	14°	12°	18°
S to 1/2"	12°	10°	16°
33/64"–3/4"	10°	8°	14°
49/64" & lgr.	8°	7°	12°

Courtesy National Twist Drill & Tool Co.

In order to grade the clearance properly along the drill lip from point to periphery, and in order to curve the back side of the cutting edge so that maximum endurance and strength, consistent with free cutting, are preserved at all points, it is necessary that every portion of the cutting lip, while being ground, rock against the grinding wheel in a bath similar to that in which the cutting lip travels when at work. If, while at work, those portions of the cutting lip nearer the point travel is shorter paths and smaller circles than the portions nearer the outer corner, this similar condition should exist while the twist drill is being ground.

The webs of twist drills increase in thickness toward the shank of the drill, in order to increase strength and rigidity. The proper web thickness should be maintained at the drill as the drills are sharpened. Thickness of the web may be further reduced for some applications to not less than 50 percent of the original web thickness (Fig. 12-10).

Automatic drill grinders are designed and constructed in accordance with geometric principles. These machines automatically locate the cone axis in relationship to the drill axis for all twist drill sizes within the capacity of the machine; they also automatically grind the correct lip clearance angle and the proper center angle.

Regrinding Tap Drills

Usually the tap drill cutting edges become dull first, and it may be necessary to grind only that portion of the tap drill. Special

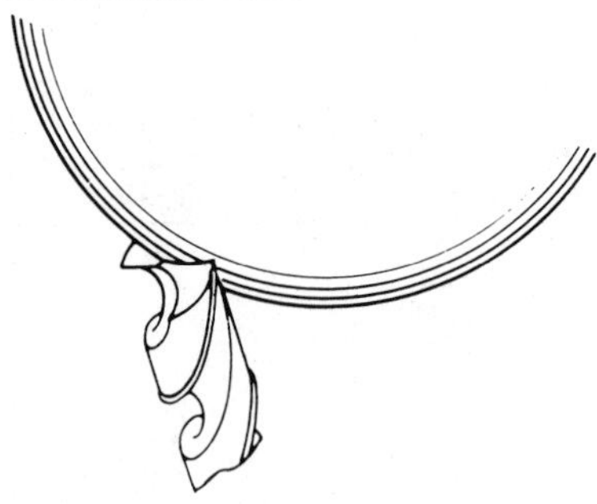

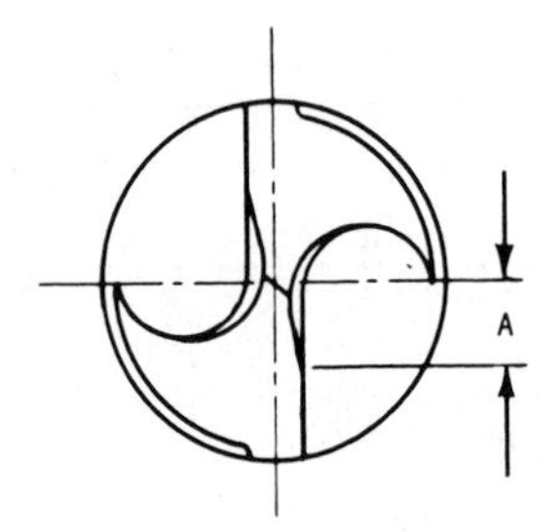

Courtesy National Twist Drill & Tool Company

(A) Using a round-faced grinding wheel to thin the web.

(B) Thinning should not extend more than half the length of the cutting edge 'A'.

Fig. 12-10. Thinning the web at the point of the drill.

machines are available for sharpening taps accurately, ensuring a uniform chamfer or taper, and correct, uniform eccentric relief.

Mounted Points and Wheels

Miscellaneous offhand grinding operations on all kinds of blanking and drop-forging dies are performed with a wide variety of mounted points and mounted wheels. They are necessary for shaping and finishing steel dies and molds that are used in the plastic and allied tool and die industries. The mounted wheels are usually driven by light, flexible-shaft, air- or motor-driven portable grinders. Surplus metal can be removed in much less time with this equipment than by scraping or filing, and a better finish may be obtained.

Cutter-Sharpening Machines

The machines designed to grind cutters may vary from single type for grinding some of the multitooth cutters, to the universal cutter and tool grinding machines, having a range for practically all toolroom requirements. Many of the cutter and tool grinders have special attachments for grinding the various milling machine cutters (Figs. 12-11 through 12-15).

Plain Milling Cutters—The cutter may be mounted on an arbor and the tooth rest adjusted to give the desired relief angle. The

Courtesy Cincinnati Milacron Company

Fig. 12-11. Long-ream grinding attachment for the cutter and tool grinder. This attachment is useful in grinding long lining reamers, extension taps, stay-bolt taps, taper reamers, and boring bars. When concentricity is important, special cutters and gages may be ground without removing them from their arbors.

Courtesy Cincinnati Milacron Company

Fig. 12-12. Clearance-setting dial on the left-hand tailstock of a cutter and tool grinder. Arbor-mounted cutters are quickly adjusted to the desired clearance angle by means of this feature. Clearance-setting dials enable the operator to set predetermined clearance angles conveniently and accurately, regardless of the cutter diameter and type of wheel used.

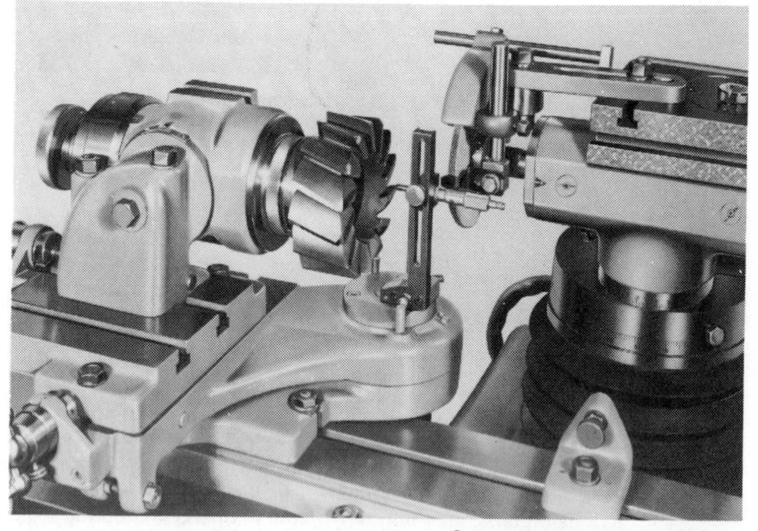

Courtesy Cincinnati Milacron Company

Fig. 12-13. Radius-grinding attachment. Medium- to large-sized milling cutters requiring an accurate 90° radius on the corner of the teeth can be ground quickly. The attachment consists of four principal elements: a base, swivel plate, adjustable table, and workhead support. A micrometer is included to determine the starting position accurately. The capacity of the attachment is 0 to 1-inch radii, 0 to 12-inch maximum cutter diameter, with 3-inch maximum width of face.

tooth rest should bear against the tooth to be ground wherever possible. The cutter should be moved slowly toward the wheel until sounds or sparks indicate contact. Holding the cutter against the tooth rest with one hand, the cutter is traversed with the other hand across the wheel face with a steady motion, either by moving the table or by sliding the cutter on a cutter bar. The cuts should not exceed 0.001 inch per pass on roughing and should be reduced to 0.0005 inch on finishing passes.

Side Milling Cutters—Cutter teeth of the side mill type are ground on the outside diameter in exactly the same manner as plain cutters. A cup wheel is usually employed in grinding the sides of the cutter teeth. The cutter is mounted on a stud arbor clamped in the universal workhead, which is swiveled to the required relief angle.

Formed Cutters—A dish-shaped grinding wheel is usually used

Courtesy Cincinnati Milacron Company

Fig. 12-14. Ample vertical range for going upward to accommodate the exceptionally large diameter face mills, or for going downward to handle complex grinding jobs, is provided by the adjustable wheelhead pile on the cutter and tool grinder.

to grind formed cutters. Either a master form or an index center should be used as a guide for the tooth rest. If these are not available, the tooth rest may be adjusted against the back of the tooth to be ground, prior to which the grinding-wheel face and the center of the cutter have been brought into the same vertical plane. Some form cutters are made with a forward rake, or undercut, tooth. In sharpening these cutters, care must be taken to offset the wheel face, so as to maintain the original rake angle.

SUMMARY

The days of sharpening a twist drill by hand have gone with the appearance of modern tool grinders in up-to-date machine shops. Machine grinding is the more accurate method of sharpening twist drills. A twist drill usually cuts faster, lasts longer, and produces a more accurate hole when ground on a machine than when ground by hand.

The cutting tips of the twist drill should form equal 59° angles with the axis of the drill. Drill manufacturers have found 59° to be

Courtesy Cincinnati Milacron Company

Fig. 12-15. Face-mill grinding attachment for sharpening face mills up to a diameter of 18 inches more quickly, more easily, and with a higher degree of accuracy. A knurled handwheel at the end of the spindle aids in indexing. The base and the swivel block are both graduated in degrees, permitting the cutter to be swiveled to the desired clearance angle.

the best angle for all-around work. If one lip of a drill is ground at 60°, and the other lip at 59°, it is easy to see that the 60° lip will do all the cutting.

Machines designed to grind cutters may vary from single types for grinding some of the multitooth cutters to the universal cutters and tool grinding machines, having a range for practically all toolroom requirements. Many of the grinding machines have special attachments for grinding various milling machine cutters.

REVIEW QUESTIONS

1. What is the recommended cutting angle of a twist drill?
2. What happens when a twist drill is not sharpened at the proper angle, or when one side is at a different angle than the other?
3. What is meant by the lip clearance angle or lip relief angle?

CHAPTER 13

Drills

A twist drill is a pointed cutting tool, usually round. It is used for cutting holes in metal or other hard substances and is driven by a machine. As with many other machinists' tools, there is a great variety of twist drills designed to meet all kinds of service.

DRILL STANDARDS

Drill diameters have become more and more standardized over the years. Standard drill diameters are classified in a decimal (inch) series that gives the machinist a wide range of twist drills from which to choose. The size of the twist drill is usually stamped on the shank. See Fig. 13-1 for parts of the drill bit. Very small drills are not identified by size, but must be measured with either a micrometer or a drill gage. These drills should be kept either in sets or in holders (Fig. 13-2). Drills are identified as to size in three ways as follows:

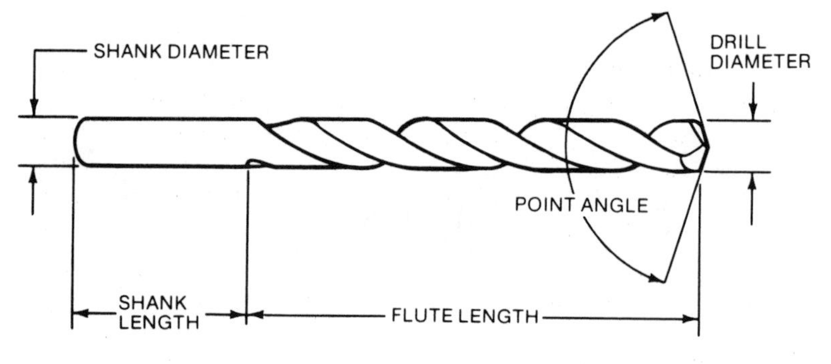

Courtesy Stanley Tools Co.

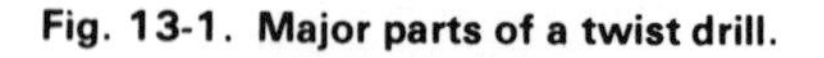

Fig. 13-1. Major parts of a twist drill.

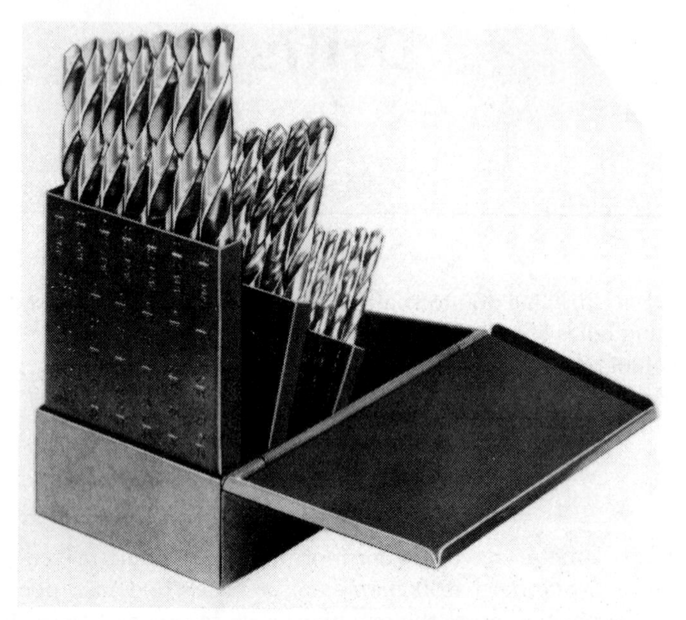

Courtesy National Twist Drill & Tool Company

Fig. 13-2. Straight-shank drill set. This is a fractional-size set (sizes 1/16 inch through 1/2 inch by 64ths).

1. *Number sizes* from No. 80 (0.0135 inch) to No. 1 (0.228 inch).
2. *Letter sizes* from "A" (0.234 inch) to "Z" (0.413 inch).
3. *Fractional sizes* from 1/64 (0.0156 inch) upward by 64ths of an inch.

The fractional sizes which come between the number and letter sizes are shown in Table 13-1. The decimal (inch) diameters of the standard twist drills in number, fraction, and letter sizes are shown in the table.

TWIST DRILL TERMINOLOGY

The parts of a twist drill can be seen in Fig. 13-1. Note the shank diameter, shank length, flute length, point angle, and drill diameter. By checking these points on the drill bit, you should be able to make sense of any table giving bit diameters and special characteristics. Some drill bits are designed for specific applications. Today the general-purpose drill bit is more or less obsolete. Take a look at the following terminology to make sure you understand the terms used today to refer to specific job-type bits.

Screw Machine Length ("Stubby")—This is a name applied to the shortest drills commonly available. They are designed to be used in automatic drilling machines primarily, and are short for maximum rigidity.

Jobbers Length—A term that applies to the length of any standard drill bit up to approximately 11⁄16 inch in diameter. It is the drill bit length most often called for by "do-it-yourselfers" and professionals.

Extended Length (Aircraft Extension)—As the name implies, these drill bits are longer than "jobbers" or "taper" length bits. They are used for drilling in hard-to-reach places, such as between studs, or for drilling through extra-thick material.

Silver and Deming—This is a name that applies only to ½-inch shank drill bits and refers to a time when drill bits were made specifically for unique machining applications. These drills feature a general-purpose flute design and are recommended for use in portable electrical drills and drill presses.

Double-Ended Drills—These are drills on which both ends are fluted for a short distance and have a solid center shank for secure chucking. These drills are designed to be reversed quickly in machine tools to reduce changeover time. They often feature a split-point design for fast starting in sheet metal.

The lengths of the drill bits therefore increase in size from screw

Table 13-1. Number, Fraction, and Letter Drill Sizes

Drill	Diameter (Inches)	Drill	Diameter (Inches)
80	.0135	42	.0935
79	.0145	3/32	.0937
1/64	.0156	41	.0960
78	.0160	40	.0980
77	.0180	39	.0995
76	.0200	38	.1015
75	.0210	37	.1040
74	.0225	36	.1065
73	.0240	7/64	.1094
72	.0250	35	.1100
71	.0260	34	.1110
70	.0280	33	.1130
69	.0292	32	.1160
68	.0310	31	.1200
1/32	.0312	1/8	.1250
67	.0320	30	.1285
66	.0330	29	.1360
65	.0350	28	.1405
64	.0360	9/64	.1406
63	.0370	27	.1440
62	.0380	26	.1470
61	.0390	25	.1495
60	.0400	24	.1520
59	.0410	23	.1540
58	.0420	5/32	.1562
57	.0430	17/64	.2656
56	.0465	H	.2660
3/64	.0469	I	.2720
55	.0520	J	.2770
54	.0550	K	.2810
53	.0595	22	.1570
1/16	.0625	21	.1590
52	.0635	20	.1610
51	.0670	19	.1660
50	.0700	18	.1695
49	.0730	13	.1850
48	.0760	3/16	.1875
5/64	.0781	12	.1890
47	.0785	11	.1910
46	.0810	10	.1935
45	.0820	9	.1960
44	.0860	8	.1990
43	.0890	7	.2010

Table 13-1. Number, Fraction, and Letter Drill Sizes (Cont'd)

Drill	Diameter (Inches)	Drill	Diameter (Inches)
13/64	0.2031	N	0.3020
6	0.2040	5/16	0.3125
5	0.2055	O	0.3160
4	0.2090	P	0.3230
3	0.2130	21/64	0.3281
7/32	0.2187	Q	0.3320
2	0.2210	R	0.3390
1	0.2280	11/32	0.3437
A	0.2340	S	0.3480
15/64	0.2344	T	0.3580
B	0.2380	23/64	0.3594
C	0.2420	U	0.3680
D	0.2460	3/8	0.3750
E	0.2500	V	0.3770
1/4	0.2500	W	0.3860
F	0.2570	25/64	0.3906
G	0.2610	X	0.3970
17/64	0.2656	Y	0.4040
H	0.2660	13/32	0.4062
I	0.2720	Z	0.4130
J	0.2770	27/64	0.4219
K	0.2810	7/16	0.4375
9/32	0.2812	29/64	0.4531
L	0.2900	15/32	0.4687
M	0.2950	31/64	0.4844
19/64	0.2969	1/2	0.5000

machine length, or stubby, to jobbers length, to taper length, to extended length.

Fractional Sizes—These are drills with diameters defined in fractions of an inch. Most drill sets start at 1/16 inch and go up to 1 inch in diameter, in increments of 1/16 inch. Fractional sizes are the most commonly specified.

Wire Gage Bits (Number Sizes)—Drill numbers correspond approximately to the Stubbs Steel Wire Gage size, an English and American standard. Number series drill bits start at No. 100, the smallest size, and increase in size up to No. 1. Number series drill bits fill the gaps between "fractional" sizes.

Letter Sizes—Letter drill sizes start where the numbers leave off and are also used to fill in the gaps between the fractional sizes. Letters run in size from larger than 7/32 (A size) up to 13/32 (Z size).

For example, the letters B, C, D, and E are between fractional sizes $^{15}/_{64}$ and $^{1}/_{4}$ inch. Letter size drills fill the need for unique hole sizes.

Metric Sizes—As the name implies, metric sizes are a series of drills in diameters corresponding to commonly used metric bolt sizes. Metric drills are sized in millimeters and tenths of a millimeter. Metric drills will become more popular as metric fasteners are more widely used.

TYPES OF DRILLS

Drills are now manufactured in a great variety of sizes and shapes for a variety of materials and special purposes. In the early days of machine work, the flat drill was used exclusively, but it has given way to the twist drill, which is much more efficient.

Twist drills are made by forging to the approximate size, and milling and grinding the forging to the finished size (Fig. 13-3).

Courtesy National Cincinnati Milacron Company

Fig. 13-3. Milling the spiral flutes on a drill.

Straight-fluted drills are sometimes used on soft metals, but most drills have spiral flutes. A twist drill is defined as a drill grooved helically along its length for the purpose of clearing itself from the waste material; the borings pass up the grooves as the drill is fed into the work. The parts of the twist drill are the *shank*, the *body*, and the *point* (Fig. 13-4).

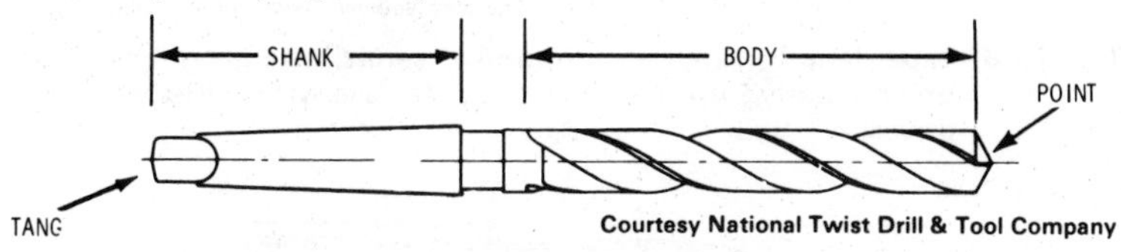

Courtesy National Twist Drill & Tool Company

Fig. 13-4. A general-purpose taper-shank twist drill.

Shank—The shank of the drill fits into the chuck of the machine that revolves the drill. Drills with straight shanks (Fig. 13-5) are held in a drill chuck, which has three jaws that grip the drill.

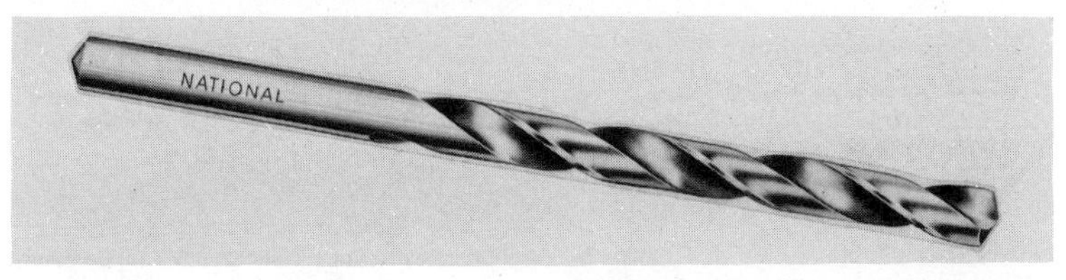

Courtesy National Twist Drill & Tool Company

Fig. 13-5. A general-purpose straight-shank twist drill.

Taper-shank drills mount directly into the taper holder of the drill press spindle, or in a sleeve. The tang of the taper-shank drill fits into a slot in the spindle to prevent the drill from slipping (Fig. 13-4). The automotive series of drills has tangs for use with split-sleeve drill drivers. (Fig. 13-6).

Body—The body of a twist drill extends from the shank to the point. The *flutes* are helical grooves running along opposite sides of the drill. A straight-fluted drill may be used for free machining brass, bronze, or other soft materials (Fig. 13-7). *Three-* or *four-fluted drills* are especially adapted for enlarging punched, cored, or drilled holes (Figs. 13-8 and 13-9). The *margin* of a drill lies along the entire length of the flute. This determines the correct size of the hole. A measurement with a micrometer across both margins gives the size of the drill.

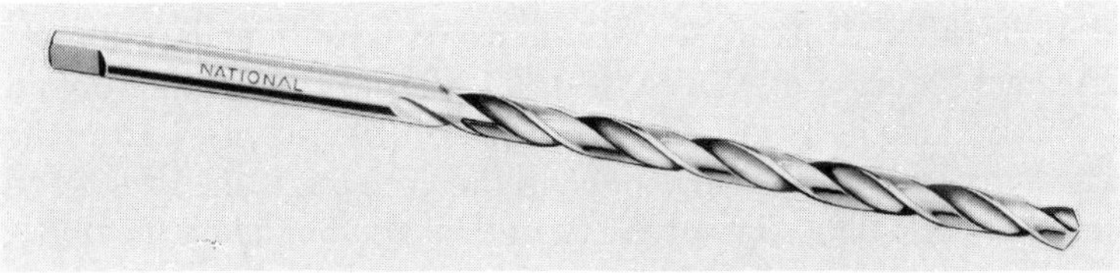

Courtesy National Twist Drill & Tool Company

Fig. 13-6. Taper-length, tanged, automotive series, straight-shank twist drill regularly furnished with tangs for use with split-sleeve drill drivers.

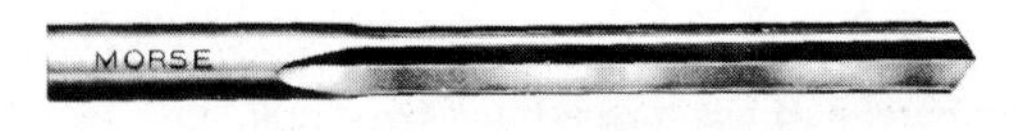

Courtesy Morse Twist Drill & Machine Company

Fig. 13-7. Straight-fluted drill for free machining brass, bronze, or other soft materials, particularly on screw machines. Also suitable for drilling thin sheet material because of lack of tendency to "hog."

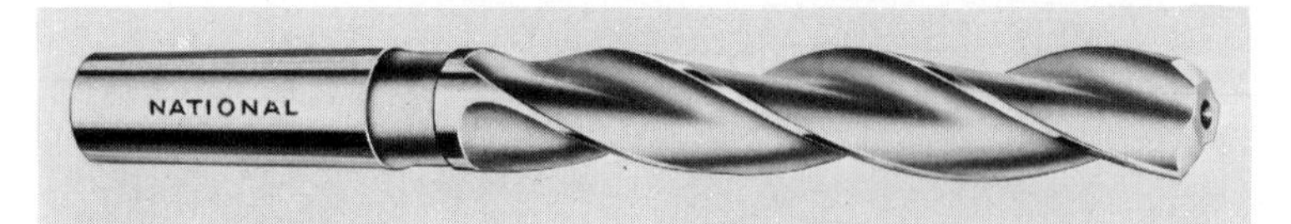

Courtesy National Twist Drill & Tool Company

Fig. 13-8. Straight-shank, three-fluted core drill.

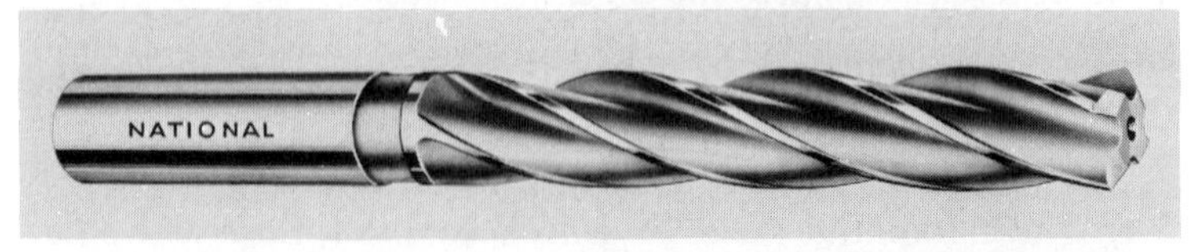

Courtesy National Twist Drill & Tool Company

Fig. 13-9. Straight-shank, four-fluted core drill.

The *body clearance* is immediately in back of the margin. This reduced diameter decreases the friction between the drill and the wall of the hole so that the drill does not bind.

The *land* is the portion of the drill body that is not cut away by the flutes. This includes both the body clearance and the margin.

Back taper allows a slight clearance for the drill in the hole. This is a slight taper in the body of the drill, from the cutting end to the tang end, made by constructing the body of the drill slightly smaller near the shank end.

In *low-helix drills* (Fig. 13-10), the flutes are made wider than those of regular drills to prevent the chips catching and clogging, or to increase the chip space. The flute core diameters are less on

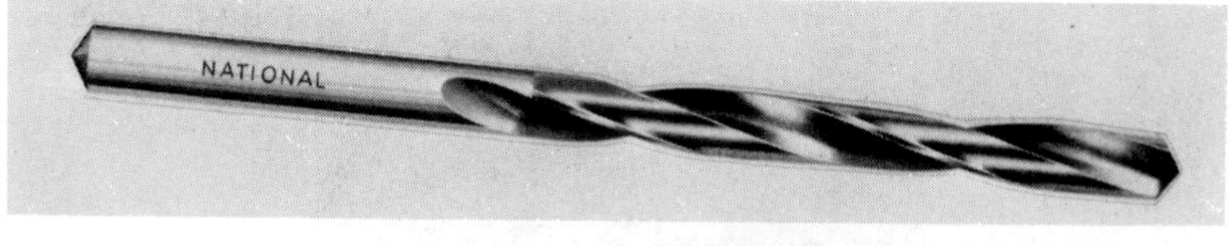

Courtesy National Twist Drill & Tool Company

Fig. 13-10. Low-helix drill used extensively in screw machines making parts from screw stock and brass.

low-helix drills; they require less pressure, give easier penetration, and develop less heat than regular drills. *High-helix drills* have higher cutting rake and improved chip-conveying properties for use in aluminum and some plastics (Fig. 13-11).

Courtesy National Twist Drill & Tool Company

Fig. 13.-11. High-helix drill designed with higher cutting rake and improved chip conveying properties for use on such materials as aluminum, diecasting alloys, and some plastics.

The length of the body of twist drills may vary. *Screw-machine length twist drills* have a shorter body (Fig. 13-12). *Center drills* (Fig. 13-13) and *starting drills* (Fig. 13-14) are also used in screw-machine operations. *Left-hand drills* are used where the spindle of the machine rotates in a left-hand direction.

Multidiameter drills of the step drill type have two or more diameters produced by successive steps on the lands of the drill. The steps are separated by square or angular shoulders. In the *oil-hole drill* (Fig. 13-15), there is a hole through the solid metal for conveying the lubricant to the point. When drilling in cast iron, air is sometimes used to blow out the chips and to keep the drill cool.

Point—This is the cone-shaped end or cutting part of the twist

Courtesy National Twist Drill & Tool Company

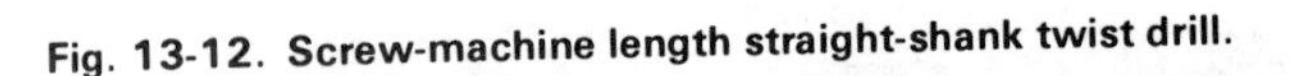

Fig. 13-12. Screw-machine length straight-shank twist drill.

Courtesy National Twist Drill & Tool Company

Fig. 13-13. Center drill.

Courtesy National Twist Drill & Tool Company

Fig. 13-14. Starting drill.

drill. The extreme tip of the drill, which forms one sharp edge, is the *dead center*, or web. The dead center, or web, acts as a flat drill, and cuts its own hole in the work. This is the reason it is common practice to drill a lead hole in the work first. The lead hole provides clearance for the dead center of a twist drill. Thus, the larger drills are kept from running off center, and less feeding pressure is required.

The cutting edges extending from the dead center to the periphery of the drill point are called *lips*. The two cutting edges, or lips, make a standard (helix) angle of 59° with the axis of the drill body. Thus, the included angle of the drill point is 118°. This angle may be varied with the type of material being drilled (Fig. 13-16)

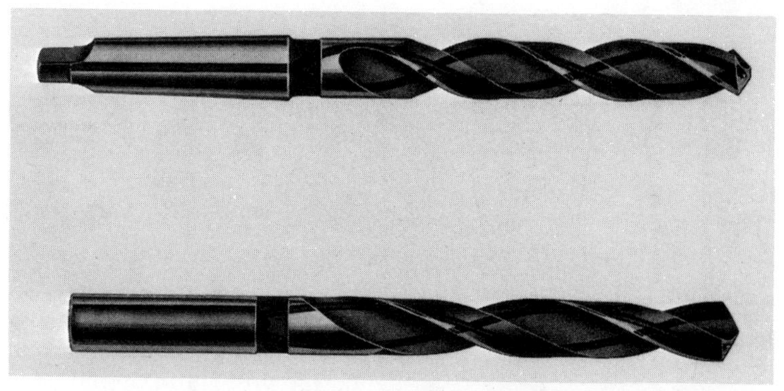

Courtesy National Twist Drill & Tool Company

Fig. 13-15. Oil-hole twist drills for production work in all types of materials on screws machines or turret lathes.

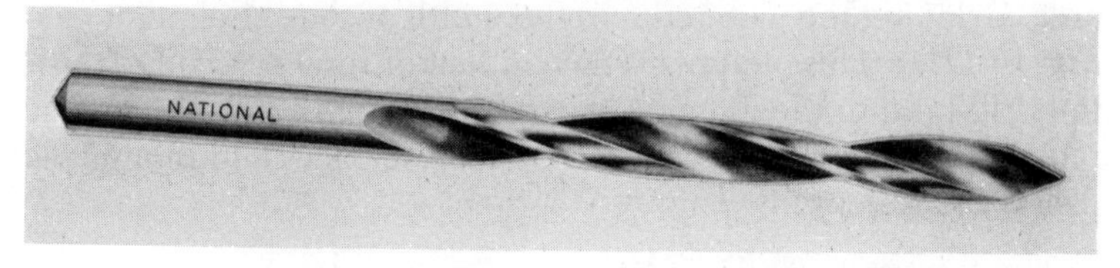

Courtesy National Twist Drill & Tool Company

Fig. 13-16. Drill used principally for drilling molded plastics. Also used for drilling hard rubber, wood, and aluminum and magnesium alloys.

Twist drills for general-purpose work usually have a *lip clearance* of 12° to 15° at the extreme diameter of the drill. Without a lip clearance, the twist drill could not cut because the metal immediately behind the lip would rub on the bottom of the newly drilled hole.

Drill-Bit Point Design

The point of a drill bit indicates its application. For example, a common drill bit has a point angle of 118° (see Fig. 13-17). Fig. 13-17A shows how the flutes have a moderate spiral going up the shank, which indicates that the drill bit is used for general drilling purposes of shallow depth in soft metals, wood, etc. If the point angle is flatter, such as 135°, as shown in Fig. 13-17B, the drill is intended to be used in tough metals where a smaller chip is taken off with each rotation. The web, the center portion of the drill bit,

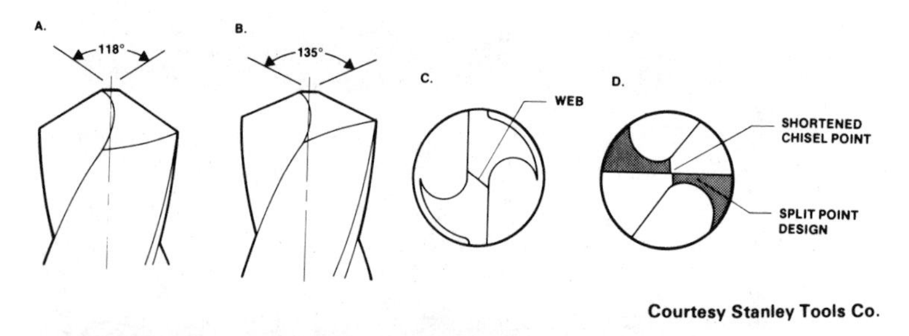

Courtesy Stanley Tools Co.

Fig. 13-17. Web and split-point-design twist drills.

is very thick (see Fig. 13-17C). If a drill bit is to be used in sheet metal, where a lot of force pushing the drill into the metal can cause denting, the point of the bit is sometimes ground in the web area. Grinding the web reduces the width of the chisel point (see Fig. 13-17D). This minimizes the amount of force needed to push the drill bit into the metal. The result is a split-point design that reduces the chance of denting the metal surfaces and minimizes drill walking.

Special-Purpose Drills

Manufacturers have developed twist drills for special purposes. The shank, body, and point have been altered to meet the various needs.

Aircraft drills (Fig. 13-18) are designed for drilling light sections of high-strength, high-temperature alloys and similar materials

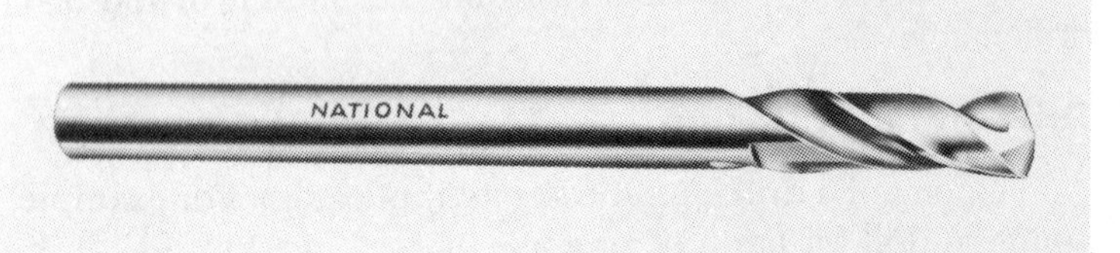

Courtesy National Twist Drill & Tool Company

Fig. 13-18. Aircraft drill used in aircraft and missile construction. These drills are effective in drilling in the higher ranges of work material hardness. Point construction produces the highest possible penetration and thrust reduction with maximum cutting lip support.

used in aircraft and missile construction. Maximum tool rigidity is built in by using heavy-duty construction and short flute lengths on regular length drills.

Die drills may be used for drilling hard steel. The drills shown in Fig. 13-19 are carbide drills. Other twist drills for special purposes

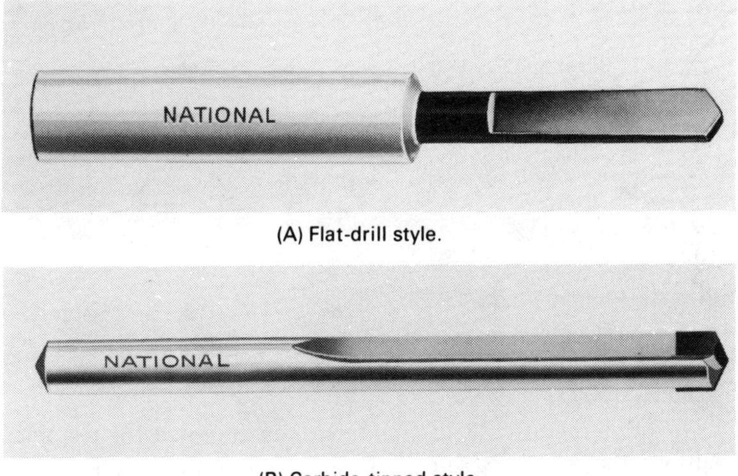

(A) Flat-drill style.

(B) Carbide-tipped style.

Courtesy National Twist Drill & Tool Company

Fig. 13-19. Die drills used for drilling hard steel.

are the *metalworking drill* (Fig. 13-20), which has a 1/4-inch-diameter shank adapted to fit the chucks of electric and hand drills, and the *Silver and Deming drill* (Fig. 13-21), with a 1/2-inch-diameter shank. The *bonding drill* (Fig. 13-22) is used for drilling holes for bonding wire in track and signal work.

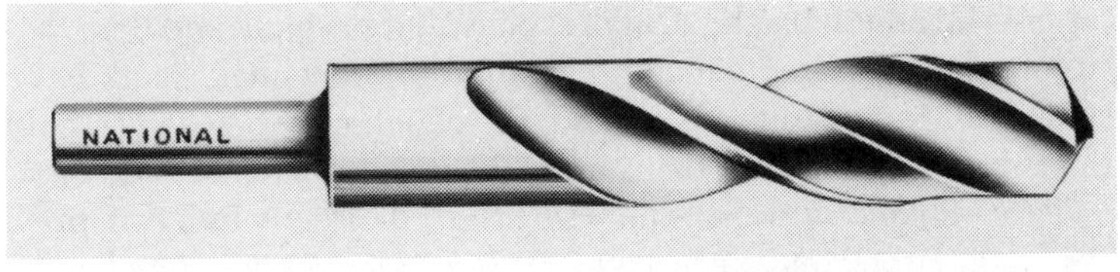

Courtesy National Twist Drill & Tool Company

Fig. 13-20. Metalworking drill with 1/4-inch diameter shank. Fits electric and hand drills with 1/4-inch chucks. Special notched point permits easy drilling in all metals.

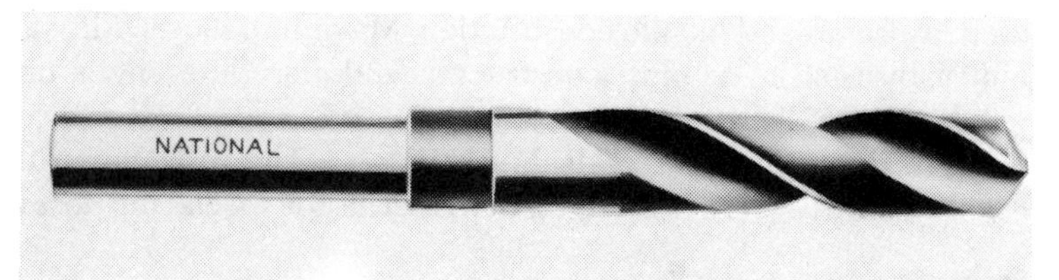

Courtesy National Twist Drill & Tool Company

Fig. 13-21. Silver and Deming drill with ½-inch diameter shank.

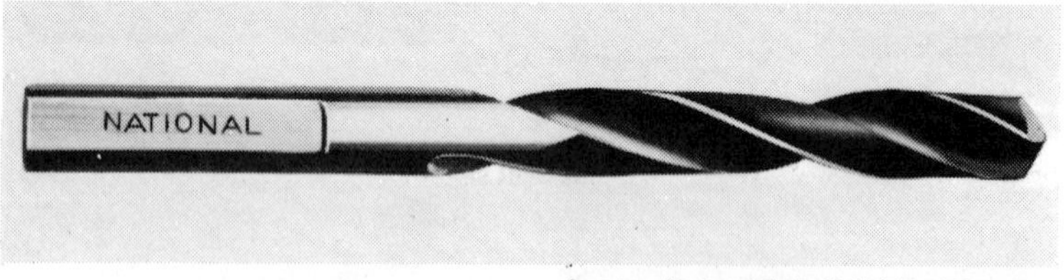

Courtesy National Twist Drill & Tool Company

Fig. 13-22. Bonding drill for drilling holes for bonding wire in track and signal work.

The *combined drill and countersink* is designed for properly finishing the center holes of work to be turned on a lathe (Figs. 13-23 and 13-24). The angle of the tapered part of the drill is 60° to conform to the taper of the lathe centers. The small extension bores out a small hole that extends beyond the point of the lathe center, protecting it from injury.

SOCKET AND SLEEVE

Most drill presses are equipped with a spindle having a Morse taper. The chuck for the drill press has a taper shank and tang that fit into the taper in the drill press spindle. The chuck shank has a standard Morse taper. Of course, the chuck may be used to hold the straight-shank twist drills up to ½ inch in size.

Taper-shank twist drills may be inserted directly into drill press spindles, or they may be inserted in various holding devices having a taper hole. The tang of the twist drill fits into a small slotted recess inside the drill press spindle, above the tapered hole, and keeps the drill from slipping. If the taper hole in the spindle is too

Courtesy National Twist Drill & Tool Company

Fig. 13-23. Plain-type combined drill and countersink.

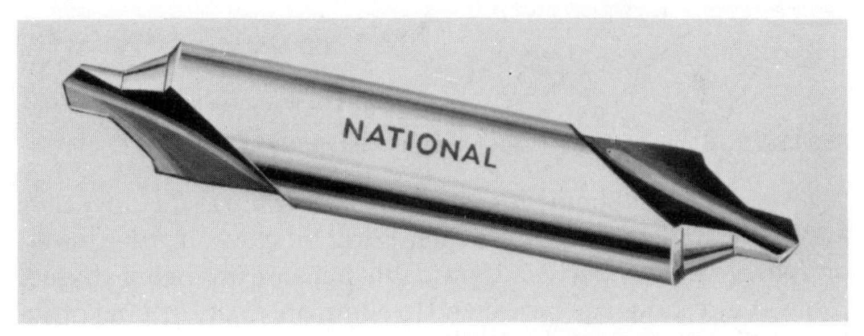

Courtesy National Twist Drill & Tool Company

Fig. 13-24. Bell-type combined drill and countersink.

large for the shank of the twist drill or chuck, the taper-shank tool is held in a steel sleeve (Fig. 13-25). The sleeve has a Morse taper on the outside so that it will fit into the tapered hole in the spindle. The tang on the end of the sleeve fits the slot in the spindle. The taper on the drill shank fits the tapered hole on the inside of the sleeve.

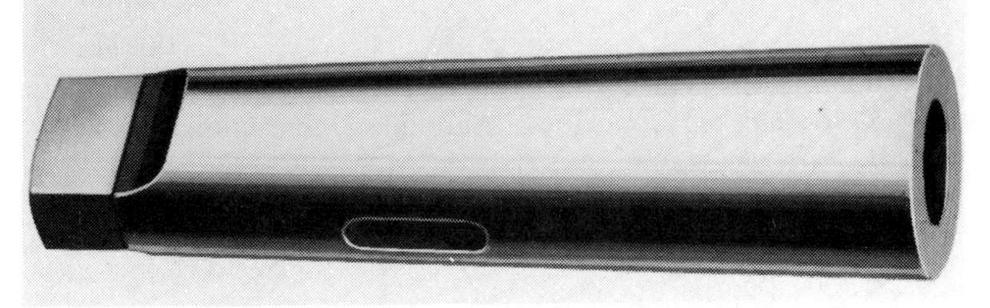

Courtesy National Twist Drill & Tool Company

Fig. 13-25. Steel sleeve for taper-shank twist drill.

The steel socket (Fig. 13-26) is different from the sleeve in that a larger taper-shank drill may be inserted in the socket than may be inserted in the spindle or sleeve.

Using the Twist Drill

The versatility of the drill bit makes it one of the most widely used power tool accessories in use today.

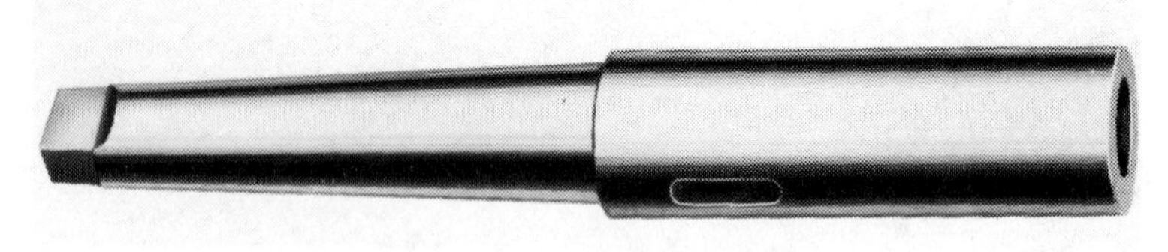

Courtesy National Twist Drill & Tool Company

Fig. 13-26. Steel socket for taper-shank twist drill.

Twist drills will stand more strain in proportion to their size than any other small cutting tool. When a drill bit chips, breaks, burns, or does not cut properly, it is probably because it is being abused. A few suggestions can be utilized to eliminate many drilling problems and ensure maximum drill life.

Secure the Work—To avoid injury or damage, the work should be securely anchored. When drilling, the bit should be perpendicular to the work. Wearing safety goggles minimizes the chance of eye injury.

Cutting Oil—The use of fluids when drilling metals increases the life of the drill bit. Fluids that aid in drilling are:

Fluid	Material
Oil	Steel
Soluble oil	Bronze, soft steel, wrought iron
Kerosene	Aluminum, aluminum alloys

SPEEDS AND FEEDS

Speed refers to the revolutions per minute (r/min) of the drill press spindle. When the speed of a twist drill is mentioned, we mean the speed calculated at the circumference of the drill, or the peripheral speed. Peripheral speed, then, is the speed of travel of a point on the largest diameter in surface feet per minute (sf/min) —not in revolutions per minute.

Extensive experimentation has determined a recommended cutting speed (sf/min) for each of the various materials as shown in Table 13-2.

Table 13-2. Recommended Drilling Speeds for Materials with High-Speed Drills

Material	Recommended Speed in Surface Feet per Minute sf/min
Aluminum and alloys	200–300
Bakelite	100–150
Plastics	100–150
Brass and bronze, soft	200–300
Bronze, high tensile	70–100
Cast iron, chilled	30–40
Cast iron, hard	70–100
Cast iron, soft	100–150
Magnesium and alloys	250–400
Malleable iron	80–90
Monel, metal	40–50
Nickel	40–60
Steel, annealed (.4 to .5 percent C)	60–70
Steel, forgings	50–60
Steel, machine (.2 to .3 percent C)	80–110
Steel, manganese (15 percent Mn)	15–25
Steel, soft	80–100
Steel, stainless (free machining)	60–70
Steel, stainless (hard)	30–40
Steel, tool (1.2 percent C)	50–60
Slate, marble, and stone	15–25
Wrought iron	50–60
Wood	300–400

After the operator has determined the recommended cutting speed for the material to be worked, he must convert the surface feet per minute (sf/min) to revolutions per minute (r/min) at which the drill press spindle must turn for the size of drill being used. The operator may either make his own calculations or consult a table.

If the operator prefers to make his own calculations, he may use the following formula:

$$V = \frac{\pi DN}{12}$$

where

V = velocity or cutting speed (sf/min) in feet per minute
D = diameter of twist drill
N = revolutions per minute (r/min)
π = 3.1416

Example: From Table 13-2, the cutting speed for cast iron is 100 sf/min, and the operator desired to use a ½-inch twist drill. Substituting in the formula:

$$100 = \frac{(3.1416)(.5)(N)}{12}$$

$$1200 = 1.5708N$$

$$\frac{1200}{1.5708} = N$$

$$764 = N, \text{ or revolutions per minute (r/min)}$$

The calculated desired speed, 764 r/min, is the same as the speed given in Table 13-3.

A simple formula for calculating the proper speed to run a twist drill for a particular metal is as follows:

$$\text{r/min} = \frac{4S}{D}$$

where

S = cutting speed or sf/min of the metal
D = diameter of twist drill

$$\text{r/min} = \frac{4(100)}{.5} = \frac{400}{.5} = 800$$

The roughly calculated speed, 800 r/min, compares favorably with the speed, 764 r/min, calculated by the formula, or obtained from the table, as a means of obtaining the desired r/min for the drill press spindle. In many instances, time can be saved by performing the rough calculation.

The feed of a twist drill refers to the downward movement into the work during each r/min of the drill press spindle. Generally, the larger the drill, the heavier the feed that may be used. The drill feed per r/min is given in Table 13-4.

The feeds should generally be less than those shown in the table for alloys and hard steels. A heavier feed may usually be used for cast iron, brass, and aluminum. Of course, a proper coolant is necessary to maintain the recommended speeds and feeds.

Table 13-3. Conversion Table for Surface Feet Per Minute (sf/min) to Revolutions Per Minute (r/min)

SIZE				SURFACE FEET PER MINUTE						
				20	25	30	35	40	45	50
Fraction	Wire Gage	Machine Screw Tap	Decimal Equivalent	REVOLUTIONS PER MINUTE						
	80		.0135	5659	7073	8488	9858	11317	12732	14146
	79		.0145	5269	6586	7903	9179	10538	11855	13172
1/64			.0156	4897	6121	7345	8531	9794	11018	12242
	78		.016	4775	5968	7162	8319	9549	10743	11937
	77		.018	4244	5306	6367	7394	8489	9550	10611
	76		.020	3820	4775	5730	6655	7639	8594	9549
	75		.021	3638	4547	5457	6338	7276	8185	9095
	74		.0225	3395	4244	5092	5915	6790	7639	8487
	73		.024	3183	3979	4775	5546	6367	7163	7958
	72		.025	3056	3820	4584	5324	6112	6875	7639
	71		.026	2938	3673	4407	5119	5876	6612	7345
	70		.028	2728	3410	4092	4753	5456	6138	6820
	69		.0293	2607	3259	3911	4542	5215	5867	6518
	68		.031	2464	3081	3697	4293	4929	5545	6161
1/32			.0312	2448	3061	3673	4266	4897	5509	6121
	67		.032	2387	2984	3581	4159	4775	5371	5968
	66		.033	2315	2893	3472	4033	4629	5208	5787
	65		.035	2183	2728	3274	3802	4365	4911	5456
	64		.036	2122	2653	3183	3697	4244	4775	5306
	63		.037	2065	2581	3097	3597	4130	4646	5162
	62		.038	2011	2513	3016	3503	4021	4524	5027
	61		.039	1959	2448	2938	3412	3917	4407	4897
	60		.040	1910	2387	2865	3327	3820	4297	4775

Table 13-3. Conversion Table for Surface Feet Per Minute (sf/min) to Revolutions Per Minute (r/min) (Cont'd)

SIZE				SURFACE FEET PER MINUTE						
Fraction	Wire Gage	Machine Screw Tap	Decimal Equivalent	20	25	30	35	40	45	50
				REVOLUTIONS PER MINUTE						
	59		.041	1863	2329	2795	3246	3726	4192	4658
	58		.042	1819	2274	2728	3169	3638	4093	4547
	57		.043	1777	2221	2665	3096	3554	3998	4442
	56		.0465	1643	2054	2465	2863	3286	3697	4108
3/64			.0469	1629	2036	2443	2837	3257	3665	4072
	55		.052	1469	1836	2204	2559	2938	3305	3673
	54		.055	1389	1736	2083	2420	2778	3125	3472
	53		.059	1295	1619	1942	2556	2590	2913	3237
		0	.060	1273	1592	1910	2219	2547	2865	3184
1/16			.0625	1222	1528	1833	2129	2445	2750	3056
	52		.0635	1203	1504	1805	2096	2406	2707	3008
	51		.067	1141	1426	1711	1987	2281	2566	2581
	50		.070	1092	1365	1638	1902	2183	2456	2729
	49	1	.073	1047	1308	1570	1823	2093	2355	2616
	48		.076	1005	1257	1508	1751	2011	2262	2513
5/64			.0781	978	1222	1467	1704	1956	2200	2445
	47		.0785	973	1217	1460	1696	1947	2190	2433
	46		.081	943	1179	1415	1644	1887	2123	2359
	45		.082	932	1165	1398	1624	1864	2097	2330
	44	2	.086	888	1111	1333	1548	1777	1999	2221
	43		.089	859	1073	1288	1496	1717	1932	2147
	42		.0935	817	1022	1226	1424	1635	1839	2044
3/32			.0938	814	1018	1222	1419	1629	1832	2036
	41		.096	796	995	1194	1387	1592	1791	1990
	40		.098	779	974	1169	1358	1558	1753	1948
		3	.099	772	964	1157	1344	1543	1736	1929
	39		.0995	768	960	1152	1338	1536	1727	1919
	38		.1015	752	941	1129	1311	1505	1693	1881
	37		.104	735	919	1102	1280	1470	1654	1837
	36		.1065	717	897	1076	1250	1435	1614	1793
7/64			.1094	698	873	1047	1216	1396	1571	1746
	35		.110	694	868	1042	1210	1389	1562	1736
	34		.111	688	860	1032	1199	1377	1549	1721
	33		.113	676	845	1014	1178	1352	1521	1690

Table 13-3. Conversion Table for Surface Feet Per Minute (sf/min) to Revolutions Per Minute (r/min) (Cont'd)

SIZE				SURFACE FEET PER MINUTE						
		Machine		20	25	30	35	40	45	50
Fraction	Wire Gage	Screw Tap	Decimal Equivalent	REVOLUTIONS PER MINUTE						
		4	.115	664	831	997	1158	1329	1495	1662
	32		.116	659	823	988	1147	1317	1482	1646
	31		.120	636	795	955	1109	1273	1432	1591
1/8		5	.125	611	764	917	1065	1222	1375	1528
	30		.1285	594	743	892	1035	1189	1337	1486
	29		.136	561	702	842	978	1123	1263	1404
		6	.138	554	692	831	965	1108	1246	1385
	28		.1405	544	680	816	948	1088	1224	1360
9/64			.1406	543	679	815	946	1086	1222	1358
	27		.144	530	663	795	924	1060	1193	1325
	26		.147	519	649	779	905	1039	1169	1299
	25		.1495	511	639	767	890	1022	1150	1278
	24		.152	503	628	754	876	1005	1131	1257
	23		.154	496	620	744	864	992	1116	1239
5/32			.1562	489	611	733	852	978	1100	1222
	22		.157	487	608	730	848	973	1095	1217
	21		.159	481	601	721	837	961	1081	1201
	20		.161	474	593	712	826	949	1067	1186
		8	.164	466	583	699	812	932	1049	1165
	19		.166	460	575	690	801	920	1035	1150
	18		.1695	451	563	676	785	901	1014	1127
11/64			.172	444	555	666	773	888	999	1110
	17		.173	442	552	662	769	883	994	1104
	16		.177	432	540	647	752	863	971	1079
	15		.180	425	531	637	740	850	956	1062
	14		.182	419	524	629	731	839	944	1049
	13		.185	413	517	620	720	827	930	1033
3/16			.1875	407	509	611	709	814	916	1018
	12		.189	404	505	606	704	808	909	1010
		10	.190	402	502	603	700	804	904	1005
	11		.191	400	500	600	697	801	901	1001
	10		.1935	395	494	592	688	790	889	987
	9		.196	390	487	584	679	779	877	974
	8		.199	384	480	576	669	769	865	961
	7		.201	380	476	571	663	761	856	951

Table 13-3. Conversion Table for Surface Feet Per Minute (sf/min) to Revolutions Per Minute (r/min) (Cont'd)

SIZE				SURFACE FEET PER MINUTE						
				20	25	30	35	40	45	50
Fraction	Wire Gage	Machine Screw Tap	Decimal Equivalent	REVOLUTIONS PER MINUTE						
13/64			.2031	376	470	564	655	752	846	940
	6		.204	374	468	561	652	749	842	936
	5		.2055	372	465	558	648	744	837	930
	4		.209	365	456	548	636	730	822	913
	3		.213	358	448	537	624	717	806	896
		12	.216	354	442	531	616	707	796	884
7/32			.2188	349	436	524	608	698	786	873
	2		.221	345	432	518	602	691	777	863
	1		.228	335	419	503	584	671	755	838
			.234	326	408	489	568	652	734	816
15/64			.2344	326	408	489	568	652	734	816
			.238	321	401	481	559	642	722	802
		14	.242	316	394	473	550	631	710	789
	D		.246	311	389	466	542	622	700	777
1/4	E		.250	306	382	458	532	611	688	764
	F		.257	297	371	446	518	594	669	743
	G		.261	293	366	439	510	585	658	731
17/64			.2656	288	360	432	502	576	648	720
	H		.266	287	359	431	500	574	646	718
	I		.272	281	351	422	490	562	633	703
	J		.277	276	345	414	480	552	621	689
	K		.281	272	340	408	474	544	612	680
9/32			.2815	271	339	407	472	542	610	678
	L		.290	264	329	395	459	527	593	659
	M		.295	259	324	388	451	518	583	647
19/64			.2969	257	322	386	449	515	579	644
	N		.302	253	316	379	441	506	569	632
5/16			.3125	244	306	367	426	489	550	611
	O		.316	241	302	362	421	483	543	604
	P		.323	237	296	355	413	474	533	592
21/64			.3281	233	291	350	406	466	524	583
	Q		.332	230	287	345	401	460	517	575
	R		.339	225	282	338	393	451	507	563
11/32			.3438	222	278	333	387	445	500	556

Table 13-3. Conversion Table for Surface Feet Per Minute (sf/min) to Revolutions Per Minute (r/min) (Cont'd)

SIZE				SURFACE FEET PER MINUTE						
		Machine		20	25	30	35	40	45	50
Fraction	Wire Gage	Screw Tap	Decimal Equivalent	REVOLUTIONS PER MINUTE						
	S		.348	219	274	329	382	439	493	548
	T		.358	213	266	320	371	426	480	533
23/64			.3594	212	265	319	370	425	478	531
	U		.368	208	260	312	362	416	468	519
3/8			.375	204	255	306	355	408	459	510
	V		.377	202	253	304	353	405	456	506
	W		.386	198	247	297	345	396	445	495
25/64			.3906	196	244	293	341	391	440	489
	X		.397	193	241	289	335	385	433	481
	Y		.404	189	237	284	330	379	426	474
13/32			.4062	188	235	282	327	376	423	470
	Z		.413	185	231	277	322	370	416	462
7/16			.4375	175	219	262	305	350	394	437
15/32			.4688	163	203	244	283	325	366	407
1/2			.500	153	191	229	266	306	344	382
9/16			.5625	136	170	204	237	272	306	340
5/8			.625	122	153	183	213	244	275	306
11/16			.6875	111	138	166	193	221	249	277
3/4			.7500	105	127	152	177	203	229	254
13/16			.8125	94	117	141	164	188	211	235
7/8			.875	87	109	131	152	174	196	218
15/16			.9375	82	102	123	142	163	184	204
1			1.000	76	95	115	133	153	172	191
1 1/8			1.125	68	85	102	118	136	153	170
1 1/4			1.250	61	76	92	106	122	138	153
1 3/8			1.375	56	70	84	97	112	125	139
1 1/2			1.500	51	64	77	89	102	115	128
1 5/8			1.625	47	59	71	83	95	107	118
1 3/4			1.750	44	54	65	76	87	98	109

CLEARANCE DRILLS

A clearance drill is used to drill a hole of sufficient size so that a bolt or screw will pass through it. This drill creates a hole with a clearance for the outside (major) diameter of the bolt or screw.

Table 13-3. Conversion Table for Surface Feet Per Minute (sf/min) to Revolutions Per Minute (r/min) (Cont'd)

SIZE				SURFACE FEET PER MINUTE						
		Machine		20	25	30	35	40	45	50
Fraction	Wire Gage	Screw Tap	Decimal Equivalent	REVOLUTIONS PER MINUTE						
1⅞			1.875	40	51	61	71	81	91	101
2			2.000	38	48	57	67	76	86	95
2½			2.500	31	38	46	53	61	69	76
3			3.000	25	32	38	44	50	57	63
3½			3.500	22	28	33	39	44	50	55
4			4.000	19	24	29	33	38	43	48
4½			4.500	17	21	25	29	34	38	42
5			5.000	15	19	23	27	31	34	38
5½			5.500	14	17	21	24	28	31	34
6			6.000	13	16	19	23	26	29	32
6½			6.500	11	14	17	20	23	26	29
7			7.000	12	13	16	19	21	24	27
8			8.000	10	12	15	17	20	22	25
9			9.000	08	11	13	15	17	19	21
10			10.000	08	10	11	13	15	17	19
11			11.000	07	09	10	12	14	15	17
12			12.000	06	08	09	11	12	14	15

The difference between the clearance drill size and the thread is referred to as the clearance. From Table 13-5, clearance drill equals the diameter of the bolt or screw plus the clearance:

Clearance drill = diameter of thread + clearance
CD = 1″ + 1/64
CD = 1 1/64″

Table 13-4. Drill Feed in Inches Per Revolution

Reference Symbol	Diameter of Drill—Inches			
	Under ⅛	⅛ to ¼	¼ to 1	Over 1
L—Light	.001	.002	.003	.006
M—Medium	.0015	.003	.006	.012
H—Heavy	.0025	.005	.010	.025

Table 13-5. Clearance Drill Sizes

Thread Sizes				Clearance Drill (Inches)		
UNC NC (ISS)	UNF NF (SAE)	Outside Diameter (Inches)	Root Diameter (Inches)	Size	Decimal Equivalent	Clearance (Inches)
	#0-80	0.0600	0.0438	#51	0.0670	0.0070
#1-64	....	0.0730	0.0527	#47	0.0785	0.0055
	#1-72	0.0730	0.0550	#47	0.0785	0.0055
#2-56	....	0.0860	0.0628	#42	0.0935	0.0075
	#2-64	0.0860	0.0657	#42	0.0935	0.0075
#3-48	....	0.0990	0.0719	#36	0.1065	0.0075
	#3-56	0.0990	0.0758	#36	0.1065	0.0075
#4-40	....	0.1120	0.0795	#31	0.1200	0.0080
	#4-48	0.1120	0.0849	#31	0.1200	0.0080
#5-40	....	0.1250	0.0925	#29	0.1360	0.0110
	#5-44	0.1250	0.0955	#29	0.1360	0.0110
#6-32	....	0.1380	0.0974	#25	0.1495	0.0115
	#6-40	0.1380	0.0155	#25	0.1495	0.0115
#8-32	....	0.1640	0.1234	#16	0.1770	0.0130
	#8-36	0.1640	0.1279	#16	0.1770	0.0130
#10-24	....	0.1900	0.1359	13/64	0.2031	0.0131
	#10-32	0.1900	0.1494	13/64	0.2031	0.0131
#12-24	....	0.2160	0.1619	7/32	0.2187	0.0027
	#12-28	0.2160	0.1696	7/32	0.2187	0.0027
1/4"-20	....	0.2500	0.1850	17/64	0.2656	0.0156
	1/4"-28	0.2500	0.2036	17/64	0.2656	0.0156
5/16"-18	....	0.3125	0.2403	21/64	0.3281	0.0156
	5/16"-24	0.3125	0.2584	21/64	0.3281	0.0156
3/8"-16	....	0.3750	0.2938	25/64	0.3906	0.0156
	3/8"-24	0.3750	0.3209	25/64	0.3906	0.0156
7/16"-14	....	0.4375	0.3447	29/64	0.4531	0.0156
	7/16"-20	0.4375	0.3725	29/64	0.4531	0.0156
1/2"-13	....	0.5000	0.4001	33/64	0.5156	0.0156
	1/2"-20	0.5000	0.4350	33/64	0.5156	0.0156
9/16"-12	....	0.5625	0.4542	37/64	0.5781	0.0156
	9/16"-18	0.5625	0.4903	37/64	0.5781	0.0156
5/8"-11	....	0.6250	0.5069	41/64	0.6406	0.0156
	5/8"-18	0.6250	0.5528	41/64	0.6406	0.0156
3/4"-10	....	0.7500	0.6201	49/64	0.7656	0.0156
	3/4"-16	0.7500	0.6688	49/64	0.7656	0.0156
7/8"-9	....	0.8750	0.7307	57/64	0.8906	0.0156
	7/8"-14	0.8750	0.7822	57/64	0.8906	0.0156
1"-8		1.0000	0.8376	1 1/64	1.0156	0.0156
	1"-14	1.0000	0.9072	1 1/64	1.0156	0.0156

DRILL PROBLEMS AND CAUSES

Problem	*Cause*
Drill will not cut	Dull drill bit; work piece too hard; drill speed too fast.
Hole rough	Too much pressure on the drill; work not secure; dull or improperly sharpened drill; cutting oil needed.
Hole oversize	Drill bit bending; improper size drill; work not secure; drill not sharpened properly.
Drill bit walks	No center punch; no pilot hole to lead drill into place; improper drill point.
Drill bit grabs work	Improper drill point; work not secure; too much speed or pressure.
Drill bit burning	Too much speed or pressure; flutes clogged; work too hard; cutting oil needed; dull drill bit.
Drill breaks when drilling	Flutes clogged with chips; work not secure; improper drill type; too much pressure or pressure not applied on centerline of bit.

SUMMARY

Drills are manufactured in many sizes and shapes for a variety of materials and special purposes. Twist drills are made by forging to the approximate size, and then milling and grinding to the finished size. Straight fluted drills are sometimes used on soft metals, but most drills have spiral flutes.

The parts of the twist drill are the shank, the body, and the point. The shank of the drill fits into the chuck of the machine that revolves the drill. This part can be straight or tapered, depending on the type of drill. The body of the drill extends from the shank to the point. The flutes or grooves run along opposite sides of the body. The point is the cone-shaped end or cutting part of the twist drill. The cutting edges extending from dead center to the periphery of the drill point are called lips. The lips are formed at 59°, which is determined as the best cutting angle.

REVIEW QUESTIONS

1. What are the shank, body, and point of a twist drill?
2. What is an oil-hole drill, and why is it used?
3. Explain the purpose of the multidiameter drill.
4. What are the lips of the drill?
5. How are drill sizes identified?

CHAPTER 14

Reamers

A reamer is a precision tool designed to finish, to a specified diameter, a hole that has been produced either by drilling or by other means. It is quite impossible to drill a hole to an exact standard diameter. Where precision is required, a hole is first drilled to a few thousandths of an inch less than the desired size and then reamed to exact size.

A reamer is both fluted and slightly tapered in its construction. The blades are worked out of the solid metal by planing or milling on a machine. The flutes are then backed off to give a cutting edge. The size of a reamer can be determined with a micrometer, measuring across two opposite cutting edges.

TYPES OF REAMERS

There are many types of reamers in use. A reamer has its place in industry as a finishing and sizing tool. The use of reamers has been somewhat reduced by the introduction of the internal grinder, but work that must undergo machining in the lathe, for example, may be finished and brought to size with a reamer. Reamers may be used on work in the drill press without removing the work from the drill press, and on milling machines. Of course, hand reamers are used extensively.

Hand Reamers

Hand reamers have a square end to engage the hand wrench. The fluted or cutting part on hand reamers is slightly tapered on the end to facilitate starting the reamer properly. The shank is 0.005 inch under the size of the reamer. Hand reamers are made with either straight (Fig. 14-1) or helical (Fig. 14-2) flutes.

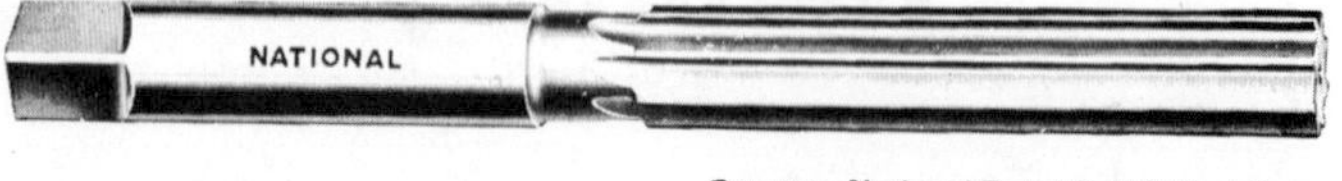

Courtesy National Twist Drill & Tool Company

Fig. 14-1. Hand reamer with straight flutes.

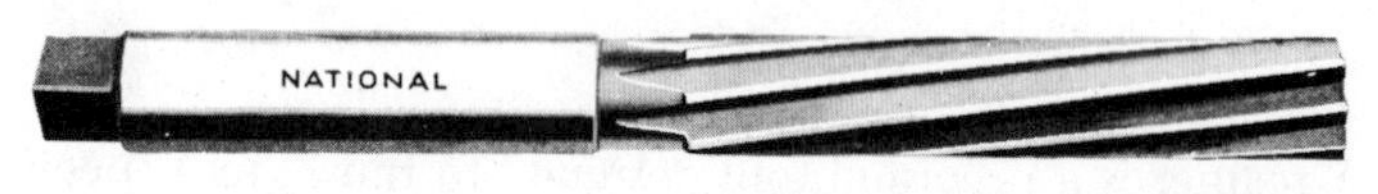

Courtesy National Twist Drill & Tool Company

Fig. 14-2. Hand reamer with helical flutes.

The hand reamer is used to give the hole a smooth finish and correct diameter. Taper reamers are available both in Morse taper (Fig. 14-3) and in Brown and Sharpe taper (Fig. 14-4) for hand finishing.

Other hand reamers are the *adjustable hand reamer* (Fig. 14-5) and the *expansion hand reamer* (Fig. 14-6). The expansion reamer with the helical flute is suited for use in holes where the cut is

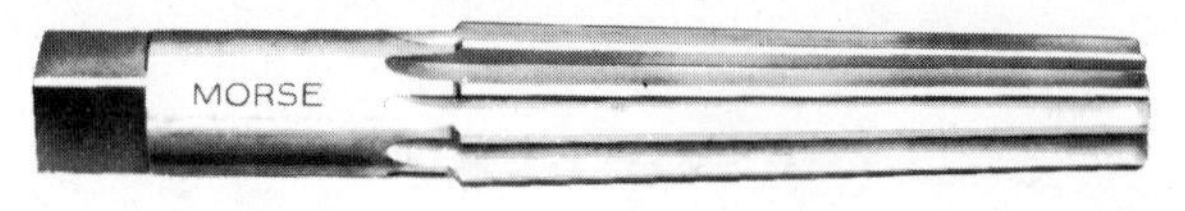

Courtesy Morse Twist Drill & Tool Company

Fig. 14-3. Taper reamer for hand-finishing Morse taper holes in sockets, sleeves, and spindles.

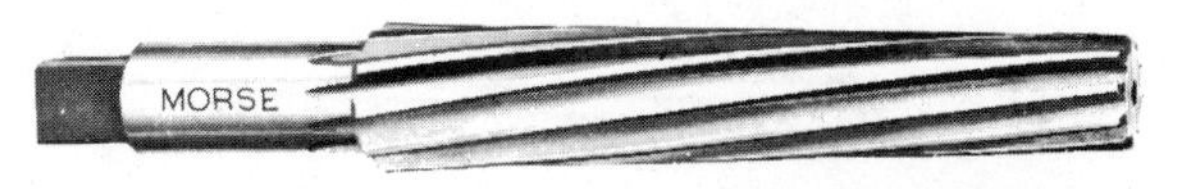

Courtesy Morse Twist Drill & Tool Company

Fig. 14-4. Taper reamer for hand reaming Brown & Sharpe taper holes.

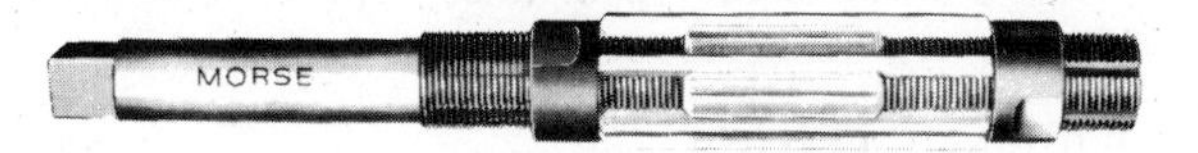

Courtesy Morse Twist Drill & Tool Company

Fig. 14-5. Adjustable hand reamer.

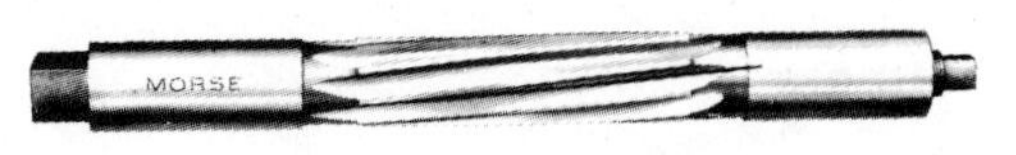

Courtesy Morse Twist Drill & Tool Company

Fig. 14-6. Expansion hand reamer with helical flutes. Especially suited for use in holes where the cut is interrupted by a longitudinal slot or keyway.

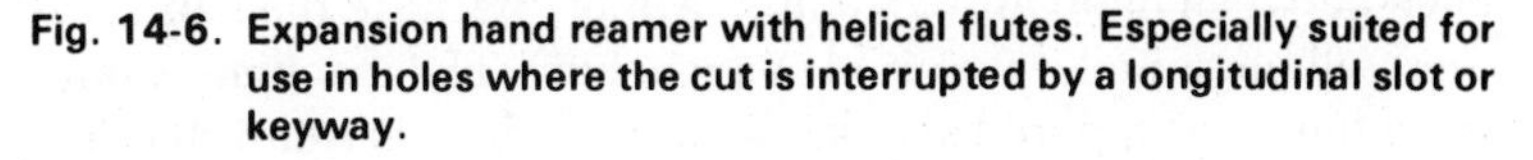

interrupted by a longitudinal slot or keyway. *Taper pin reamers* (Figs. 14-7 and 14-8) in both straight and spiral flutes are available for hand operations.

If possible, all hand reaming should be done in a vertical position. The hand reamer should not be expected to remove a considerable amount of stock. The largest amount of stock to expect the hand reamer to remove is 0.005 inch. Holes should be drilled with this in mind when the hand reamer is to be used. These reamers are slightly tapered at the end for a distance of 3/8 inch to 1/2 inch, and

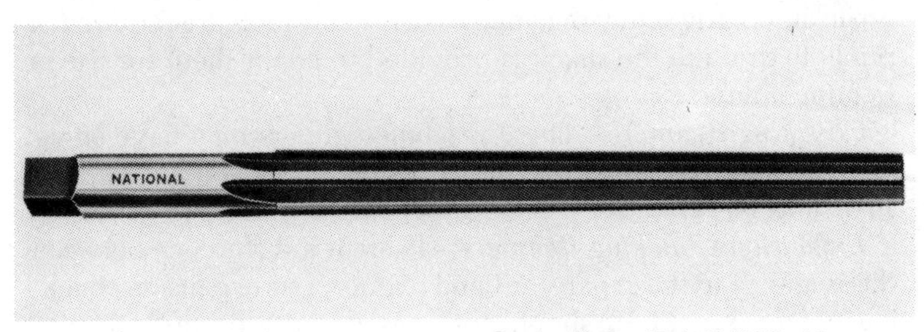

Courtesy National Twist Drill & Tool Company

Fig. 14-7. Taper pin reamer with straight flutes. Taper is 1/4 inch per foot.

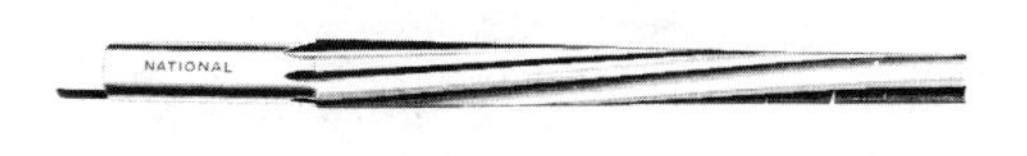

Courtesy National Twist Drill & Tool Company

Fig. 14-8. Taper pin reamer with spiral flutes. Taper is 1/4 inch per foot.

they are 0.010 inch to 0.012 inch smaller at the end. This taper facilitates the entrance of the reamer and enables it to make a good start.

Machine Reamers

The beginning students should be able to recognize the difference between hand and machine reamers because the hand reamer will be ruined if it is used in a machine. The machine reamer has a tang and tapered shank, like a twist drill. It is inserted in a sleeve for use in the drill press spindle. Machine reamers are sometimes followed by hand reamers.

Stub Screw-Machine Reamers—These reamers are free-cutting production tools, economical to use, as their short length practically eliminates breakage (Fig. 14-9). Stub reamers are particularly desirable on production jobs where close tolerances must be maintained without lost time in gaging small parts, sharpening tools, and making machine adjustments. They are manufactured in decimal sizes to provide for accurate production where close tolerances are desired. Stub reamers are designed primarily for use in automatic screw machines where short tools are required; a pin hole through the shank is provided to adapt them for use in floating holders.

Chucking Reamers—These machine reamers may have either straight or taper shanks, and either straight or helical flutes (Figs. 14-10 and 14-11).

Expansion Chucking Reamers—There is a difference between this reamer and the expansion hand reamer. The expansion chucking reamer (Fig. 14-12) should not be used as an adjustable reamer to cut oversize holes. This reamer is adapted to light finishing cuts in all types of materials.

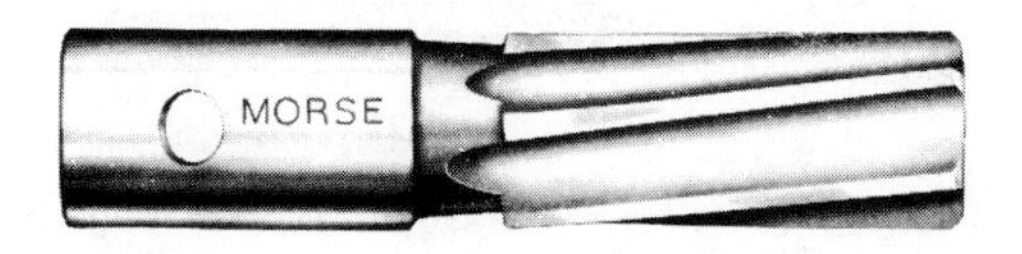

Courtesy Morse Twist Drill & Tool Company

Fig. 14-9. Stub screw-machine reamer. Available in decimal sizes from 0.0600 inch to 1.010 inclusive. Regularly furnished with right-hand cut and 7° left-hand helix.

Courtesy Morse Twist Drill & Tool Company

Fig. 14-10. Straight-shank, chucking reamer with straight flutes.

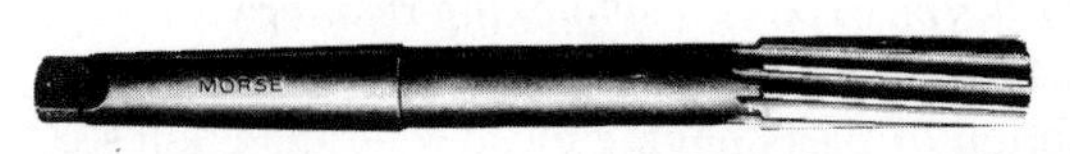

Courtesy Morse Twist Drill & Tool Company

Fig. 14-11. Taper-shank chucking reamer with helical flutes.

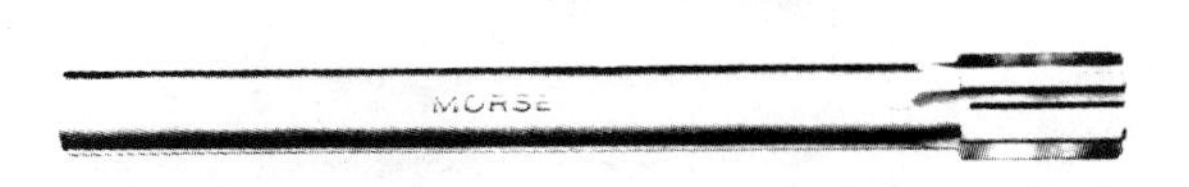

Courtesy Morse Twist Drill & Tool Company

Fig. 14-12. Expansion chucking reamer for light finishing cuts in all types of materials. It should not be used to cut oversize holes.

Rose Chucking Reamer—This machine reamer is made primarily for roughing cuts (Fig. 14-13). The leading edge of the rose reamer is given a back-off chamfer of 45°. The reamer gets its name from the fact that the end view has a rose-like appearance.

Shell Reamers—The shell reamer is used for sizing and finishing operations. It is a reamer head which fits either a straight-shank or

Courtesy Morse Twist Drill & Machine Company

Fig. 14-13. Rose chucking reamer. Used primary for roughing cuts.

a taper-shank arbor. Several different sizes of shell reamers may be fitted to the same arbor, which results in a saving in parts.

Shell reamers are available in either straight flutes or in helical flutes (Figs. 14-14 and 14-15). These reamers also are available either with straight shanks or tapered shanks for machine use. The roughing reamers have lateral grooves across the lands, while the finishing reamers have straight, tapered lands. The fluted shell reamer is made to remove small amounts of metal, and cuts along the edges. The *rose shell reamer* has no clearance along the lands and is made to cut on the end. It is usually a few thousandths of an inch undersize so that the hole can be finish reamed with either a hand reamer or a fluted chucking reamer. The rose shell reamer is not intended to be a finishing reamer.

A *shell-type expansion chucking reamer* with replaceable shells (Fig. 14-16) is designed for high-precision reaming in the mass production of parts having close-tolerance holes. The expansion feature is to compensate for loss of size due to wear.

Courtesy Morse Twist Drill & Machine Company

Fig. 14-14. Straight-fluted shell reamer.

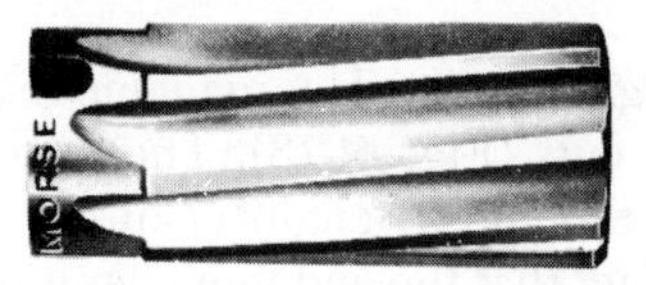

Courtesy Morse Twist Drill & Machine Company

Fig. 14-15. Helical fluted shell reamer.

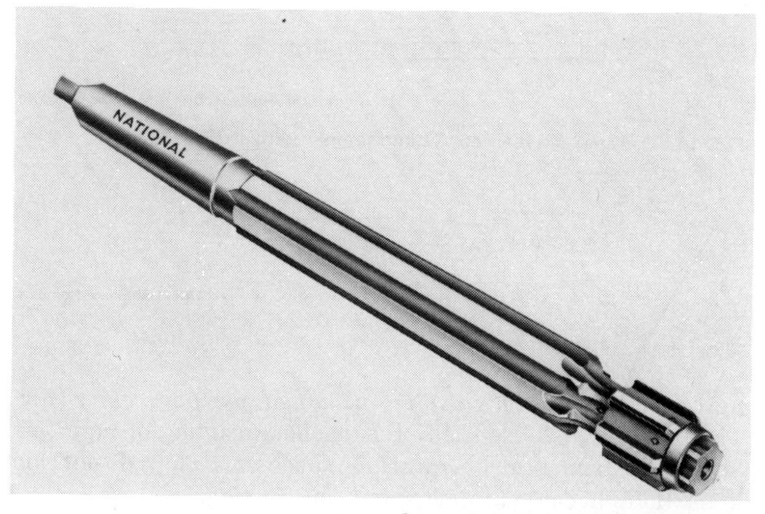

Courtesy National Twist Drill & Tool Company

Fig. 14-16. Shell-type expansion chucking reamer designed for high-precision reaming of parts having close tolerance holes.

Carbide-tipped reamers are used in many production setups and especially where abrasive material such as foundry sand and scale are encountered. The carbide-tipped helical flute chucking reamer (Fig. 14-17) is recommended for highly abrasive materials, heat treated steels, and other hand materials.

Tapered Reamers—Several types of tapered reamers are available for use both in hand reaming and in machine reaming operations.

Taper pin reamers (Figs. 14-18 and 14-19) are all of the same taper; the point of each reamer will enter the hole reamed by the

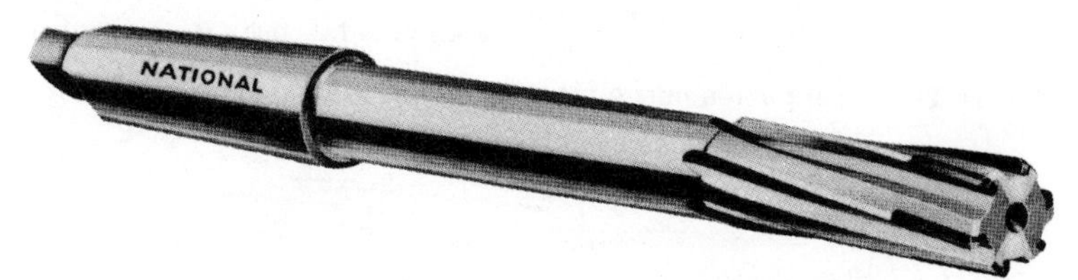

Courtesy National Twist Drill & Tool Company

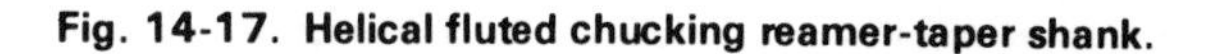

Fig. 14-17. Helical fluted chucking reamer-taper shank.

Courtesy Morse Twist Drill & Machine Company

Fig. 14-18. Taper pin reamer. A hand reamer with straight flutes.

Courtesy Morse Twist Drill & Machine Company

Fig. 14-19. Taper pin reamer. A hand reamer with spiral flutes.

next smaller size. A *high-spiral machine taper pin reamer* (Fig. 14-20) is designed especially for machine reaming of taper pin holes. This reamer can be run at high speeds, and chips do not clog in the flutes because of their spiral construction.

Morse taper reamers are available for both hand reaming and for machine reaming of standard Morse-taper holes in sleeves, sockets, and spindles. A machine Morse taper is shown in Fig. 14-21.

Reamers of the tapered type used for special purposes are the *bridge reamer* (Fig. 14-22) and the *car reamer* (Fig. 14-23). The bridge reamer is used for reaming the rivet and bolt holes in structural iron and steel, boiler plate, and similar work. The deep flutes provide adequate space for chips sheared out by the reamer, and the tapered point facilitates entering holes that are out of alignment.

Courtesy Morse Twist Drill & Machine Company

Fig. 14-20. Taper pin reamer. A machine reamer with high spirals.

Courtesy Morse Twist Drill & Machine Company

Fig. 14-21. Morse taper reamer for machine finishing of Morse taper holes.

Courtesy Morse Twist Drill & Machine Company

Fig. 14-22. Bridge reamer. Used for reaming the rivet and bolt holes in structural iron and steel.

Courtesy Morse Twist Drill & Machine Company

Fig. 14-23. Car reamer.

The *repairman's taper reamer* (Fig. 14-24) is used by automobile and bicycle repairmen, blacksmiths, electricians, machinists, plumbers, and carpenters for enlarging holes in thin metals, etc. The straight-shank repairman's taper reamer (Fig. 14-25) may be used in an electric drill.

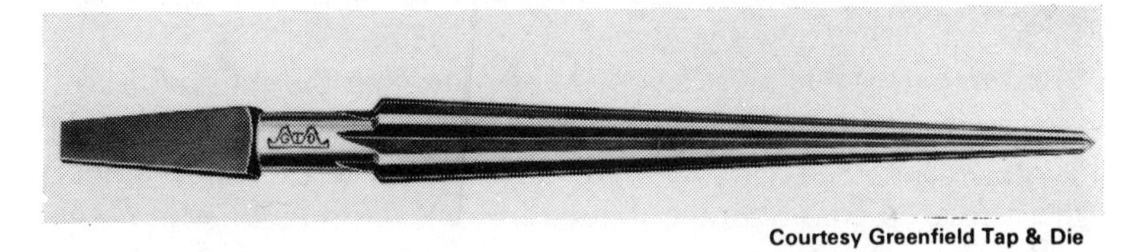

Courtesy Greenfield Tap & Die

Fig. 14-24. Repairman's taper reamer. Used for enlarging holes in thin metals.

The *diemakers' reamer* (Fig. 14-26) is another special-purpose reamer. In laying out a die, holes can be drilled close together, outlining the shape desired, then reamed with the diemakers' reamer until the holes run together and the central piece drops out. The reamer can then be run along the outline of the edge of the die as a spiral mill; the resulting clearance is correct for the finished die. Rapid action and freedom from clogging are two valuable features of this reamer. Another valuable feature is its freedom from breakage, as the easy shearing cut imposes little strain upon the cutting edges.

Pipe reamers are tapered reamers for reaming holes to be

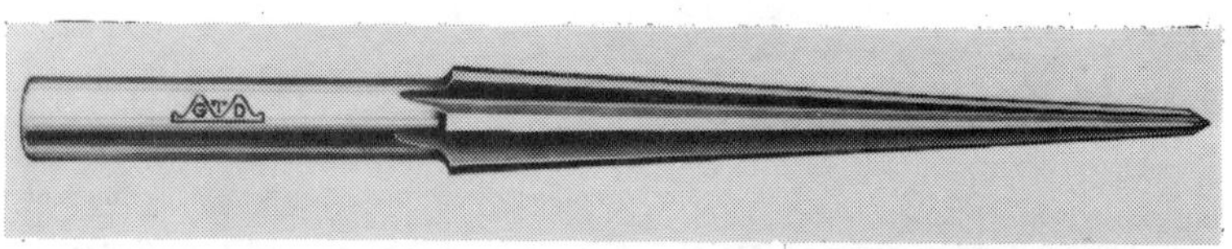

Courtesy Greenfield Tap & Die

Fig. 14-25. Repairman's taper reamer with straight shank for use in an electric drill.

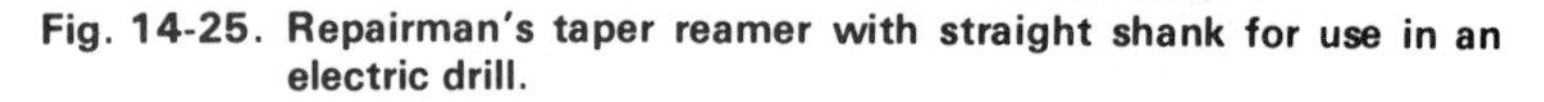

Courtesy Morse Twist Drill & Machine Company

Fig. 14-26. Diemaker's reamer. Taper is 0.013 inch per inch.

tapped with American National Standard Taper Pipe Taps (Fig. 14-27). The pipe reamer is also made with inserted lands (Fig. 14-28). All sizes of these reamers are tapered ¾ inch to the foot. The nominal size refers to the pipe size for which the reamer is intended, rather than the actual size of the reamer.

Burring reamers are designed for removing the internal burrs caused by cutting pipe (Figs. 14-29 and 14-30). The straight-shank burring reamer may be used in an electric drill. Burring reamers are also used for countersinking and for enlarging holes in sheet metal. A *ratchet burring reamer* (Fig. 14-31) may also be used for enlarging holes in sheet metal, countersinking, etc.

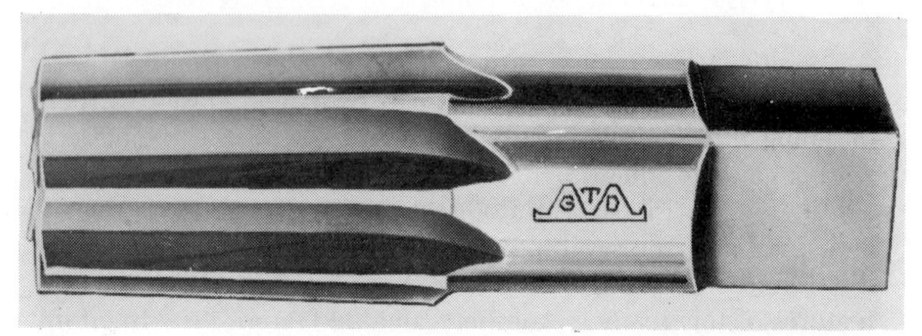

Courtesy Greenfield Tap & Die

Fig. 14-27. Taper pipe reamer with straight flutes for reaming holes to be tapped with American National Standard Taper Pipe Taps.

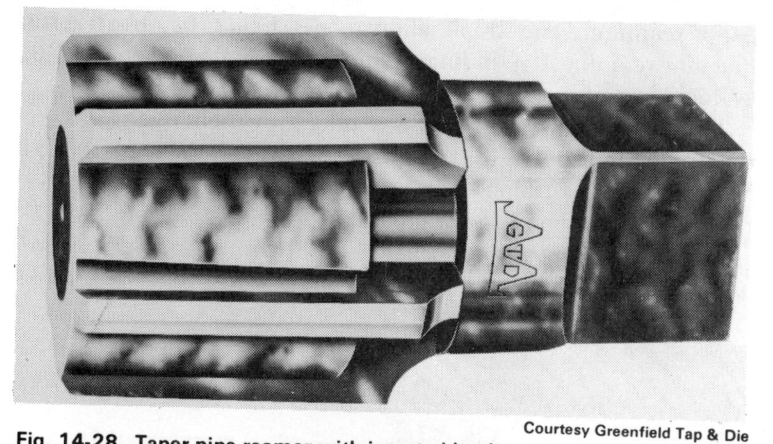

Courtesy Greenfield Tap & Die

Fig. 14-28. Taper pipe reamer with inserted lands.

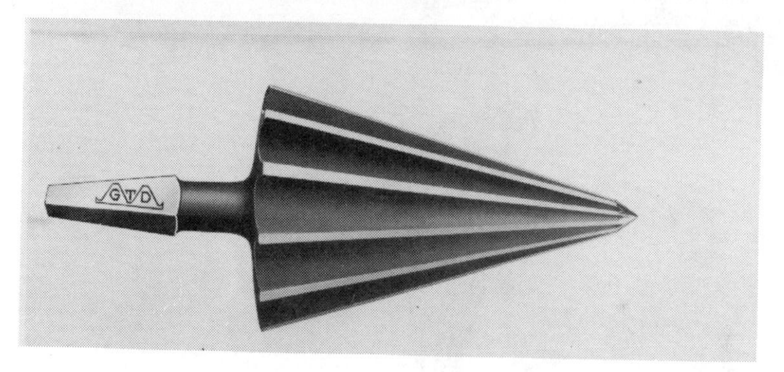

Courtesy Greenfield Tap & Die

Fig. 14-29. Burring reamer for removing interior burrs caused by cutting pipe.

USE AND CARE OF REAMERS

Proper use of reamers determines the accuracy of drilled holes and the quality of their finish. Proper resharpening of reamers, suitable cutting fluids, rigidity of machine tools and fixtures, and correct feeds and speeds are all factors related to the proper use of reamers.

An adequate amount of stock should be left for the reamer to cut. About 0.001 inch to 0.003 inch of stock should be allowed for hand reaming. The stock allowances should be greater for machine reaming, depending on the size: 1/4-inch diameter, 0.010; 1/2-inch diameter, 0.015; and 1-inch diameter, 0.020.

The recommended speeds for reamers are about two-thirds the recommended speeds for twist drills. However, the feeds can be greater than for drilling. Between 0.0015 inch and 0.004 inch per flute per revolution is recommended for reamers. Chatter, oversize holes, and poor finish usually result from lack of alignment and rigidity in machine tools and fixtures.

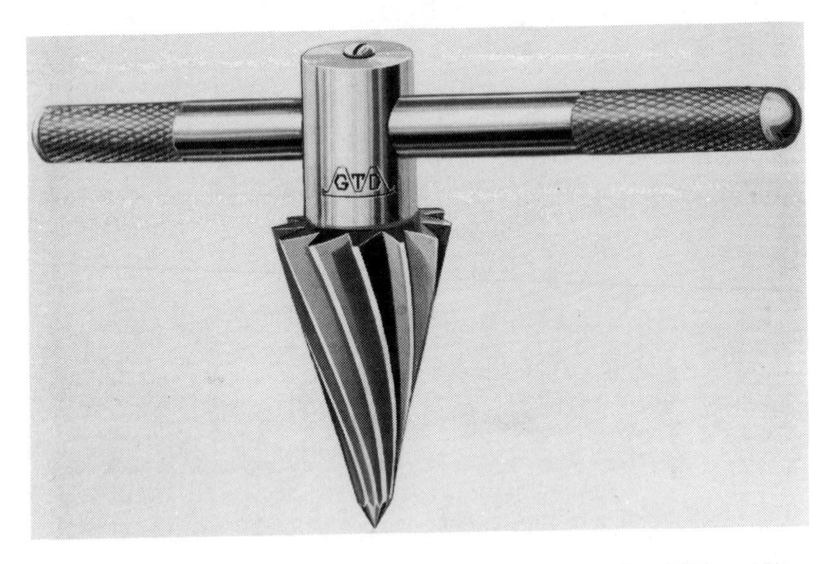

Courtesy Greenfield Tap and Die

Fig. 14-30. Burring reamer with T-handle shank.

Reamers should not be permitted to become excessively dull before resharpening. Proper equipment should be used for sharpening, and great care should be exercised.

Reamers should be transported and stored in containers having separate compartments for each reamer. These tools are delicate and easily damaged. When not in use, they should be covered with a rust preventive.

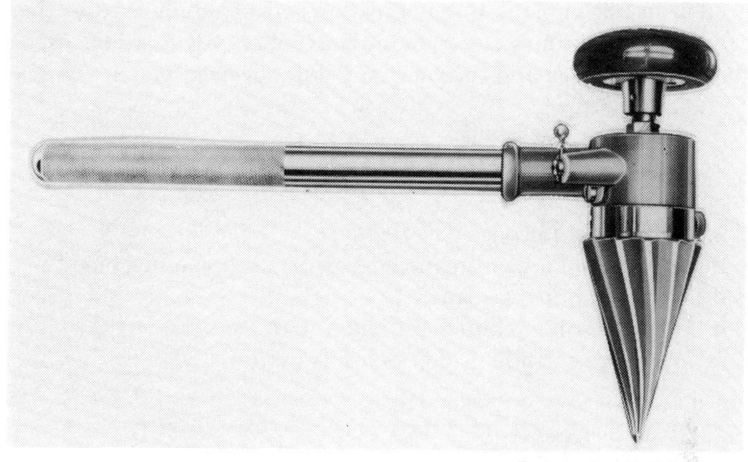

Courtesy Greenfield Tap & Die

Fig. 14-31. Ratchet burring reamer.

SUMMARY

A reamer is a precision tool designed to finish, to a given diameter, a hole that has been made by drilling. Where precision is required, the hole is first drilled to within a few thousandths of an inch of the correct size and then is reamed to the exact size.

There are various types of hand and machine reamers. Hand reamers have a square end on the shank to engage a hand wrench. The flute or cutting part on hand reamers is slightly tapered on the end to start the reamer properly. Hand reamers are made with either straight or helical flutes.

There is a difference between machine and hand reamers. If a hand reamer is used in a machine, it will be ruined. A machine reamer has a shank similar to a twist drill and is made for high-speed operation. Stub reamers are designed primarily for use in automatic screw machines where short tools are required. A pin hole through the shank is provided to adapt the reamer for use in floating holders.

Burring reamers are designed to remove the internal burrs caused by cutting pipe. Straight-shank burring reamers may be used in an electric drill. Ratchet reamers are generally used on the job by plumbers and electricians. Burring reamers are also used for countersinking and for enlarging holes in sheet metal.

REVIEW QUESTIONS

1. What is the purpose of a reamer?
2. Why should a hand reamer never be used in a machine?
3. What is a burring reamer?
4. Why are some reamers designed with a taper?
5. When are carbide tipped reamers used?

CHAPTER 15

Taps

A tap is a tool used for cutting internal threads. Taps are threaded accurately and fluted—the flutes extend the length of the threaded portion, forming a series of cutting edges. Taps are made of carbon steel and high-speed steel. Carbon steel taps are used for hand-tapping of cast iron and general work that is not too severe. High-speed steel taps are used for either hand or machine tapping on tough or abrasive materials, such as aluminum, brass, Bakelite, malleable iron, die castings, fiber, hard rubber, low-carbon steels, and other materials with similar characteristics.

The threads of taps are either *cut* or *ground*. Precision work requires taps that have ground threads. Taps are available with two, three, or four flutes. The flutes may be straight, angular, or helical.

TYPES OF TAPS

The object of the tapping operation is to make helical grooves, or threads, in holes so that they may hold bolts, studs, screws, etc. The most common taps are the hand tap and the machine-screw tap. Other types of taps are pipe, pulley, nut, tapper, and special taps of various kinds.

Hand Taps

Originally, hand taps were intended only for hand operation (Figs. 15-1 and 15-2), but they are now widely used on machine production work. Standard hand taps, starting with machine-screw size (No. 0), are made in sets of three (Fig. 15-3). These sets of three are standard in the Unified Screw Thread form as follows:

1. The *taper* tap is used to start the thread (Fig. 15-3A). At least six threads are tapered or chamfered before full diameter of the tap is reached.

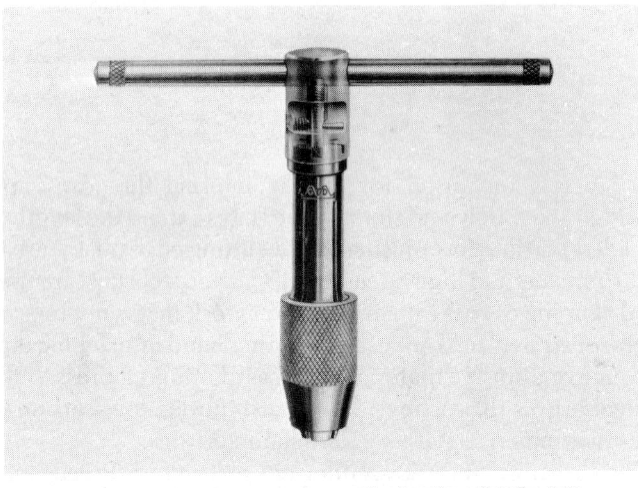

Courtesy Greenfield Tap & Die

Fig. 15-1. T-handle tap wrench with slip handle designed for use in hand tapping. This wrench can also be used on twist drills, reamers, screw extractors, etc.

2. The *plug* tap is used to cut the threads as fast as possible after the taper tap has been used (Fig. 15-3B). Three to five threads of the plug tap are chamfered or tapered.
3. The *bottoming* tap is used last to drive the thread to the bottom of a blind hole (Fig. 15-3C). This tap has only 1½ threads chamfered.

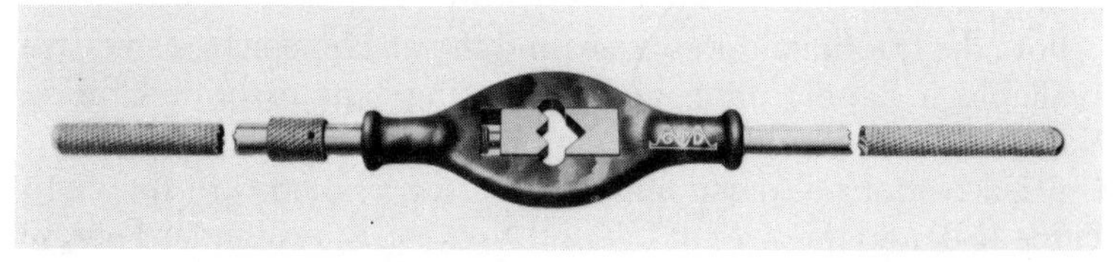

Courtesy Greenfield Tap & Die

Fig. 15-2. Tap wrench with solid handles. One handle is movable and can be turned to adjust the center.

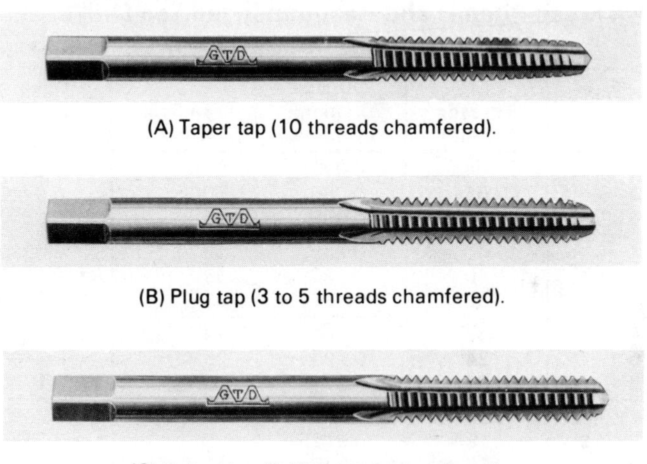

(A) Taper tap (10 threads chamfered).

(B) Plug tap (3 to 5 threads chamfered).

(C) Bottoming (1 -½ threads chamfered).

Courtesy Greenfield Tap & Die

Fig. 15-3. Set of three hand taps used in succession for tapping a blind hole. These taps are identical in general dimensions, their only difference being in the chamfered threaded portion at the point.

These taps are used in order when tapping a blind hole (a hole which does not extend entirely through the work) and it is desired to cut threads to the bottom of the hole. For a hole extending entirely through the work, of course, only the taper tap and plug tap are necessary.

Taps are available in other than the Unified Thread form. Other thread forms are: Metric, Whitworth, Acme, and Square threads.

Size of Taps—The size of hand taps ¼ inch in diameter and larger is indicated in fractional dimensions. Machine-screw taps are really small standard hand taps. Their size is indicated by the machine-screw system of sizes, ranging from No. 0 (0.060 inch) up to No. 12 (0.216 inch).

Both the fractional screw sizes and the machine-screw sizes are available in taper, plug, and bottoming taps in both Unified National Coarse and Unified National Fine Threads. The sizes of Unified National Coarse Standard Screw Threads are shown in Table 15-1, and the sizes of Unified National Fine Standard Screw Threads are shown in Table 15-2.

Table 15-1. Unified National Coarse Standard (UNC) Screw Thread Pitches and Recommended Tap Drill Sizes

Formerly American National Form Thread

Sizes	Threads Per Inch	Outside Diameter of Screw	Tap Drill Sizes	Decimal Equivalent of Drill
1	64	0.073	53	0.0595
2	56	0.086	50	0.0700
3	48	0.099	46	0.0810
4	40	0.112	43	0.0890
5	40	0.125	38	0.1015
6	32	0.138	33	0.1130
8	32	0.164	29	0.1360
10	24	0.190	25	0.1495
12	24	0.216	16	0.1770
1/4	20	0.250	7	0.2010
5/16	18	0.3125	F	0.2570
3/8	16	0.375	5/16	0.3125
7/16	14	0.4375	U	0.3680
1/2	13	0.500	27/64	0.4219
9/16	12	0.5625	31/64	0.4843
5/8	11	0.625	17/32	0.5312
3/4	10	0.750	21/32	0.6562
7/8	9	0.875	49/64	0.7656
1	8	1.000	7/8	0.875
1⅛	7	1.125	63/64	0.9843
1¼	7	1.250	1 7/64	1.1093

Courtesy South Bend Lathe, Inc.

Size of Tap Drill to Use—As tap size is determined by the major diameter of its threads, it is evident that it is necessary to drill the hole for tapping smaller than the tap size by nearly twice the depth of thread. The drilled hole must be small enough to leave sufficient stock in which to cut the screw threads. Generally, a tap drill that will give approximately a 75 percent thread is used. Tap drill size can be determined either from Table 15-1 or Table 15-2, or by use of the following formula:

$$\text{Tap drill size} = \underset{\text{(of thread)}}{\text{Major diameter}} - \frac{(0.75 \times 1.299)}{\text{No. threads per inch}}$$

Table 15-2. Unified National Fine Standard (UNF) Screw Thread Pitches and Recommended Tap Drill Sizes

Sizes	Threads Per Inch	Outside Diameter of Screw	Tap Drill Sizes	Decimal Equivalent of Drill
0	80	0.060	3/64	0.0469
1	72	0.073	53	0.0595
2	64	0.086	49	0.0730
3	56	0.099	44	0.0860
4	48	0.112	42	0.0935
5	44	0.125	37	0.1040
6	40	0.138	32	0.1160
8	36	0.164	29	0.1360
10	32	0.190	21	0.1590
12	28	0.216	14	0.1820
1/4	28	0.250	7/32	0.2187
5/16	24	0.3125	I	0.2720
3/8	24	0.375	R	0.3390
7/16	20	0.4375	25/64	0.3906
1/2	20	0.500	29/64	0.4531
9/16	18	0.5625	33/64	0.5156
5/8	18	0.625	37/64	0.5781
3/4	16	0.750	11/16	0.6875
7/8	14	0.875	13/16	0.8125
1	14	1.000	15/16	0.9375
1 1/8	12	1.125	1 3/64	1.0468
1 1/4	12	1.250	1 11/64	1.1718

Courtesy South Bend Lathe, Inc.

Example: For a hole ¾ inch in diameter, 10 threads per inch, substitute in above formula.

$$\text{Tap drill size} = 0.750 - \frac{(0.75 \times 1.299)}{10}$$

$$= 0.750 - 0.097$$

$$= 0.653 \text{ (decimal equivalent of size of tap drill)}$$

$$= 21/32 \text{ (correct tap drill size nearest 0.653)}$$

For all practical purposes a simpler formula can be used as follows:

$$\text{Tap drill size} = \text{Major diameter} - \frac{1}{N \text{ (No. threads per inch)}}$$

$$= 3/4 - 1/10$$

$$= 0.750 - 0.1$$

$$= 0.650 \text{ or nearest decimal to } 21/32$$

Other Types of Hand Taps—In addition to the standard taps, consisting of taper, plug, and bottoming taps mentioned, other styles of hand taps are available in which the points and flutes are designed differently.

Spiral-pointed or "gun" hand taps with plug chamfer are designed primarily for use in through holes (Fig. 15-4A). The spiral point forces the chips ahead of the tap. This prevents the chips from clogging in the flutes and causing tap breakage. It also eliminates possible damaging of the threads by the chips when the tap is reversed. The bottoming chamfered spiral-pointed tap is used in blind holes where a space is provided at the bottom of the hole for chip collection (Fig. 15-4B).

Spiral-fluted hand taps are recommended for tapping blind holes in such ductile materials as aluminum and magnesium where chip removal is a problem (Fig. 15-5). These taps are most effective when the material being tapped produces long, stringy, curling chips. The taps cut freely while ejecting chips from the tapped hole, which prevents clogging and damage to both the

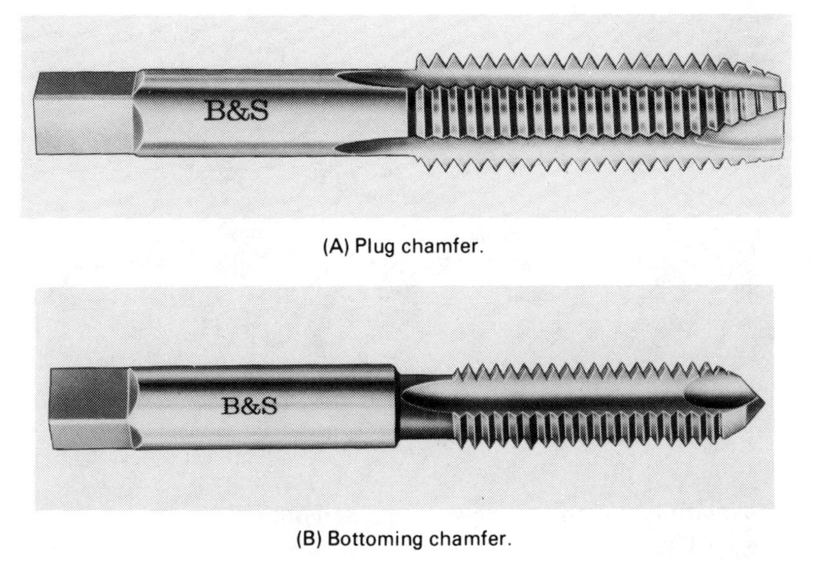

(A) Plug chamfer.

(B) Bottoming chamfer.

Courtesy American Twist Drill Company

Fig. 15-4. Spiral-pointed hand taps.

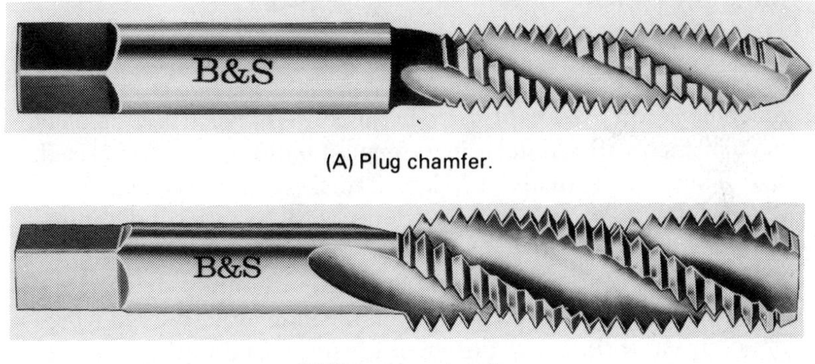

(A) Plug chamfer.

(B) Bottoming chamfer.

Courtesy American Twist Drill Company

Fig. 15-5. Spiral-fluted hand taps.

threaded parts and the taps. *Fast spiral-fluted hand taps* are recommended for deep blind holes where chip removal is a problem (Fig. 15-6).

Spiral-pointed hand taps with a short flute are used for through

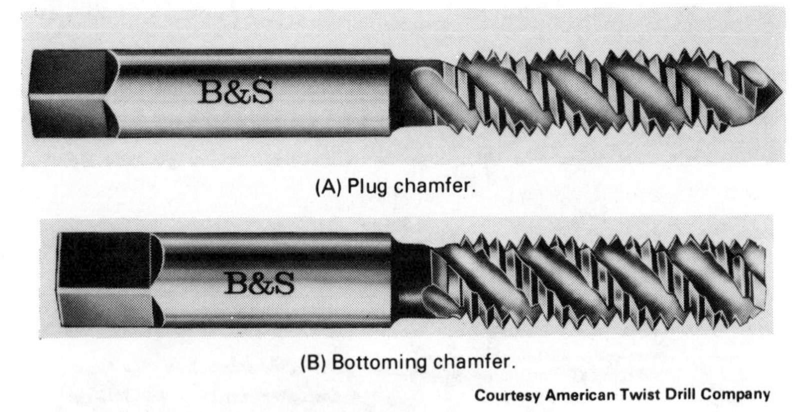

(A) Plug chamfer.

(B) Bottoming chamfer.

Courtesy American Twist Drill Company

Fig. 15-6. Spiral-fluted hand taps.

hole tapping of holes in sheet metal and other thin sections (Fig. 15-7). This design is suitable where the tapped hole is not deeper than the diameter of the tap.

Machine-Screw Taps

There is in reality no difference in standard hand taps and machine-screw taps, except that the latter are made in the machine-screw system of sizes. In this system, No. 0 is equivalent of 0.060 inch with a regular increment of 0.013 inch between each size. Thus, No. 1 equals 0.073 inch, No. 2 equals 0.086 inch, No. 3 equals 0.099 inch, etc. Machine-screw taps are used for tapped holes below 1/4 inch in diameter. They are available in the Unified Screw form of thread in taper, plug, and bottoming styles.

Machine-screw taps are also available in several styles in which the points and flutes are designed for improved performance in certain applications and materials. The *fast spiral-fluted machine-*

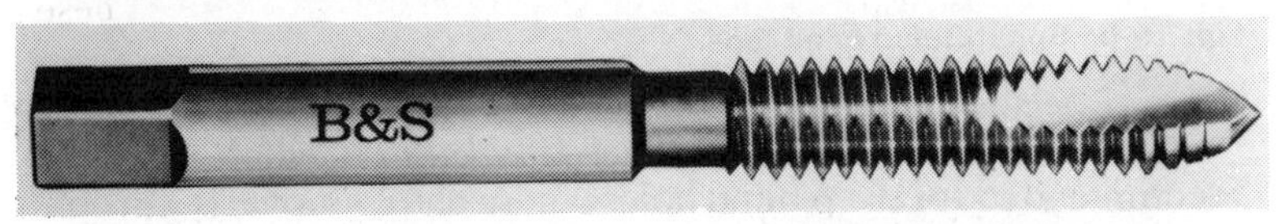

Courtesy American Twist Drill Company

Fig. 15-7. Spiral-pointed hand tap with short flute (plug chamfer).

Table 15-3. Metric Tap Drill Sizes

TAPPING DRILL DIAMETERS FOR
METRIC SCREW THREADS

Nominal Dia. of Thread (In mm)	Depth of External Thread (In mm)	Tapping Drill Dia. (In mm)	Radial Engagement with Ext. Thread %
1.6	0.2147	1.30	70
1.8	0.2147	1.50	70
2	0.2454	1.65	71.3
2.2	0.2760	1.80	72.5
2.5	0.2760	2.10	72.5
3	0.3067	2.60	65.2
3.5	0.3681	3.00	68
4	0.4294	3.40	70
4.5	0.4601	3.90	65.3
5	0.4908	4.30	71.3
6	0.6134	5.20	65.4
7	0.6134	6.20	65.4
8	0.7668	7.00	65.2
9	0.7668	8.00	65.2
10	0.9202	8.80	65.1
11	0.9202	9.80	65.1
12	1.0735	10.60	65.2
14	1.2269	12.40	65.2
16	1.2269	14.25	71
18	1.5336	16.00	65.2
20	1.5336	18.00	65.2
22	1.5336	20.00	65.2
24	1.8403	21.50	68
27	1.8403	24.50	68
30	2.1470	27.00	69.8

screw tap is shown in Fig. 15-8. It is recommended for tapping deep blind holes in aluminum and magnesium.

Thread forming (roll) taps (Fig. 15-9) are fluteless taps that do not cut threads in the same manner as conventional taps. These taps are forming tools, and the threading action is similar to the rolling process used to produce external threads. Thread rolling taps produce a strong thread, and due to the forming action the thread surface is somewhat work-hardened. Tap drills for roll taps must be larger than those used for the size diameter thread with conventional taps.

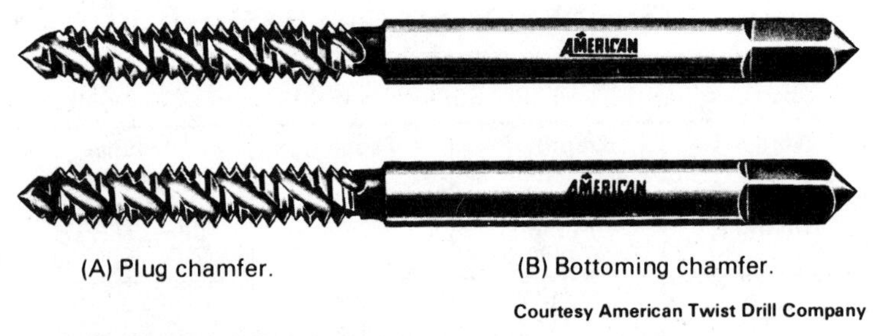

(A) Plug chamfer. (B) Bottoming chamfer.

Courtesy American Twist Drill Company

Fig. 15-8. Fast spiral-fluted machine-screw tap.

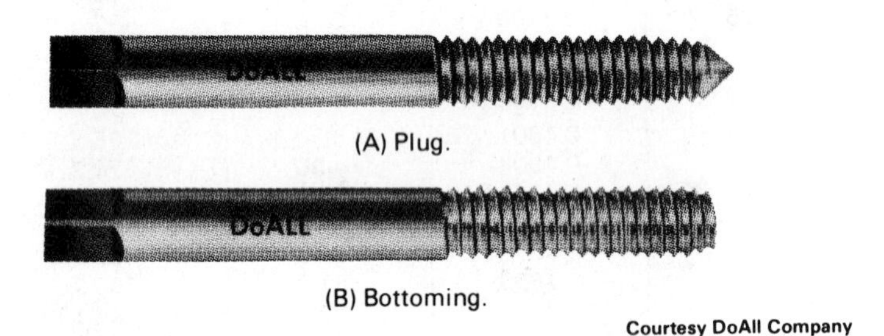

(A) Plug.

(B) Bottoming.

Courtesy DoAll Company

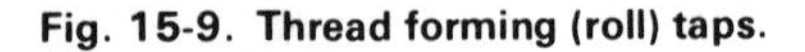

Fig. 15-9. Thread forming (roll) taps.

Pipe Taps

Pipe taps are made in two thread forms: straight and taper. The nominal size of a pipe tap is the same as that of the fitting to be tapped—not the actual size of the tap. *Straight pipe taps* (Fig. 15-10) with American Standard Pipe Thread (NPS) form are intended for tapping holes or couplings for low-pressure work to

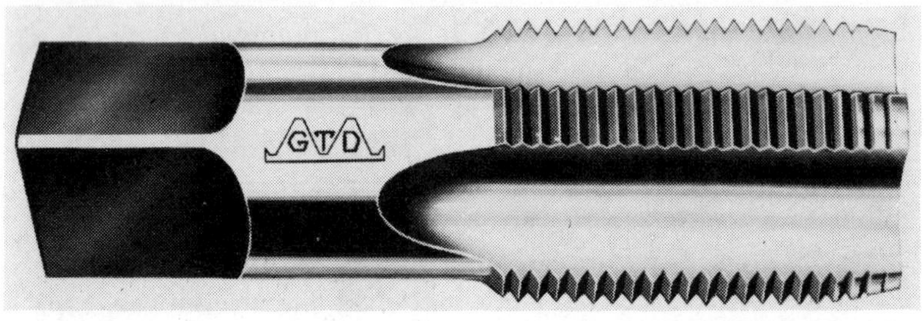

Courtesy Greenfield Tap & Die

Fig. 15-10. Straight pipe tap.

assemble with taper-threaded pipe or fittings to secure a tight joint with the use of a lubricant or sealer.

Taper pipe taps are available with both regular and interrupted threads (Figs. 15-11 and 15-12). These taps are furnished with American Standard Pipe (NPT) form of thread having a taper of ¾ inch per foot. The interrupted-thread taps have every other thread removed, except for a few threads at the point. These taps are used for tapping tough metals, which have a tendency to "load" the teeth of the tap and should be used only when the regular thread taps fail.

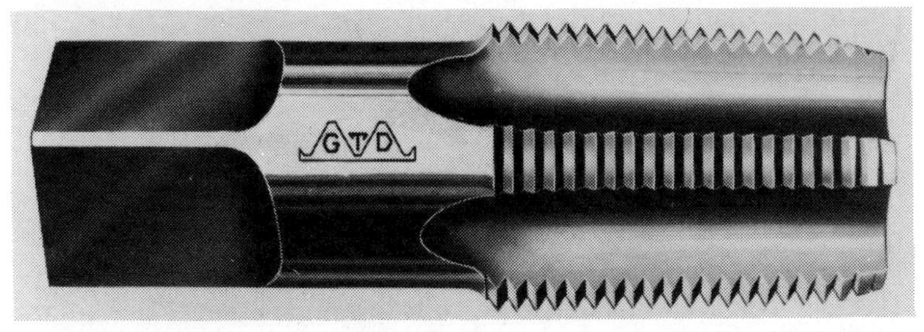

Courtesy Greenfield Tap & Die

Fig. 15-11. Taper pipe tap with regular thread.

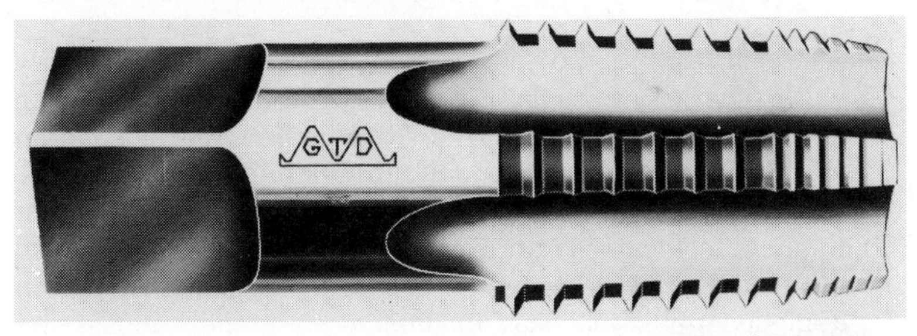

Courtesy Greenfield Tap & Die

Fig. 15-12. Taper pipe tap with interrupted thread. Every other thread is removed except for the first few threads at the point.

Nut Taps

These taps are used where the work requires a tap with a long shank, or where the hole is of a greater depth than can be handled by a hand tap (Fig. 15-13). Nut taps are used for tapping small

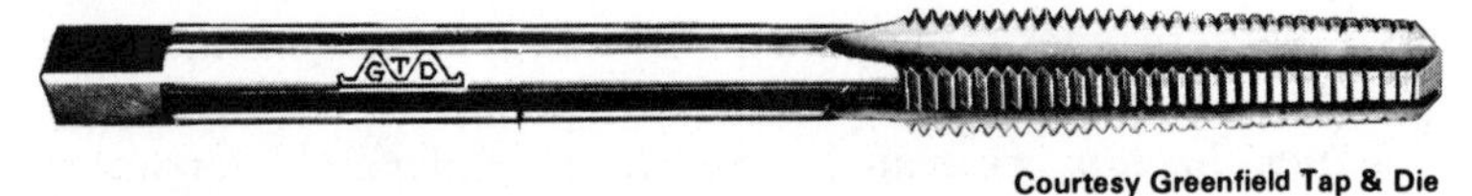

Courtesy Greenfield Tap & Die

Fig. 15-13. Nut taps.

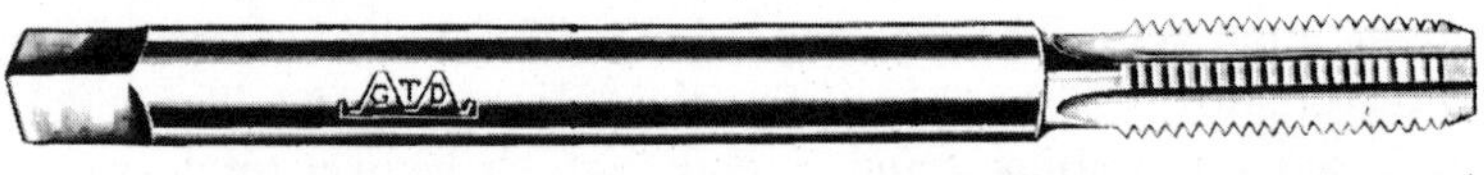

Courtesy Greenfield Tap & Die

Fig. 15-14. Pulley taps used for tapping setscrew and oil cup holes in the hubs of pulleys.

quantities made from tough material, such as stainless steel and similar alloys.

Pulley Taps

The setscrew and oilcup holes in the hubs of pulleys are tapped with pulley taps (Fig. 15-14). These taps are available in several different overall lengths because of the variation in diameter of pulleys. The long shank also permits tapping in places that might be inaccessible to hand taps.

Tapper Taps

These taps are used by nut manufacturers. They are regularly furnished with long shanks and short threads, and as succeeding nuts are tapped, they run up on the shank until it is full; then the tap is removed from the holder, and the nuts slide off (Fig. 15-15). The threaded section is made as short as practical, and a smaller number of teeth are chamfered than on the nut taps.

Bent-shank tapper taps (Fig. 15-16) are designed for use in automatic tapping machines manufactured by the National Machinery Co. Tapping in these machines is continuous—the nuts are fed to the tap automatically. There are different sizes of the machine; each machine requires a given shank length and radius in the bend.

Special-Purpose Taps

Spark-plug taps are available with the International form of thread in plug style only.

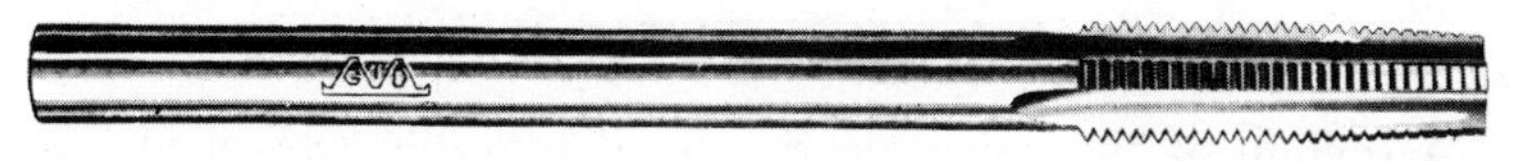

Courtesy Greenfield Tap & Die

Fig. 15-15. Tapper tap with straight shank.

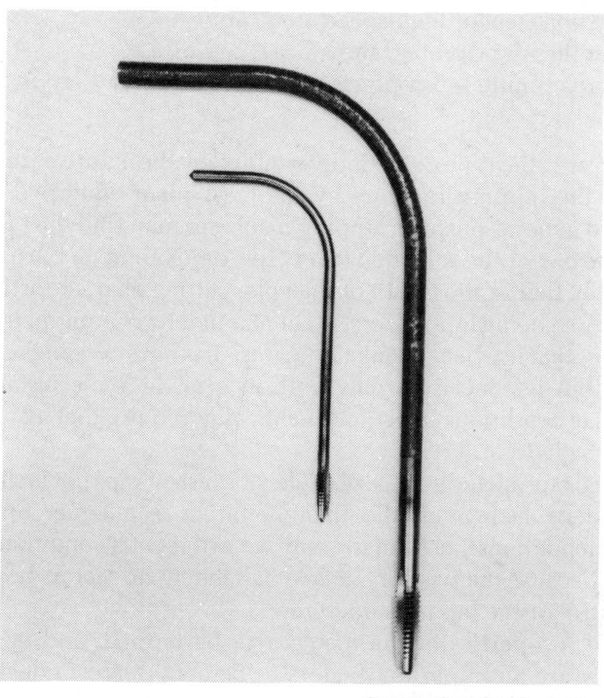

Courtesy Greenfield Tap & Die

Fig. 15-16. Bent-shank taper tap.

Acme threads are used extensively for transmitting and controlling power in lead screws on machine tools, and on valves, jacks, and other mechanisms. Acme tap sets usually consist of one or more roughing taps and a finishing tap. The ground-thread taps are usually used on Acme screw and nut assemblies, because there is less wear when the lead of the screw and nut are matched accurately.

TAP SELECTION

A review of the principles involved may be helpful in tap selection. The following tap features must be considered in selecting the proper tap for a given job:

1. The type of tap.
2. Carbon steel or high-speed steel tap.
3. Cut thread or ground thread.
4. Thread limits—if a precision or ground-thread tap is to be used.

Generally, the type of tap is determined by the nature of the job and by the tapping facilities available. A plant equipped with standard general-purpose tapping machines may find the proper tap to be one of the so-called hand taps, depending on the nature of the job, that is, the depth of the hole, cutting characteristics of the material being tapped, etc. A nut manufacturer equipped with machines that use bent-shank tapper taps has no choice in the type of tap, but is concerned only with its size. In some instances, particular conditions or requirements may require that a special tap be made to order.

Special taps include any size, style, or finish of taps not included in standard listings in the manufacturers' catalogue. Special recommendations can be made only when the exact conditions are known because the type of tap is so dependent on factors beyond the control of the tap manufacturers.

Speed, properties of the material to be tapped, and desired accuracy are all important factors to be considered in deciding whether to use carbon steel or high-speed steel taps. Carbon steel taps may be run about one-half as fast as high-speed steel taps.

Carbon steel taps may be used efficiently in brass and ferrous metals. Most nonferrous metals and abrasive materials, such as Bakelite, fiber, or hard rubber, turn cutting edges quickly and require the use of high-speed steel taps. If accuracy is required, the ground-thread taps necessary for this type of service are regularly available only in high-speed steel.

Accuracy is the only important factor to be considered in determining whether to use cut-thread or ground-thread taps. Cut-thread taps are formed by milling or "cutting." The manufacturing tolerances or variations in size and form of cut threads are not held to such uniformity or to such close "limits" as are the ground-thread taps. Ground-thread taps are finish ground with abrasive wheels to extremely close tolerances. Accuracy of the required thread is determined by the class of thread desired in the threaded parts. Thus, if a tapped hole and a threaded stud are required to "fit" to a very close tolerance, threads of "closer limits" will be required than if a "loose fit" is acceptable.

To select correctly the size of tap for a particular job, an understanding of the relationship between *tap pitch diameter tolerance* and *gage limits* is necessary. Pitch diameter tolerances are indicated as follows:

1. For taps *through 1 inch diameter*:
   - L1 = Basic to Basic minus 0.0005 inch.
   - H1 = Basic to Basic plus 0.0005 inch.
   - H2 = Basic plus 0.0005 inch to Basic plus 0.001 inch.
   - H3 = Basic plus 0.001 inch to Basic plus 0.0015 inch.
   - H4 = Basic plus 0.0015 inch to Basic plus 0.002 inch.
   - H5 = Basic plus 0.002 inch to Basic plus 0.0025 inch.
   - H6 = Basic plus 0.0025 inch to Basic plus 0.003 inch.
2. For taps *over 1 inch diameter through 1 1/2 inch diameter;*
   - H4 = Basic plus 0.001 inch to Basic plus 0.002 inch.

CLASSES OF THREADS

The various degrees of snugness of fit are expressed in terms of "class of threads." The classes of threads are given in Table 15-4.

The six classes include three for screws and three for nuts. The external classes of threads are identified as 1A, 2A, and 3A; 1B, 2B, and 3B refer to the internal classes of threads. Class 2A and 2B threads are the free-fitting type found on a majority of the commercial threaded fasteners. Any unified class of external thread may be mated with any internal class of thread as long as the product meets the specified tolerance and allowance.

After the class of threads is determined, the standard gage limits of threads that correspond to that fit must be considered, and a tap selected to produce threads conforming to these predetermined

Table 15-4. Classes of Threads—Unified Thread

Class 1A and 1B	The combination of Class 1A for external threads and Class 1B for internal threads is intended to covef the manufacture of threaded parts **where quick and easy assembly is necessary or desired** and an allowance is required to permit ready assembly.
Class 2A and 2B	The combination of Class 2A for external threads and Class 2B for internal threads is designed **for screws, bolts, and nuts.** It is also suitable for a wide variety of other applications. A similar allowance is provided which minimizes galling and seizure encountered in assembly and use. It also accommodates, to a limited extent, platings, finishes or coatings.
Class 3A and 3B	The combination of Class 3A for external threads and Class 3B for internal threads is provided for those applications **where closeness of fit and accuracy of lead and angle of thread are important.** These threads are obtained consistently only by use of high quality production equipment supported by a very efficient system of gaging and inspection. No allowance is provided.

Courtesy American Twist Drill Co.

limits of size. If the "limits" of cut-thread taps are not close enough to assure the class of fit or quality of thread desired, the choice, of course, is a ground-thread tap.

Determination of either too fast or too slow tapping speed is essential to efficient tapping. There are certain speeds at which taps operate efficiently in specific materials and are shown in Table 15-5. The composition of the material to be threaded, the kind of steel from which the tap is made, and the design of both the tap and tapping machine are all important factors in determining the proper tapping speed for a given job. High or maximum speeds may be determined by gradual stages of experimentation.

SUMMARY

A tap is a precision tool used for cutting internal threads. The threads of taps are either cut or ground. Precision work requires taps that have ground threads. Taps are available with two, three, or four flutes, which may be straight, angular, or helical. Hand taps were intended only for hand operation, but are widely used on machine production work. Standard hand taps are generally made

Table 15-5. Recommended Cutting Speeds and Lubricants for Machine Tapping

Material	Speeds In Feet Per Minute (ft/min)	Lubricant
Aluminum	90-100	Kerosene and Light Base Oil
Brass	90-100	Soluble Oil or Light Base Oil
Cast Iron	70-80	Dry or Soluble Oil
Magnesium	20-50	Light Base Oil Diluted with Kerosene
Phosphor Bronze	30-60	Mineral Oil or Light Base Oil
Plastics	50-70	Dry or Air Jet
Steels		
Low Carbon	40-60	Sulfur Base Oil
High Carbon	25-35	Sulfur Base Oil
Free Machining	60-80	Sulfur Base Oil
Molybdenum	10-35	Soluble Oil
Stainless	10-35	Soluble Oil

in sets of three: taper tap, plug tap, and bottoming tap. The taper tap is used to start the thread, the plug tap is used to cut the threads as far as possible, and the bottoming tap is used to drive the thread to the bottom of the hole. These taps are used when tapping a blind hole (a hole that does not extend entirely through the work).

Other types of hand taps are spiral-point, which are used generally in through holes, and spiral-fluted, which are recommended for tapping blind holes in ductile materials such as aluminum and magnesium.

REVIEW QUESTIONS

1. How many flutes are generally on a tap?
2. Can hand taps be used in machine work?
3. Standard taps are generally made in sets of three. Name them.
4. What must be considered when selecting a tap for a given job?

5. How much stock should be left for threads in a drilled hole?
6. What is the object of the tapping operation?
7. For what is the taper tap used?
8. For what is the plug tap used?
9. How is the bottoming tap used?
10. How is the size of hand taps ¼ inch in diameter and larger indicated?
11. List types of hand taps *other than* the taper, plug, and bottoming taps.
12. How do fluteless taps cut threads?
13. Where do you use a nut tap?
14. Where do you use bent-shank tapper taps?
15. What four tap features do you need to know in order to select a tap properly?

CHAPTER 16

Threading Dies

A threading die is a tool with an internal thread, similar to the thread in a nut, used to cut *external threads* on bolts and round stock. The thread is cut by the teeth of the die as the die advances, while being turned or screwed onto round stock.

TYPES OF DIES

Several types of thread-cutting dies are available to meet varied conditions. The various threads have been covered in the preceding discussion of taps.

Solid Dies

A special holder is not needed for these dies. A large wrench may be used to turn them. Solid dies are used principally for rethreading bruised or rusty threads, but they may be used for cutting new threads.

Solid square bolt dies are one-piece nonadjustable dies (Fig. 16-1). They are used for cutting threads on bolts and pipes, as well as for repair work on threads.

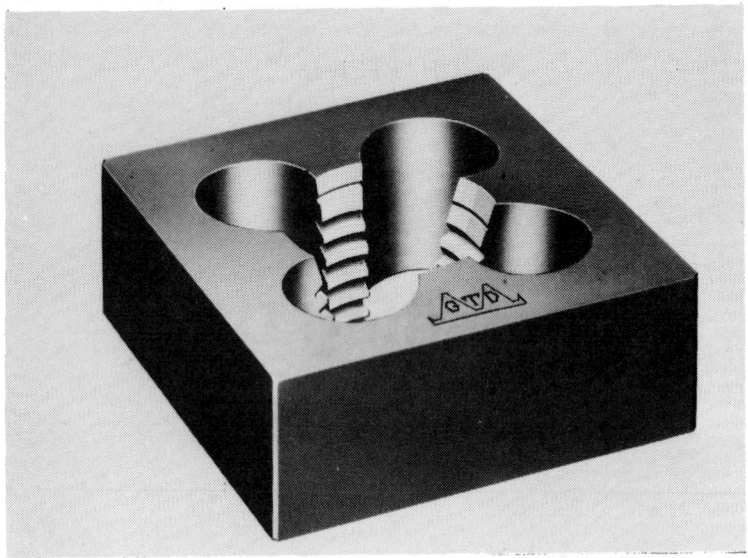

Courtesy Greenfield Tap & Die

Fig. 16-1. Solid square bolt dies.

Solid hexagon rethreading dies, as the name implies, are used principally for repairing bruised or rusty threads (Fig. 16-2). They can be turned with any size wrench that will fit.

Round Split Dies

Adjustable dies of the split type are made in all standardized thread sizes, have limited adjustment for size, and cut threads in easy stages. The round split die is split on one side and can be adjusted to the desired class of thread. Most split dies are set to cut the thread either slightly over or slightly under the designated size.

Round split dies are of two types: (1) screw adjusting, and (2) open adjusting. Adjustment of the screw-adjusting type is by means of a fine-pitch screw which forces the sides of the split die apart, or allows them to spring together (Fig. 16-3). This adjust-

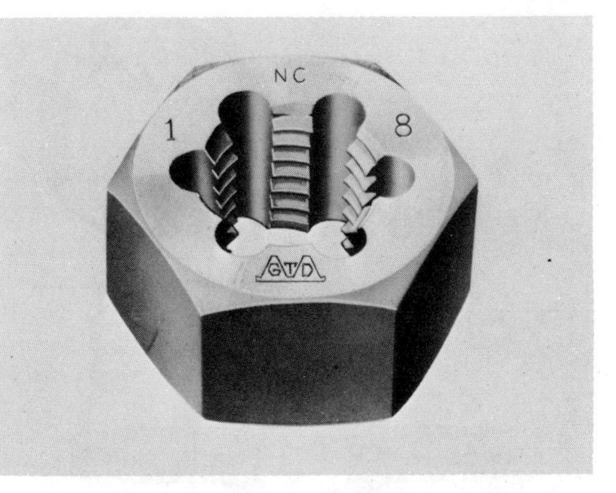

Courtesy Greenfield Tap & Die

Fig. 16-2. Hexagon rethreading dies.

Fig. 16-3. Screw-adjusting type of round adjustable die. A fine-pitch screw forces the sides of the die apart or allows them to spring back together.

Courtesy Morse Twist Drill & Machine Company

ment remains positive when the die is removed from the machine holder or from the hand stock, so that each time the die is used, a new adjustment is not necessary. The slot in these dies is beveled so that the adjusting screw can be removed, if necessary, when the dies are used in a machine holder, and adjustment can be made by the adjusting screw in the holder. The open type of die (Fig. 16-4), designed for use in both screw machines and diestocks, is adjusted by means of three screws in the holder, one screw for expanding and two screws for compressing the dies (Fig. 16-5).

Since the range of adjustment of round dies is limited, only slight adjustments are possible. Adjustments several thousandths of an inch larger than the nominal size of the die result in poor perfor-

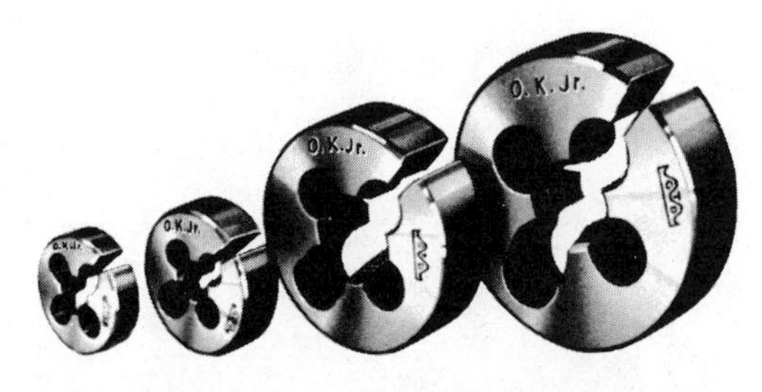

Courtesy Greenfield Tap & Die

Fig. 16-4. The open type of round die.

Courtesy Greenfield Tap & Die

Fig. 16-5. The three-screw type of round diestock for holding round adjustable dies.

mance because of drag on the heel of the thread sections. Excessive expansion may cause the die to break in two. If it is necessary to cut a thread that requires more than a slight adjustment of the die, a two-piece die should be used.

Two-Piece Adjustable Dies

This type of adjustable die is made in two separate halves (Fig. 16-6). The halves are usually held in a collet, which consists of a cap and a guide (Fig. 16-7). The die is adjusted by setscrews at either end of the slot. Pipe threading dies may also be of the two-piece type. These are regularly available with the standard pipe taper of ¾ inch per foot.

"Acorn" Dies

These dies are used in specially designed holders. The "Acorn" die has a tapered nose at the end of the lands, accurately ground concentric with the thread (Fig. 16-8). A solid adjusting cap (Fig.

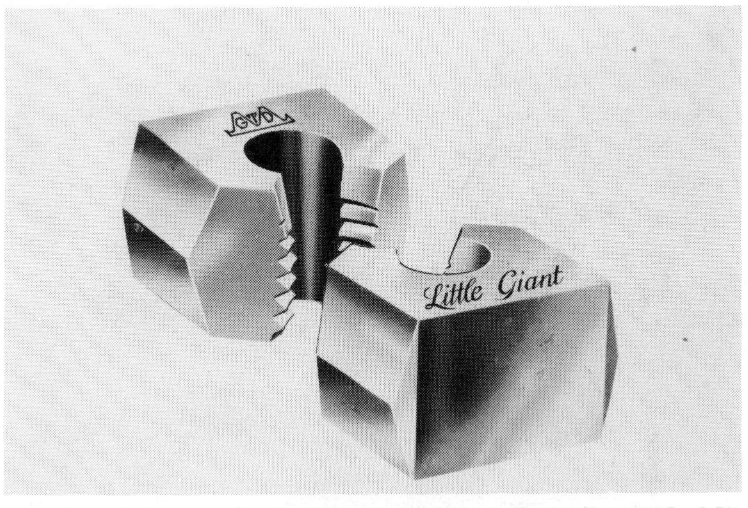

Courtesy Greenfield Tap & Die

Fig. 16-6. Two-piece adjustable die.

16-9) effects a uniform closing in of the lands when it is tightened. The die is driven by lugs on the holder which fit two slots in its base. The adjustment is governed by screwing the cap more or less tightly onto the holder. Proper adjustment is maintained by a lock nut.

"Acorn" die holders (Fig. 16-10) are designed for use on automatic screw and other machines that provide for automatically reversing the die, or the rod, at the instant the desired length of thread has been cut. The body has a longitudinal "float" that allows the die to follow its own lead independently of any lag in the machine.

USE OF DIES TO CUT THREADS

A diestock with two-piece adjustable dies is used to cut pipe threads. The die and ways should be absolutely clean of chips, dust, etc., before placing the die halves in the ways of the diestock. Adjust the two-piece die by loosening the holding bolts, and turn each adjusting bolt until the reference mark on each half die

(A) Cap.

(B) Guide.

(C) Collet.

Courtesy Greenfield Tap & Die

Fig. 16-7. Collet for use with two-piece dies, consisting of a cap and a guide.

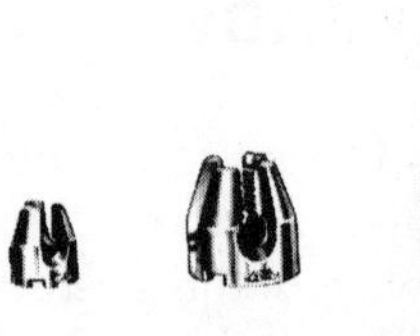

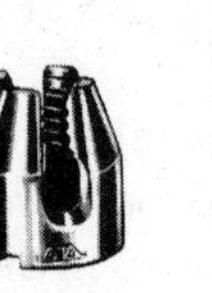

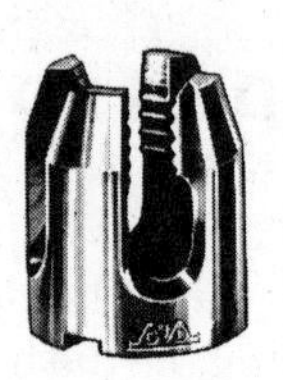

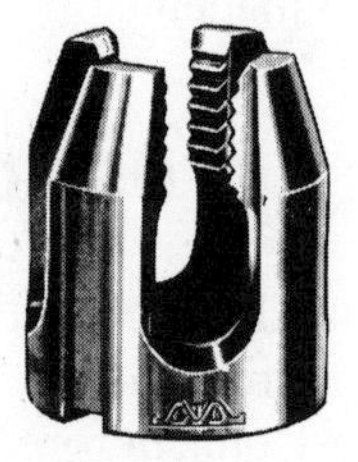

Courtesy Greenfield Tap & Die

Fig. 16-8. Acorn dies for use in specially designed holder.

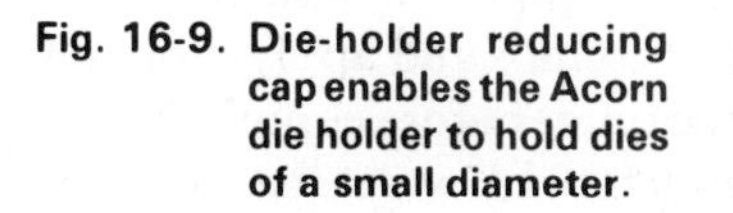

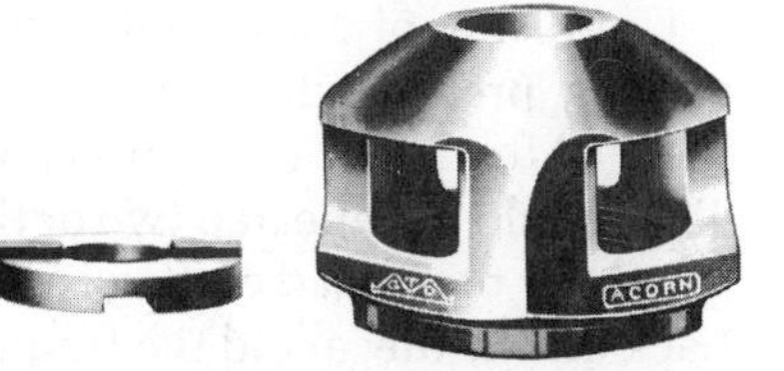

Fig. 16-9. Die-holder reducing cap enables the Acorn die holder to hold dies of a small diameter.

Courtesy Greenfield Tap & Die

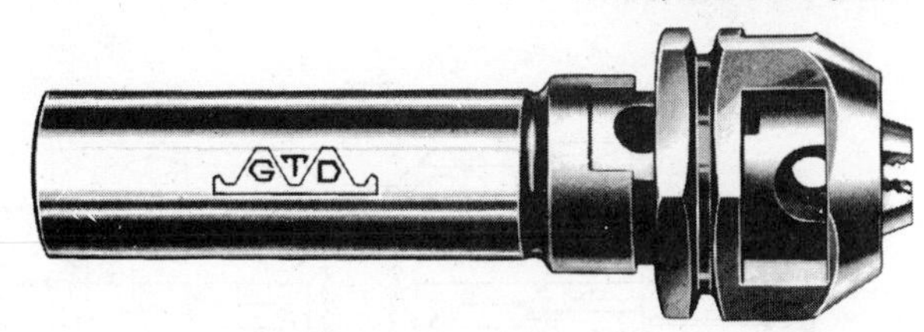

Courtesy Greenfield Tap & Die

Fig. 16-10. Acorn die holder designed for use on automatic machines.

registers with the mark "S" on the diestock (Fig. 16-11). When these are in position, tighten the holding bolts firmly. Use only the wrench provided with the stock for this purpose.

The die will cut a standard thread when the reference marks register properly (Fig. 16-12). Adjusting is done only when irregularity or variations in fittings (elbows, T's, etc.) make it necessary. After the die halves are properly placed and all adjustments are made, select the proper guide collar, place it in the sleeve, and secure it by tightening the thumb bolt. Screw the two arms into place, and the tool is ready to cut threads. The pipe to be threaded should be clamped in a pipe vise (Fig. 16-13).

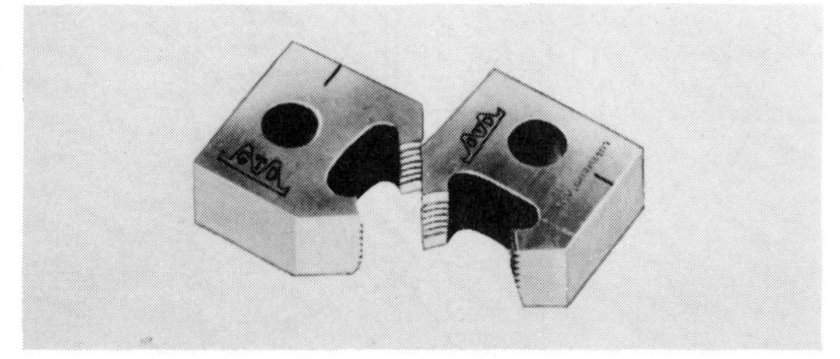

Courtesy Greenfield Tap & Die

Fig. 16-11. Two-piece adjustable pipe dies.

Use plenty of cutting oil in starting and cutting the thread. In starting, press the dies firmly against the pipe end until they "take hold." After a few turns, blow out the chips and apply more oil. This should be repeated two or three times before completing the cut. After the thread cutting is completed, blow out the chips and back off the die; avoid the frequent reversals which are made by most pipe fitters.

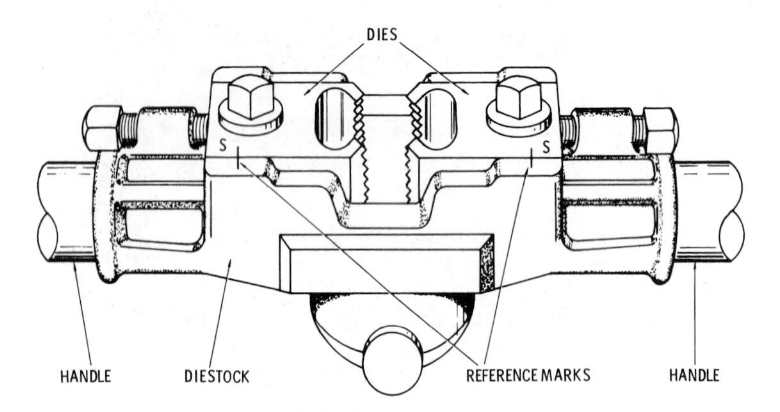

Fig. 16-12. Pipe dies and diestock.

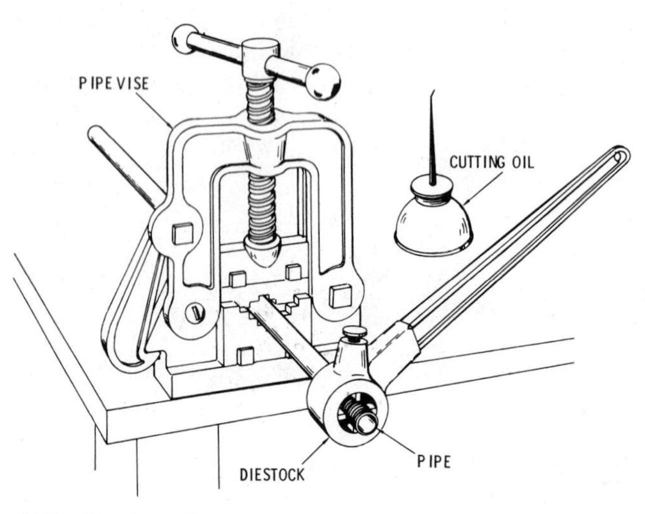

Fig. 16-13. Pipe threading.

Soluble oil has been found preferable to lubricating oil as a cutting oil. The heat generated when cutting threads is dissipated by the water in the emulsion which flows to the cutting edge of the die, giving continuous lubrication, rather than spasmodic flooding, as with lubricating oil.

SUMMARY

Threading dies are tools with internal threads similar to the threads on a nut. These tools are used to cut threads on a bolt or almost any round material, such as steel. The threads are cut by the teeth of the die as the die advances while being turned or screwed onto round stock.

There are various types of dies, such as solid, round split, two-piece adjustable, and Acorn dies. The solid die is one-piece and nonadjustable and is used for cutting threads on standard bolts and pipe. This type of die is also used to repair bruised or rusty threads. The round split die is split on one side and can be closed or opened gradually to the desired size of the thread. Most split dies are set to cut the thread slightly under the designated size.

The two-piece adjustable dies are made in two separate halves and are usually held together in a collet, which consists of a cap and a guide. The die is adjusted by setscrews at either end of the slot. The Acorn dies are used in specially designed holders, and have a tapered nose at the end of the lands accurately ground concentric with the thread. Acorn die holders are designed for use on automatic screws and other machines which provide for automatically reversing the die.

REVIEW QUESTIONS

1. Name the four types or designs of threading dies.
2. How are the two-piece adjustable dies adjusted?
3. Why are cutting fluids used?
4. What tool is used to hold a die?
5. How much can a split die be adjusted?

CHAPTER 17

Milling Machine Cutters

Milling cutters are usually referred to as multitooth, cylindrical, rotary cutting tools designed for mounting on milling machine arbors. The principal dimensions of the most commonly used types of milling cutters have been adopted as standard by the cutting manufacturers, and these standards have been approved by the American Standards Association. The most commonly used cutters in milling operations are made of high-speed steel, cemented carbide, or cast alloys.

MILLING OPERATION

For many years, *up milling*—rotating the cutter opposite the direction of feed (Fig. 17-1)—was considered the only practical way to use milling cutters. In recent years, however, *down milling*—rotating the cutter in the direction of feed (Fig. 17-2)—has been recognized.

In the up milling operation, the cutter tooth has a tendency to slide along the surface for a short distance. This sliding action under pressure tends to dull the cutter tooth. The cutter revolution marks so familiar on milled surfaces are caused by the alternate sliding action and breaking through of the cutter teeth. Down milling is not practical on all milling machines. This method should not be used unless the nature of the job permits that both the work and the cutter be held rigidly, and the milling machine is equipped with an anti-backlash device. If down milling can be used, however, a better surface finish, larger feeds per tooth, and longer cutter life without regrinding can be expected.

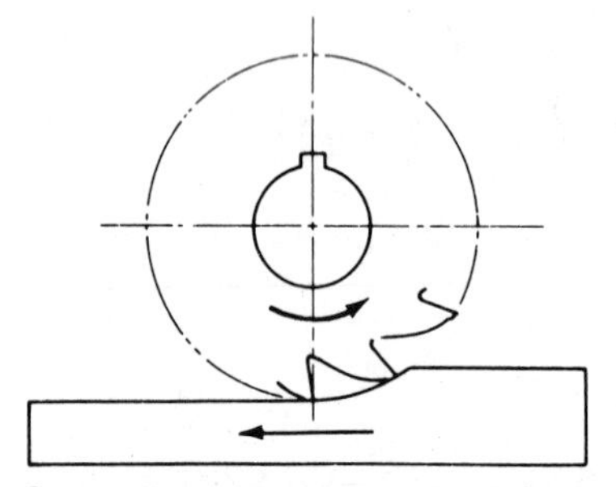

Fig. 17-1. Up milling action.

Courtesy National Twist Drill and Tool Company

In down milling, full engagement of the tooth with the work occurs practically instantaneously. Thus, gradual building up of peripheral pressures and the resulting sliding action and dulling of the cutter are prevented. Also, gradual disengagement of the teeth with the work largely eliminates feed marks.

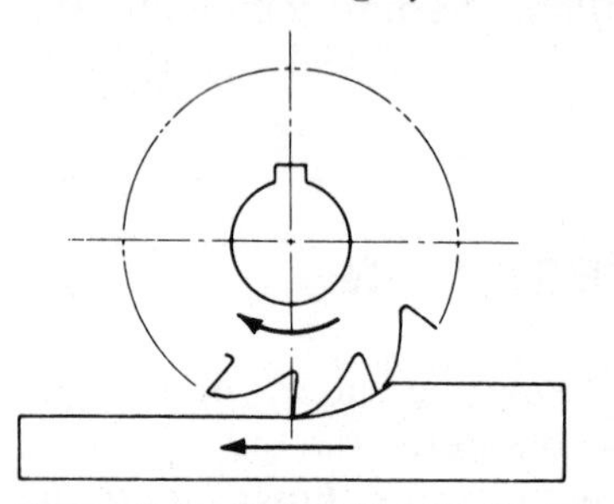

Fig. 17-2. Down milling action.

Courtesy National Twist Drill and Tool Company

CLASSIFICATION OF MILLING CUTTERS

Milling cutters may be classified as to clearance (or relief) of teeth. *Profile cutters* are sharpened by grinding on the periphery of the teeth. The clearance is obtained by grinding a narrow land back of the cutting edge. The cutters are called shaped profile cutters if the cutting edges are curved or irregular in shape. *Formed cutters* are sharpened by grinding the face of the teeth. The eccentric relief, or clearance, back of the cutting edge has the same contour as the cutting edge itself.

Another classification of cutters is based on the method of mounting. Cutters with a hole for mounting on an arbor are designated *arbor cutters*. Cutters having either a straight shank or a tapered shank integral with the cutter are called *shank cutters*. Those cutters that can be attached directly to a spindle end or a stub arbor are called *facing cutters*.

Milling cutters are either *right-hand* or *left-hand* cutters. A right-hand cutter rotates counterclockwise, and a left-hand cutter rotates clockwise, as viewed from the front when mounted on the spindle.

GENERAL TYPES OF MILLING CUTTERS

The milling machine operator should be familiar with each cutter by name and with the operation that it can perform. The different milling cutters are designed for a specific purpose.

Plain Milling Cutters

The plain milling cutter (commonly called a mill) is used to mill flat surfaces parallel to the axis of rotation. The cutter teeth have a $12\frac{1}{2}°$ rake. Cutters with less than a $\frac{3}{4}$-inch face have straight (axial) teeth, and the larger sizes have left-hand helical or "spiral" teeth. The left-hand helical tooth causes the cutting thrust which tends to keep the spindle tight in its bearings.

Plain mills are adapted for work of a slabbing nature, when the work is narrower than the cutter face. When used on flat work where there is a shoulder, the lead end of helical teeth should work in the corner, and shallow end teeth or side chip grooves should be ground into that end for best results.

Light-duty plain cutters are best suited for moderate cuts in malleable iron, steel, and cast iron (Fig. 17-3). Heavy-duty plain cutters have a large heavy rake, coarse teeth, deep flutes, and a steep helix (Fig. 17-4).

A helical cutter is a plain milling cutter with an extra-steep helical angle, usually 52° (Fig. 17-5). The helix is generally opposite the direction of rotation, thereby utilizing the end thrust to

Courtesy National Twist Drill and Tool Company

Fig. 17-3. Plain milling cutter. Used for light-duty milling.

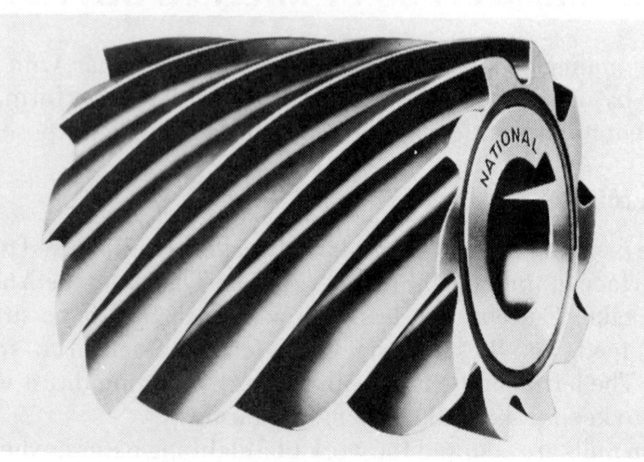

Courtesy National Twist Drill and Tool Company

Fig. 17-4. Plain milling cutter. Used for heavy-duty milling.

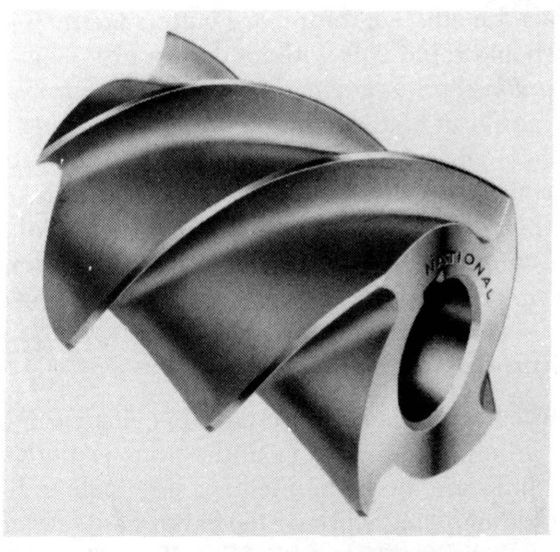

Courtesy National Twist Drill and Tool Company

Fig. 17-5. Helical plain milling gear.

keep the spindle tight in its bearings. This is not a general-purpose cutter. It can be run with a light cut at high speeds and fast feeds on either brass or soft steels. It cuts with a shearing action, forces the chips off sidewise, does not show revolution marks, and does not spring away from the work. This feature makes it especially adaptable for use on thin work or for intermittent cuts where the amount of stock to be removed varies. For extremely wide surfaces, cutters can be made in sections with the spiral angle reversed in each succeeding section.

The term "spiral" should not be used interchangeably for helix. A *helix* is a curve generated by a point which both rotates and advances axially on a cylindrical surface. A *spiral* is a curve generated by a point having three motions: (1) rotation about an axis, (2) increase in distance from the axis, and (3) advancement parallel with the axis. Plain cutters with a helix angle of 25° to 45° are commonly, but incorrectly, called *spiral mills.*

Coarse-tooth cutters are capable of removing a considerable quantity of metal in a given time without overloading the cutter or machine. The wide spaces between the teeth permit the cutting edges to be well backed up, which is not always possible with closely spaced teeth. Therefore, the cutters are well adapted to handle deep and rapid cuts without danger of failing.

Nicked milling cutters are made with nicked or grooved teeth to enable the cutter to take deeper cuts, as for roughing work. The effect of the nicks is to reduce the power required to drive the cutter. The nicks are arranged so that a cutting edge (of the next tooth) will be behind a nick. Thus, instead of a continuous chip, a number of chips are made by each cutting edge. Nicked cutters having a very wide (long) face are known as slabbing cutters.

Side Milling Cutters

These cutters are plain milling cutters of cylindrical form which have teeth around the periphery and on one or both sides (Fig. 17-6). Side mills are recommended for side milling, for slotting, and for straddle milling work. If the cutter has teeth on only one side, it is a *half side milling cutter* (Fig. 17-7). Half side mills are used for heavy-duty straddle mill work.

Courtesy National Twist Drill and Tool Company

Fig. 17-6. Side milling cutter. Used for slotting and light-duty milling. Cutters can be ganged and used as straddle mills.

The *staggered-tooth side milling cutter* (Fig. 17-8) is designed for deep slotting and for heavy-duty side milling. The shear cutting action, alternately right and left, eliminates side thrust. The

Courtesy National Twist Drill and Tool Company

(A) Right-hand. (B) Left-hand.

Fig. 17-7. Half-side milling cutters. Used for heavy-duty straddle milling.

alternate right- and left-hand spirals of the teeth, with considerable angle of undercut, enable this cutter to remove a large amount of metal without destructive vibration and chatter, permit taking deep cuts, and leave a good finish. Free cutting action makes increased speed and feed possible. Cuts that would stall an ordinary cutter can be taken easily. Although they were intended primarily for deep cuts in steel, staggered-tooth side mills can be used for shallow cuts, which is an advantage if the work requires cuts of varying depths. Ganging two or more of these cutters, instead of using a wide cutter with a wide tooth space, is recommended when wide slots are to be milled.

Interlocking side cutters are useful where a slot width must be held to extremely accurate limits. These cutters can be separated by spacing collars of the required thickness to obtain the correct width of face.

Courtesy National Twist Drill and Tool Company

Fig. 17-8. Staggered-tooth side milling cutter. Used for deep slotting and for heavy-duty side milling. Cutters can be ganged and used as straddle mills.

The shearing action, alternately right and left, eliminates side thrust; the cutting action is very smooth and rapid. The cutters can also be made with inserted teeth, in the larger sizes. The teeth have a positive rake on all cutting faces. These cutters have the added advantage of being suitable for finishing the bottom of a slot, while having an adjustment for holding the slot to a required width.

Inserted-tooth cutters (Fig. 17-9) are commonly used for the larger cutters. Inserted teeth (sometimes called blades) are generally used because this construction is cheaper, and all teeth can be replaced easily if necessary. Inserted-tooth construction avoids the danger of cracking while being hardened. The teeth may be made of either high-speed steel, cemented carbide, or cast alloy.

Various methods of holding the inserted teeth are used. They are generally made long enough to permit sharpening a great number of times. These cutters are used for heavy-duty side milling and for face milling jobs where long life under severe working conditions is desirable.

End Mills

An end mill, by strict definition, is a milling cutter that has cutting teeth only on its end. However, in addition to the end teeth,

Courtesy National Twist Drill and Tool Company

Fig. 17-9. Milling cutter with solid carbide indexible throwaway inserts.

end mills may have teeth along the periphery or cylindrical surface.

Because they have cutting teeth on the end of the mill, end mills are usually held by shanks. They may have either a straight shank (Fig. 17-10) or a taper shank (Fig. 17-11) to fit various collets and adapters.

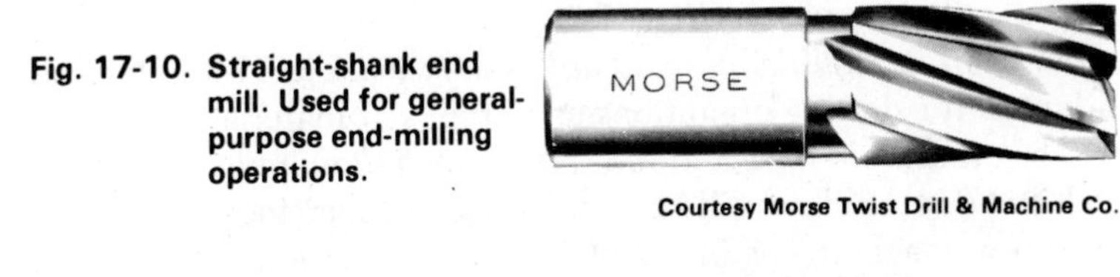

Fig. 17-10. Straight-shank end mill. Used for general-purpose end-milling operations.

Courtesy Morse Twist Drill & Machine Co.

Courtesy Morse Twist Drill & Tool Co.

Fig. 17-11. Taper-shank end mill.

End mills may be made for either right-hand or left-hand rotation. The helix (right-hand or left-hand) may be in either the same or the opposite direction as the cutter rotation. When the helix and cutter rotation are the same (either right-hand or left-hand), the teeth have a positive rake angle. For some specific purposes, end mills are available with the cutter rotation and the helix opposite —for example, left-hand cutter rotation with right-hand helix. Generally, end mills having both a right-hand helix and a right-hand cutter rotation are preferred.

Several types of cutting ends for end mills are produced to further adapt them to a wide variety of uses, such as profiling, end milling, slotting, surface milling, and many other milling machine operations. The different types of cutting ends available on end mills are:

1. Two flute.
2. Multiple flute (three-flute, four-flute, six-flute).
3. Single end.
4. Double end.
5. Hollow.
   a. Solid.
   b. Adjustable.
6. Ball end.
7. Carbide-tipped.
8. Shell end.

Two-flute single-end end mills are adapted to slotting operations in all kinds of materials. This mill can cut to center, which permits plunge cutting (Fig. 17-12).

A *two-flute double-end end mill* is shown in Fig. 17-13. It is also adapted for slotting operations in all types of materials.

The *adjustable-type* and the *solid-type hollow mills* are shown in Fig. 17-14. Hollow mills are used for sizing bar stock of all types in screw machines or in turret lathes. They are available with undercut teeth for use in steel or with straight teeth for use in brass. Compensation for internal wear can be made on the adjustable type of hollow mill.

Ball end mills are used for machining fillets and slots with corner radii. They are also used extensively for die sinking and machining dies (Fig. 17-15).

Carbide-tipped end mills are available in most types and shapes (Fig. 17-16). The advantage of carbide-tipped cutters is the increased cutting speeds. The surface feet per minute for carbide-tipped cutters is double that of high-speed steel, as indicated in Table 17-2.

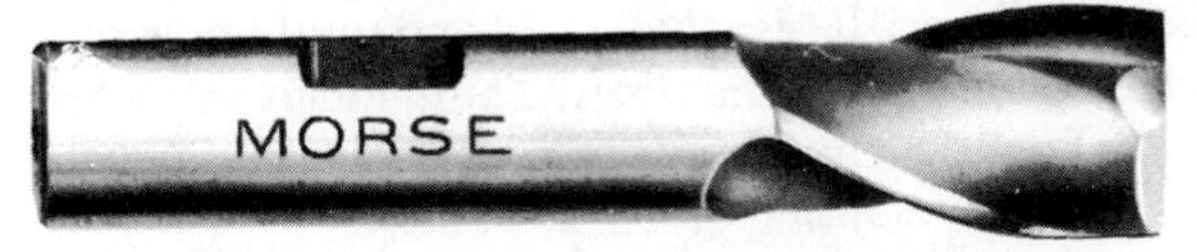

Courtesy Morse Twist Drill & Machine Company

Fig. 17-12. Two-fluted single-end end mill.

Courtesy Morse Twist Drill & Machine Company

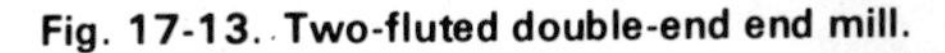

Fig. 17-13. Two-fluted double-end end mill.

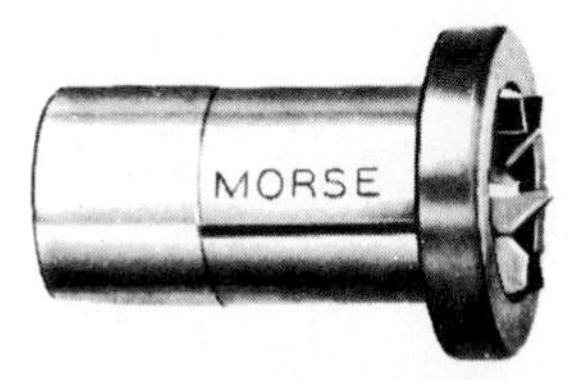

(A) Adjustable type.

(B) Solid type.

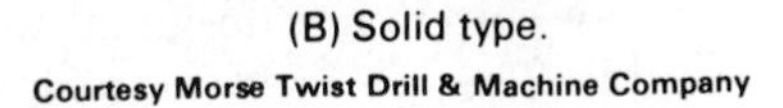

Courtesy Morse Twist Drill & Machine Company

Fig. 17-14. Hollow mills. Used in screw machines or turret lathes for sizing bar stock of all types.

Courtesy National Twist Drill and Tool Company

Fig. 17-15. Four-flute ball end mill.

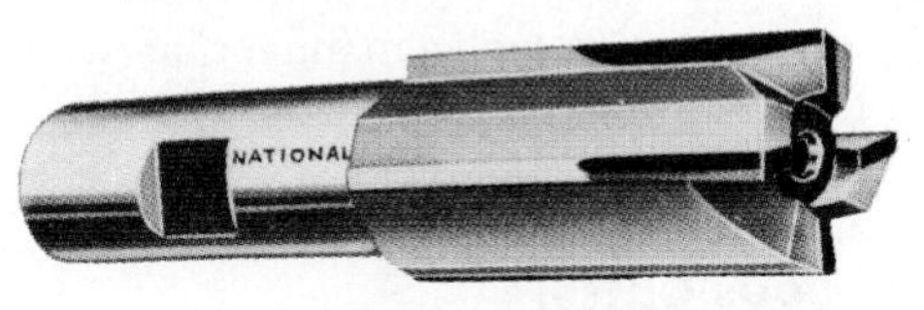

Courtesy National Twist Drill and Tool Company

Fig. 17-16. Carbide-tipped straight flute end mill.

Shell end mills are larger than solid end mills and a range from 1¼ inches to 6 inches in diameter. These mills have a hole in the center to mount the cutter on an arbor (Fig. 17-17). Cutters of this type are used for slabbing or surface cuts.

Courtesy National Twist Drill and Tool Company

Fig. 17-17. High-speed steel shell mill.

Angle Milling Cutters

These cutters are designed to mill at an angle to the axis of rotation. They are used for milling surfaces at various angles to the axis of rotation, and are often used in making other milling cutters.

Angle cutters are made for right-hand rotation and for left-hand rotation. *Single-angle milling cutters* (Fig. 17-18) are used to mill ratchet teeth or to mill dovetails. The common single-angle cutters vary from 40° to 80°. *Double-angle milling cutters* (Fig. 17-19) are available with included angles of 45°, 60°, or 90°.

Single-angle cutters have one side at an angle of 90° to the axis of rotation, and the other side at, usually, either 45° or 60°—only one side cuts at an angle other than 90° to the rotation axis. A double-angle cutter is constructed in such a manner that two angles cut at an angle other than 90° to the rotation axis.

Slitting Saws, Slotting Saws, and Miscellaneous Cutters

Thin, straight-toothed, plain milling cutters are generally called slitting saws. A *plain metal-slitting saw* is shown in Fig. 17-20. This

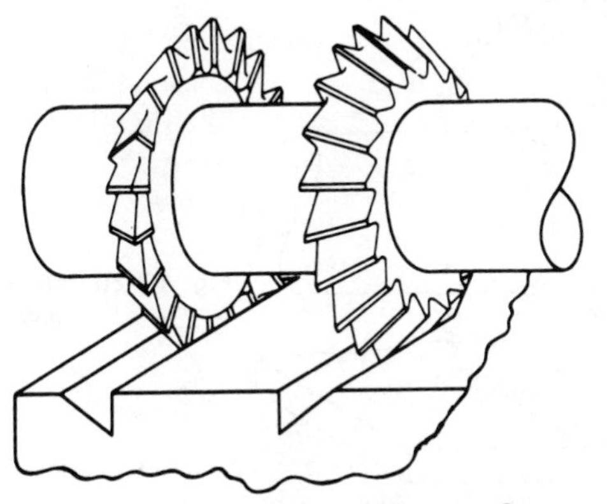

Courtesy Morse Twist Drill & Machine Company

Fig. 17-18. Single-angle milling cutter used for milling ratchet teeth or for milling dovetails.

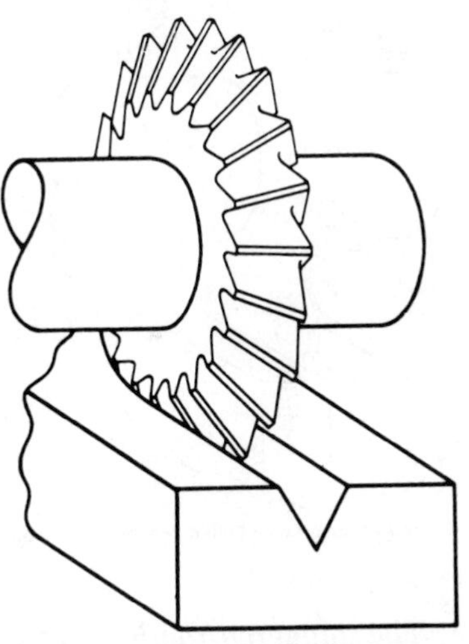

Fig. 17-19. Double-angle milling cutter. Used for milling grooves, notches, serrations, or threads.

Courtesy Morse Twist Drill & Machine Company

saw is used for general-purpose slotting, parting, or cutting-off operations of moderate depth in both ferrous and nonferrous materials.

Side-tooth metal-slitting saws have side chip-clearance gashes (Fig. 17-21). They are used for cutting-off operations in all types of materials.

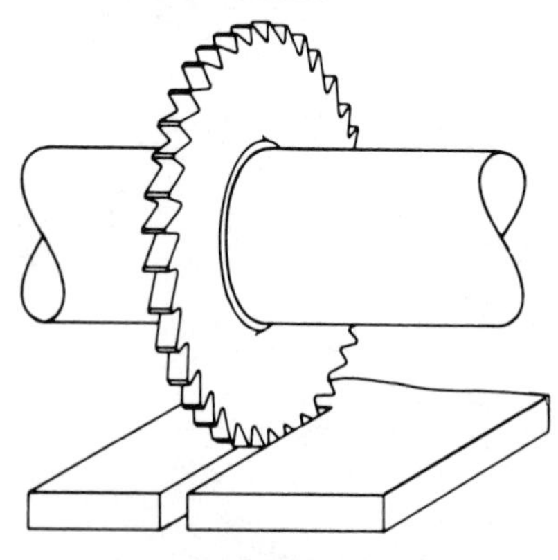

Courtesy National Twist Drill & Machine Company

Fig. 17-20. Plain metal-slitting saw.

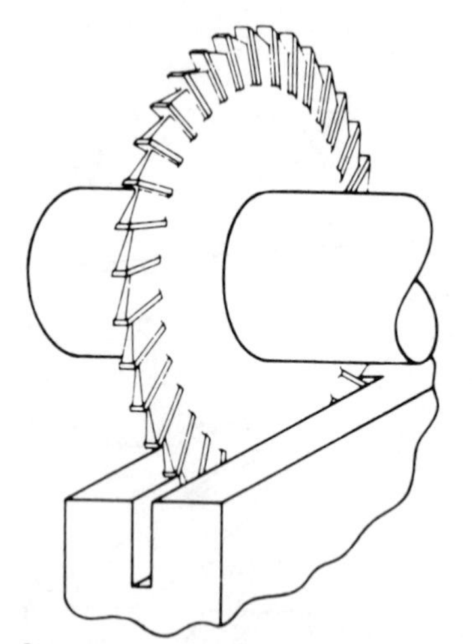

Courtesy Morse Twist Drill & Machine Company

Fig. 17-21. Side-tooth metal-slitting saw. Note the chip-clearance gashes on the side.

The *staggered-tooth metal-slitting saw* (Fig. 17-22) is manufactured with alternate helical teeth for shear cutting action and chip clearance between the side teeth. These saws are used for heavy-duty slotting in all types of materials and can make deeper cuts under coarser feeds than other types of saws.

A *screw-slotting cutter* is shown in Fig. 17-23. This cutter is ground with side clearance and is used for slotting both ferrous and nonferrous screwheads, sheet, or tubing. Screw-slotting cut-

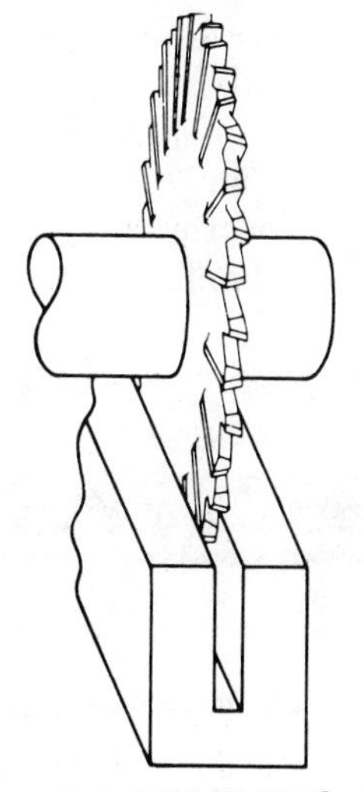

Fig. 17-22. Staggered-tooth metal-slitting saw. Used for heavy-duty slotting in all types of materials.

Courtesy Morse Twist Drill & Machine Company

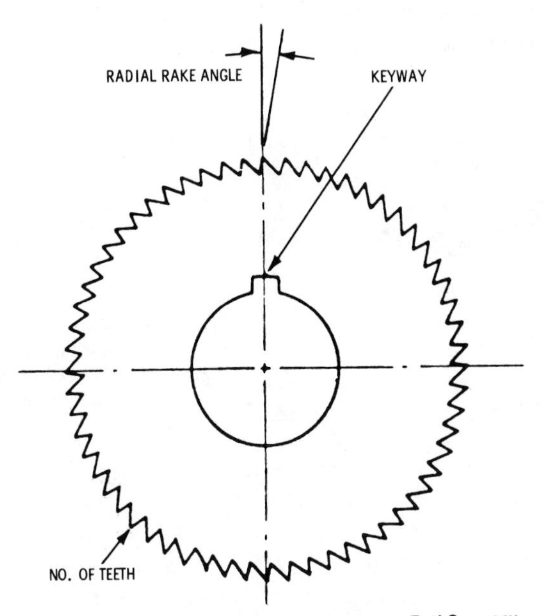

Courtesy National Twist Drill and Tool Company

Fig. 17-23. Screw-slotting cutter. Used for slotting ferrous and nonferrous screwheads, sheet, or tubing.

ters are used only for making shallow, short slots similar to those in screwheads.

A *T-slot milling cutter* is a special form of end mill for making T-slots (Fig. 17-24). This cutter is designed to mill the wide bottom of a T-slot after the narrow portion has been milled with a side mill or an end mill.

Courtesy Morse Twist Drill & Machine Company

Fig. 17-24. T-slot milling cutter.

Woodruff key-seat cutters are made in both shank and arbor types. The shank type (Fig. 17-25) is used for cutting the smaller Woodruff key seats. The larger Woodruff key seats are cut with

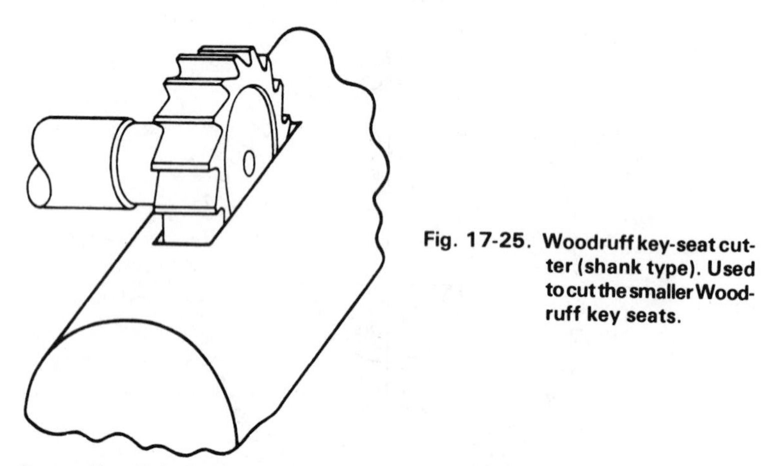

Fig. 17-25. Woodruff key-seat cutter (shank type). Used to cut the smaller Woodruff key seats.

Courtesy Morse Twist Drill & Machine Company

the arbor type (Fig. 17-26) of key-seat cutter. These cutters have profile teeth and are used for cutting semicircular keyways in shafts.

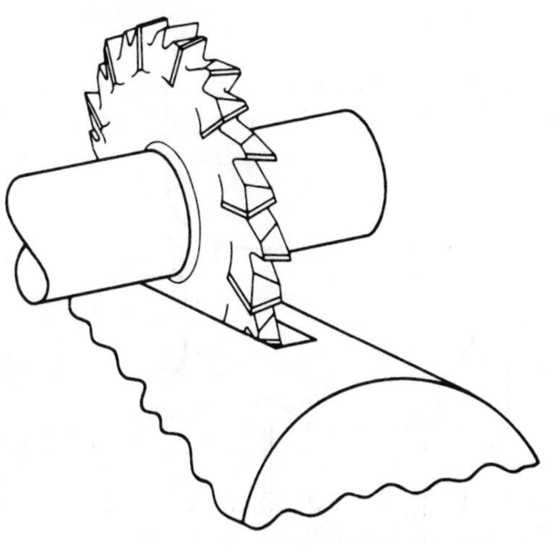

Courtesy Morse Twist Drill & Machine Company

Fig. 17-26. Woodruff key-seat cutter (arbor type). Used to cut the larger Woodruff key seats.

The *threaded-hole single-angle milling cutter* (Fig. 17-27) may be used for dovetail milling in all types of materials. The cutter with a 60° included angle is generally used for milling dovetails.

Form-Relieved Cutters

These cutters are special milling cutters shaped to meet various milling specifications. This cutter is sharpened by grinding only the faces of the teeth; the outline of the teeth corresponds to the required form, and should not be ground when resharpening.

Convex-cutters (Fig. 17-28) are used to mill half-circles in all types of materials. If convex half-circles are desired, a *concave cutter* (Fig. 17-29) must be used. Convex quarter-circles are milled with *corner-rounding cutters* (Fig. 17-30).

Gear cutters are available in many different diameters and hole sizes, so that they can be used on almost any regular milling machine or arbor. *Finishing gear milling cutters* (Fig. 17-31) with a 14½° pressure angle regular involute form are most commonly used. These cutters can be sharpened on the face of the teeth without changing their form. Finishing gear milling cutters are

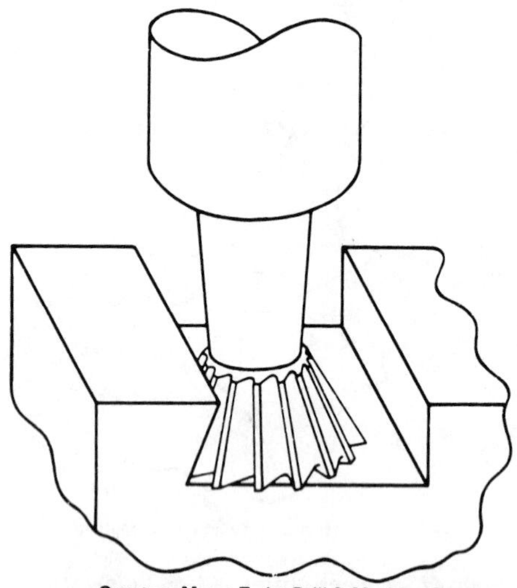

Courtesy Morse Twist Drill & Machine Company

Fig. 17-27. Threaded-hole single-angle cutter. Used for dovetail-milling.

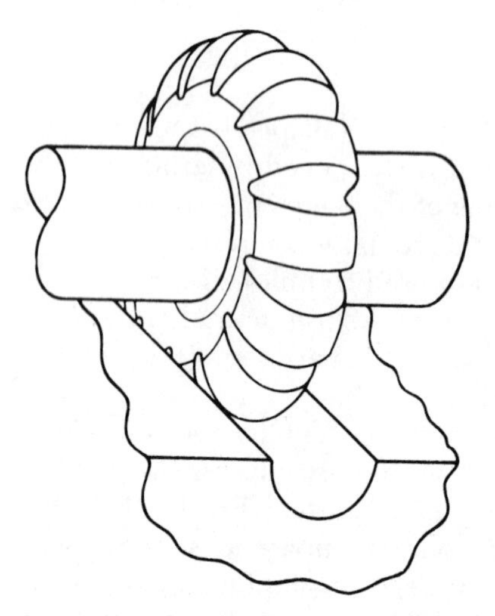

Fig. 17-28. Convex cutter. A form-relieved cutter used to mill concave half-circles in all types of materials.

Courtesy Morse Twist Drill & Machine Company

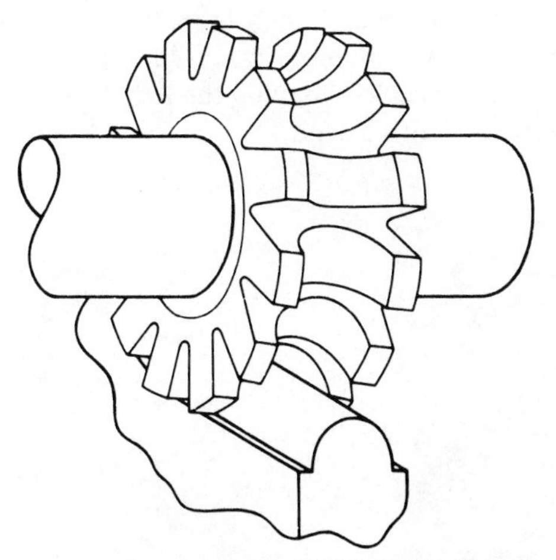

Courtesy Morse Twist Drill & Machine Company

Fig. 17-29. Concave cutter. A form-relieved cutter used to mill convex half-circles in all types of materials.

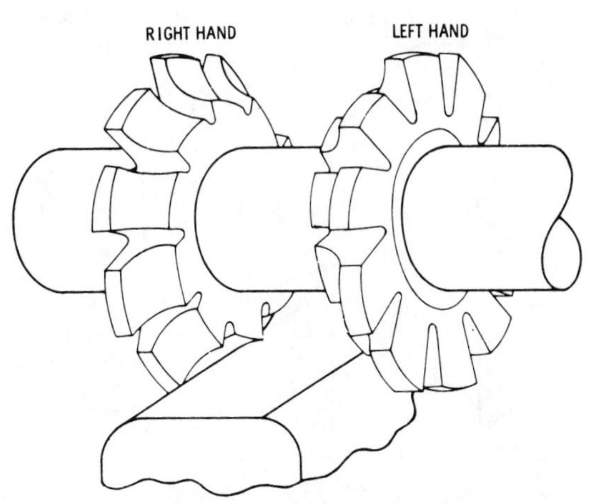

Courtesy Morse Twist Drill & Machine Company

Fig. 17-30. Corner-rounding cutter. A form-relieved cutter used for milling convex quarter-circles on the edges of all types of materials.

made with eight cutters for each pitch-eight cutters being required to produce a full set of gears, ranging from 12 teeth to a rack.

A *sprocket cutter* is shown in Fig. 17-32. This cutter is used for cutting roller chain sprockets having the American National Standard Roller Chain Tooth Form.

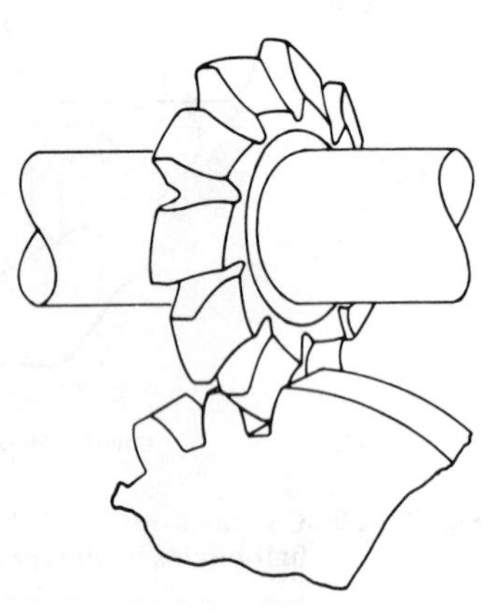

Courtesy National Twist Drill and Tool Company

Fig. 17-31. Finishing gear milling cutter. A form-relieved cutter with 14 ½° pressure angle.

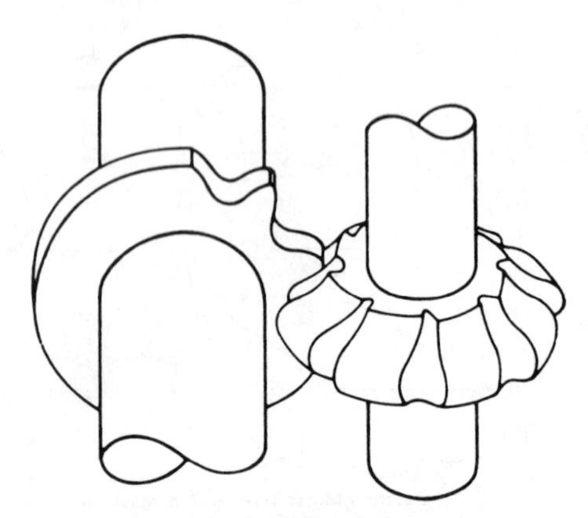

Courtesy Morse Twist Drill & Machine Company

Fig. 17-32. Sprocket cutter. Used for cutting roller chain sprockets.

Hobs

A hob is a hardened, threaded cutter, formed like a worm (Fig. 17-33). It is commonly used for cutting teeth in spur and helical gears, herringbone gears, worm gears, worms, splined shafts, ratchets, square shafts, and sprockets for silent chain, roller chain, and block chain.

Courtesy National Twist Drill and Tool Company

Fig. 17-33. Hob. Used to hob spur gears and helical gears.

Hobs are made in either the ground or unground form. For average hobbing operations, unground forms are more economical and more commonly used. Ground-form hobs are better for extreme accuracy.

Spur-gear hobs can be used to hob either spur gears or helical gears. A right-hand hob can be used to cut either a right-hand helical gear or a left-hand helical gear. However, it is recommended that a right-hand hob be used to cut a right-hand helical gear, and a left-hand hob be used to cut a left-hand helical gear.

In general, the hobbing process can be employed to produce any form that is repeated regularly on the periphery of a circular part.

CARE OF MILLING CUTTERS

Every precaution should be taken to prevent milling cutters from becoming nicked and dulled. In making milling machine setups, the cutters should not be bumped against the workpiece, the tools, or the machine. It is common practice to coat the cutting edges of cutters with plastic to prevent nicking and dulling. The cutters should be stored carefully when they are not in use.

SPEEDS AND FEEDS

Individual experience and judgment are extremely valuable in selecting the correct milling speeds and feeds. Even though suggested rate tables are given, it should be remembered that these are suggestions only. The lower figure in the table for a particular material should always be used until sufficient practical experience has been gained. Then, the speed can be increased until either excessive cutter wear or chatter indicates that the practical limit has been exceeded.

Speed and feed rates are governed by many variable factors. Some of these are: material, cutter, width and depth of cut, required surface finish, machine rigidity and setup, power and speed available, and cutting fluid.

Speeds

The speed at which the circumference of the cutter passes over the work is always given in surface feet per minute (sf/min). However, the spindle speed of a milling machine is always given in revolutions per minute (r/min). Table 17-1 can be used to convert surface feet per minute (sf/min) to revolutions per minute (r/min), for making speed adjustments on the milling machine. The table gives the r/min for cutters of different diameters at various surface speeds. Speeds not listed in the table can be found by simple calculation—for example, for 200 sf/min, 1 ¾-inch-diameter cutting must run twice as fast as for 100 sf/min ($2 \times 509 = 1018$). If a table of speeds is unavailable, a formula can be used for spindle speed as follows:

$$\text{r/min} = \frac{\text{sf/min} \times 12}{\pi \times \text{diameter}}$$

The cutting speeds for the different materials to be milled are given in Table 17-2. These speeds are based on average conditions for high-speed steel tools. The cutting speeds may be doubled for carbide-tipped cutters. Note that the cutting speeds in Table 17-2 are given in surface feet per minute (sf/min). If a table of cutting speeds is unavailable, sf/min may be calculated by the formula:

$$\text{sf/min} = \frac{\pi \times \text{diameter} \times \text{r/min}}{12}$$

Milling cutters with helical teeth can be run faster than those with straight teeth because the shearing action enables them to cut more freely. Coarse-tooth cutters can be run faster than fine-tooth cutters because fewer teeth are in contact with the work at any time. Roughing cuts are usually made at slow speeds and heavy cuts; finishing cuts are made at high speeds and fine feeds. However, there are exceptions to these practices.

Feeds

The feed rate is the rate at which the work advances past the cutter. Feed rate is commonly given in inches per minute (in/min). Generally, the rule in production work is to use all the feed that the machine and the work can stand. However, it is a problem to know where to start with the feed. Tables 17-3 and 17-4 give the suggested speed per tooth for high-speed steel and carbide-tipped milling cutters, respectively.

Note that the feeds are given in thousandths of an inch per tooth for the various cutters. Multiply the feed per tooth by the number of teeth, and multiply that product by the r/min to determine the feed rate in in/min as follows:

$$\text{in/min} = \text{feed per tooth} \times \text{number of teeth} \times \text{r/min}$$

In actual practice, it is better to start the feed rate at a somewhat lower figure than that indicated in the table and work up gradually until the most efficient removal rates are reached. Too high a feed

Table 17-1. Revolutions Per Minute

	Surface Feet Per Minute						
Diameter	40	50	60	70	80	90	100
1/4	611	764	917	1070	1222	1375	1528
5/16	489	611	733	8566	978	1100	1222
3/8	407	509	611	713	815	917	1019
7/16	349	437	524	611	698	786	873
1/2	306	382	458	535	611	688	764
5/8	244	306	367	428	489	550	611
3/4	204	255	306	357	407	458	509
7/8	175	218	262	306	349	393	437
1"	153	191	229	267	306	344	382
1 1/8	136	170	204	238	272	306	340
1 1/4	122	153	183	214	244	275	306
1 3/8	111	139	167	194	222	250	278
1 1/2	102	127	153	178	204	229	255
1 5/8	94	117	141	165	188	212	235
1 3/4	87	109	131	153	175	196	218
1 7/8	81	102	122	143	163	182	204
2"	76	95	115	134	153	172	191
2 1/4	68	85	102	119	136	153	170
2 1/2	61	76	92	107	122	137	152
2 3/4	52	69	83	97	111	125	139
3"	51	64	76	89	102	115	127
3 1/2	44	55	65	76	87	98	108
4"	38	48	57	67	76	86	95
4 1/2	34	42	51	59	68	77	85
5"	31	38	46	54	61	69	76
5 1/2	28	35	42	49	56	63	70
6"	25	32	38	45	51	57	64
7"	22	27	33	38	44	49	55
8"	19	24	29	33	38	43	48
9"	17	21	25	30	34	38	42
10"	15	19	23	27	31	34	38
11"	14	17	21	24	28	31	35
12"	13	16	19	22	25	29	32
13"	12	15	18	21	24	27	29
16"	10	12	14	17	19	22	24
18"	8	11	13	15	17	19	21

Courtesy Cincinnati Inc.

Table 17-2. Cutting Speeds

(Surface Feet Per Minute)

MATERIAL	High-Speed Steel Rough	Finish	Carbide-Tipped Rough	Finish	COOLANT
Cast Iron	50-60	80-110	180-200	350-400	Dry
Semi-steel	40-50	65-90	140-160	250-300	Dry
Malleable Iron	80-100	110-130	250-300	400-500	Soluble, Sulfurized, or Mineral Oil
Cast Steel	45-60	70-90	150-180	200-250	Soluble, Sulfurized, Mineral, or Mineral Lard Oil
Copper	100-150	150-200	600	1000	Soluble, Sulfurized or Mineral Lard Oil
Brass	200-300	200-300	600-1000	600-1000	Dry
Bronze	100-150	150-180	600	1000	Soluble, Sulfurized, or Mineral Lard Oil
Aluminum	400	700	800	1000	Soluble or Sulfurized Oil, Mineral Oil and Kerosene
Magnesium	600-800	1000-1500	1000-1500	1000-1500	Dry, Kerosene, Mineral Lard Oil
SAE Steels					
1020 (coarse feed)	60-80	60-80	300	300	Soluble, Sulfurized, Mineral, or Mineral Lard Oil
1020 (fine feed)	100-120	100-120	450	450	" " "
1035	75-90	90-120	250	250	" " "
X-1315	175-200	175-200	400-500	400-500	" " "
1050	60-80	100	200	200	" " "
2315	90-110	90-110	300	300	" " "
3150	50-60	70-90	200	200	" " "
4340	40-50	60-70	200	200	Sulfurized or Mineral Oil
Stainless Steel	100-120	100-120	240-300	240-300	" " "

Note: Feeds should be as much as the work and equipment will stand, provided a satisfactory surface finish is obtained.

Table 17-3. Suggested Feed per Tooth for High-Speed Steel Milling Cutters

Material	Face Mills	Helical Mills	Slotting and Side Mills	End Mills	Form Relieved Cutters	Circular Saws
Plastics	0.013	0.010	0.008	0.007	0.004	0.003
Magnesium and Alloys	0.022	0.018	0.013	0.011	0.007	0.005
Aluminum and Alloys	0.022	0.018	0.013	0.011	0.007	0.005
Free Cutting Brasses and Bronzes	0.022	0.018	0.013	0.011	0.007	0.005
Medium Brasses and Bronzes	0.014	0.011	0.008	0.007	0.004	0.003
Hard Brasses and Bronzes	0.009	0.007	0.006	0.005	0.003	0.002
Copper	0.012	0.010	0.007	0.006	0.004	0.003
Cast Iron, Soft (150-180 B.H.)	0.016	0.013	0.009	0.008	0.005	0.004
Cast Iron, Medium (180-220 B.H.)	0.013	0.010	0.007	0.007	0.004	0.003
Cast Iron, Hard (220-300 B.H.)	0.011	0.008	0.006	0.006	0.003	0.003
Malleable Iron	0.012	0.010	0.007	0.006	0.004	0.003
Cast Steel	0.012	0.010	0.007	0.006	0.004	0.003
Low Carbon Steel, Free Machining	0.012	0.010	0.007	0.006	0.004	0.003
Low Carbon Steel	0.010	0.008	0.006	0.005	0.003	0.003
Medium Carbon Steel	0.010	0.008	0.006	0.005	0.003	0.003
Alloy Steel, Annealed (180-220 B.H.)	0.008	0.007	0.005	0.004	0.003	0.002
Alloy Steel, Tough (220-300 B.H.)	0.006	0.005	0.004	0.003	0.002	0.002
Alloy Steel, Hard (300-400 B.H.)	0.004	0.003	0.003	0.002	0.002	0.001
Stainless Steels, Free Machining	0.010	0.008	0.006	0.005	0.003	0.002
Stainless Steels	0.006	0.005	0.004	0.003	0.002	0.002
Monel Metals	0.008	0.007	0.005	0.004	0.003	0.002

Courtesy Cincinnati Inc.

Table 17-4. Suggested Feed per Tooth for Carbide-Tipped Cutters

Material	Face Mills	Helical Mills	Slotting and Side Mills	End Mills	Form Relieved Cutters	Circular Saws
Plastics	0.015	0.012	0.009	0.007	0.005	0.004
Magnesium and Alloys	0.020	0.016	0.012	0.010	0.006	0.005
Aluminum and Alloys	0.020	0.016	0.012	0.010	0.006	0.005
Free Cutting Brasses and Bronzes	0.020	0.016	0.012	0.010	0.006	0.005
Medium Brasses and Bronzes	0.012	0.010	0.007	0.006	0.004	0.003
Hard Brasses and Bronzes	0.010	0.008	0.006	0.005	0.003	0.003
Copper	0.012	0.009	0.007	0.006	0.004	0.003
Cast Iron, Soft (150-180 B.H.)	0.020	0.016	0.012	0.010	0.006	0.005
Cast Iron, Medium (180-220 B.H.)	0.016	0.013	0.010	0.008	0.005	0.004
Cast Iron, Hard (220-300 B.H.)	0.012	0.010	0.007	0.006	0.004	0.003
Malleable Iron	0.014	0.011	0.008	0.007	0.004	0.004
Cast Steel	0.014	0.011	0.008	0.007	0.005	0.004
Low Carbon Steel, Free Machining	0.016	0.013	0.009	0.008	0.005	0.004
Low Carbon Steel	0.014	0.011	0.008	0.007	0.004	0.004
Medium Carbon Steel	0.014	0.011	0.008	0.007	0.004	0.004
Alloy Steel, Annealed (180-220 B.H.)	0.014	0.011	0.008	0.007	0.004	0.004
Alloy Steel, Tough (220-300 B.H.)	0.012	0.010	0.007	0.006	0.004	0.003
Alloy Steel, Hard (300-400 B.H.)	0.010	0.008	0.006	0.005	0.003	0.003
Stainless Steels, Free Machining	0.014	0.011	0.008	0.007	0.004	0.004
Stainless Steels	0.010	0.008	0,006	0.005	0.003	0.003
Monel Metals	0.010	0.008	0.006	0.005	0.003	0.003

Courtesy Cincinnati Inc.

is indicated by excessive cutter wear. A cutter may be spoiled by too fine a feed or too heavy a feed. Rubbing, rather than a cutting action, may dull the cutting edge, and excessive heat may be generated. This fact is also true in relationship to depth of cut. The first cut on castings and rough forgings should always be made well below the surface skin. Cuts less than 0.015 inches in depth should be avoided. To obtain a good finish, take a roughing cut followed by a finishing cut, with a higher speed and lighter feed for the finishing cut.

In general, feeds are increased as speeds are reduced. Therefore, the feed is increased on abrasive, sandy, or scaly material, and for heavy cuts in heavy work. Feed should be increased if cutter wear is excessive or if there is chatter.

Feed should be decreased for better finish, when taking deep slotting cuts, or if work cannot be held rigidly. If the cutter begins to chip or to produce long, continuous chips, the feed should be decreased.

SUMMARY

For many years, up milling machine cutters rotated opposite the feed. In recent years, milling machine cutters were designed to rotate in the same direction as the feed. This type of design has eliminated the need of frequent cutter sharpening and overall maintenance. This type of cutting action is called down milling.

An end mill is a milling cutter with cutting teeth on one end only. End mills may be either right-hand or left-hand rotation. The helix may be in either the same or opposite direction as the cutter rotation. Generally, end mills having both a right-hand helix and a right-hand cutter rotation are preferred.

Several types of cutting ends for end mills are produced. The different types of cutting ends available are: two flute, multiple flute, single and double end, and hollow end, both solid and shell.

Various shapes and sizes of cutters are produced, such as angle cutters which are generally made for right-hand rotation. The common single-angle cutters include angles of 45°, 60°, or 90°. Other shapes include double angle, plain, side-tooth, convex, concave, and corner-rounding.

REVIEW QUESTIONS

1. What is the down milling action?
2. Define the term end milling.
3. What is side milling?
4. Name a few cutting ends available for end milling.
5. What is the advantage of using carbide tipped cutters?
6. How do you sharpen profile cutters?
7. How do you sharpen formed cutters?
8. What type of cutter do you use for moderate cuts in malleable iron, steel, or cast iron?
9. How are nicked milling cutters made?
10. What sizes are standard for shell-end mills?
11. What is a Woodruff key-seat cutter?
12. Where would you use a convex cutter?
13. What is a hob?
14. Explain how a sprocket cutter actually does its job.
15. How do you designate the cutter speed in relation to its circumference?

CHAPTER 18

Milling Machine Arbors, Collets, and Adapters

Milling machine operators should be familiar with the holding devices used on the machines that they operate. An inexperienced operator may mistake an arbor, which is used to hold milling cutters, for a mandrel, which is used to hold bored parts while turning the outside surface on a lathe.

ARBORS

An arbor is a tapered cylindrical shaft designed to hold milling cutters. The holding part of an arbor is cylindrical; the cutter is clamped with a nut for light work and secured further by a key for heavy cutting (Fig. 18-1). The precision and trueness of arbors influence the accuracy of milling operations. Arbors must be carefully handled both in use and in storage.

Manufacturers have attempted to standardize milling machine spindles and arbors. A uniform numbering system for standard

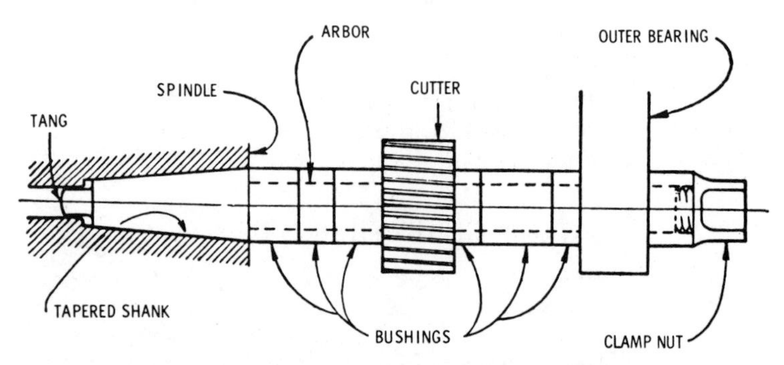

Fig. 18-1. Tang-type arbor in position in the spindle. The cutter and bushings are held in a firm frictional grip by tightening the clamp nut.

tapered arbors has been approved, and the No. 50 taper is used on most milling machines. Specifications are listed in sequence as follows: taper size, diameter, style, length from shoulder to nut, and size of bearing. For example, Arbor No. 50-1A18-4 indicates that the arbor has No. 50 standard taper, 1 inch in diameter, style A, a length of 18 inches from the shoulder to the nut, and a No. 4 bearing.

Arbors are held firmly in the hollow spindle of the milling machine by means of an arbor draw-in bar. Two drive keys on the spindle nose, which fits into corresponding slots on the arbor flanges, give driving contact and positive drive to the arbor.

Styles of Arbors

Steel taper-shank arbors to fit the national milling machine standard spindle are made in three styles as follows:

1. *Style* A has a pilot on the outer end (Fig. 18-2). The arbor is firmly supported as it turns in the arbor support bearing, suspended from the overarm.

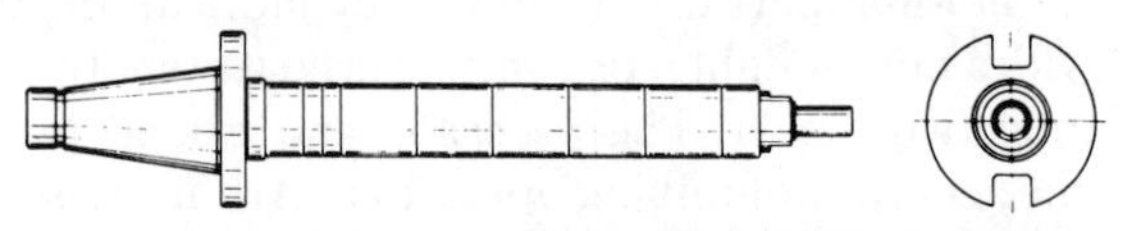

Courtesy Cincinnati Milacron Company

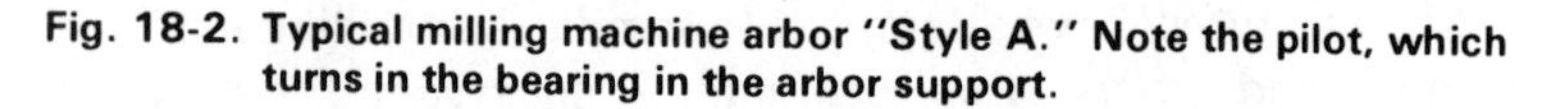

Fig. 18-2. Typical milling machine arbor "Style A." Note the pilot, which turns in the bearing in the arbor support.

2. *Style B* does not have a pilot (Fig. 18-3). A bearing sleeve fits over the arbor and is keyed to it. The sleeve revolves in the bearing in the arbor support. Support can be placed close to the cutters on the arbor for rigidity. Style B arbors are used wherever heavy cuts are made.

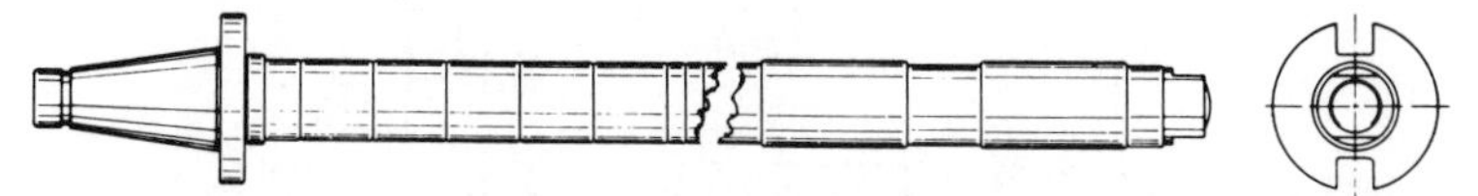

Courtesy Cincinnati Milacron Company

Fig. 18-3. Typical milling machine arbor "Style B." These arbors are longer and require support near the cutters for rigidity.

3. *Style C* is a short arbor requiring no arbor support (Fig. 18-4). It is used to hold cutters which are too small to be bolted directly to the spindle nose, such as the smaller sizes of shell-end-mill, and face-milling cutters. This style of arbor is sometimes called a shell-end-mill arbor. The cutter is driven by solid lugs on the outer end of the arbor.

Courtesy Brown & Sharpe Manufacturing Company

Fig. 18-4. Typical milling machine arbor "Style C" used for shell-type milling cutters.

Methods of Driving the Cutters

Several methods are used to prevent the cutter from slipping or turning on the arbor. The type of work to be performed and the type of cutter influence the design of the arbor.

Friction drive can be used on light work. The cutter is kept from slipping by tightening the end nut with a wrench. This forces the bushings against the cutter, and its endwise movement is prevented by the shank collar; thus, a firm frictional grip on the cutter is produced. This is by no means a positive drive (Fig. 18-5).

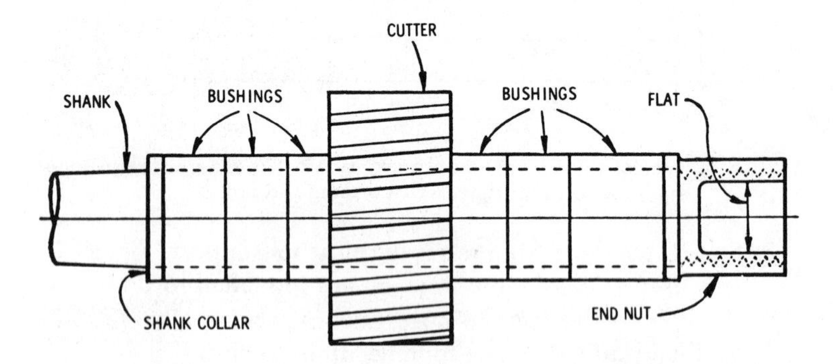

Fig. 18-5. Arbor with assembled cutter and bushings, illustrating frictional cutter drive.

A *key drive* is a positive drive and prevents slipping on any work within the capacity of the machine (Fig. 18-6). Cutters used on milling machines have machined keyways; a square key fits into the keyways of both the cutter and the arbor. Cutters can be mounted in any position along the arbor.

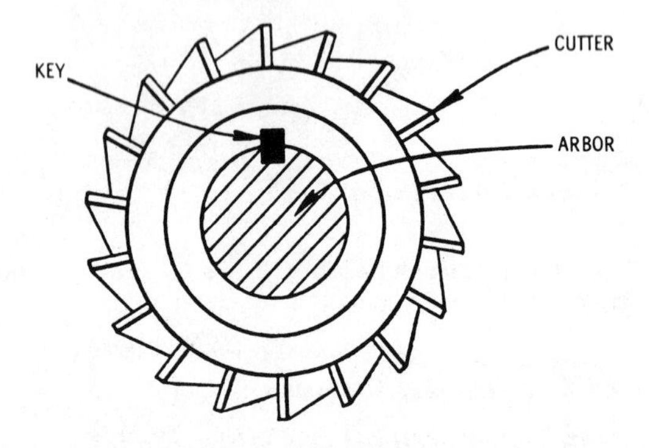

Fig. 18-6. A key-cutter drive. The cutter will not slip on the arbor for any work within the capacity of the machine.

The *arbor sleeve* has a keyway, and it may be keyed to the arbor (Fig. 18-7). The sleeve fits the bushing of the arbor support; it is used as near as possible to the cutter, or cutters, on long arbors. The arbor sleeve looks like an arbor spacing collar (Fig. 18-8), but it is different both in purpose and in diameter.

Courtesy Brown & Sharpe Manufacturing Company

Fig. 18-7. Arbor sleeve for a milling machine.

Courtesy Brown & Sharpe Manufacturing Company

Fig. 18-8. Spacing collar used on a milling machine arbor.

The *clutch drive* is used to drive gutters of the shell-end-mill type. The shell end mill is designed with a slot on each side which fits over the arbor jaws when the mill is placed on the arbor, thus forming a positive drive. Two lugs at 180° fit in the slots of the cutter to form the driving members. A tanged-end type of arbor is shown in Fig. 18-9. An arbor with a threaded end for right-hand shell end mills only is shown in Fig. 18-10.

Courtesy Morse Twist Drill & Machine Company

Fig. 18-9. A shell-end-mill arbor with tanged end.

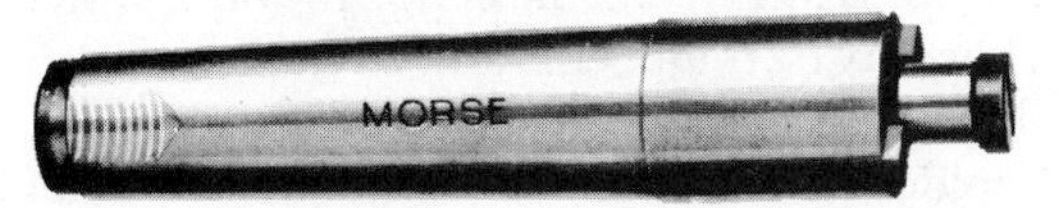

Courtesy Morse Twist Drill & Machine Company

Fig. 18-10. A shell-end-mill arbor with threaded end.

The *screw drive* is used to drive small cutters, such as the angle milling cutters which are made with a threaded hole. These cutters are screwed onto a *threaded-end arbor* (Fig. 18-11). The threaded-end arbor is made with either right-hand or left-hand threads for cutters made with either right-hand or left-hand threads, depending on the desired direction of cutter rotation. Of course, the direction of rotation must be such that cutting torque tends to screw the cutter onto the arbor.

Courtesy Morse Twist Drill & Machine Company

Fig. 18-11. An arbor used for angle milling cutters with threaded holes.

A *fly-cutter arbor* (Fig. 18-12) is used to hold a single-tooth cutter with a formed cutting edge. The arbor is slotted to receive the cutter which is secured in position by setscrews. The cutting edge of the cutter is turned to the desired profile; clearance is obtained by setting the cutter farther out from the center than the radius to which it was turned.

COLLETS

A milling machine collet is a form of sleeve bushing for reducing the size of the hole in the milling machine spindle, so that an arbor with a smaller shank can be used (Fig. 18-13). Collets are made in several forms, differing in respect to method of drive as follows:

1. Tang.
2. Draw-in.
3. Clutch.

When an arbor is inserted in the collet, the tang projects into the slot. The arbor may be removed by driving a taper key through the slot behind the tang.

A collet holder may be used to take the place of an arbor for holding the cutter. One type of collet holder is inserted in the spindle. It must be removed to change cutters. An extended type

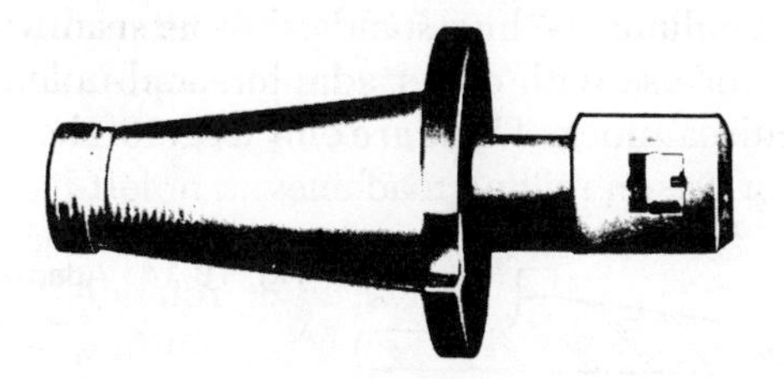

Courtesy Brown & Sharpe Manufacturing Company

Fig. 18-12. Fly-cutter arbor for milling machines with standard spindle end.

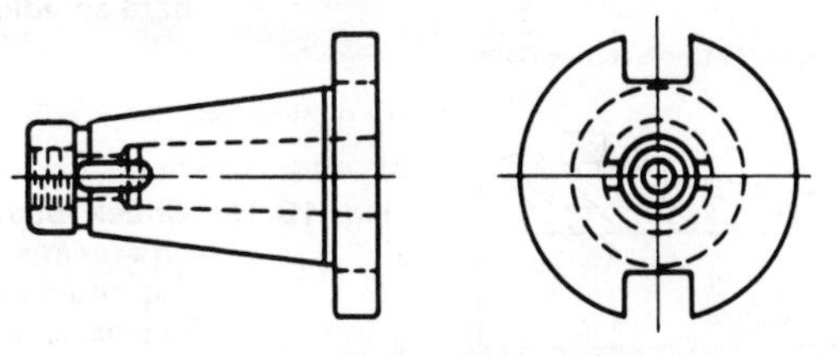

Courtesy Cincinnati Milacon Company

Fig. 18-13. Milling machines collet with draw-in bolt.

of collet holder permits the cutters to be changed without removing the holder from the spindle. A spring-collet holder is used to hold end mills.

ADAPTERS

An adapter is a form of collet having a standardized spindle end for use on milling machines. A great variety of adapters are available, rendering them suitable for holding various arbors and cutters. Arbor adapters permit the use of arbors, collets, and end mills having either Brown & Sharpe taper or Morse taper on milling machines with the standard spindle end (Fig. 18-14).

A *chuck adapter* (Fig. 18-15) has a screw end to provide a means for attaching chucks. This device is used on milling machines having a standard spindle end.

Several types of adapters are used to hold the various cutters and milling attachment spindles. A standard-taper shank adapter may be used for single-end mills. Still another adapter for end mills having milling machine standard-taper shanks available with camlock is for use with cutter adapters and milling attachment spindles with camlock. These are only a few of the many adapters available for use on milling machines.

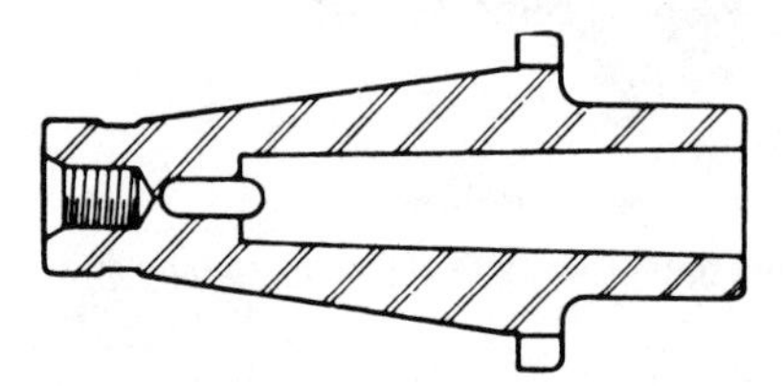

Courtesy Brown & Sharpe Manufacturing Company

Fig. 18-14. Adapter for arbors, collets, and end mills; made with either Brown & Sharpe or Morse-taper shanks for use on milling machines having a standard spindle end.

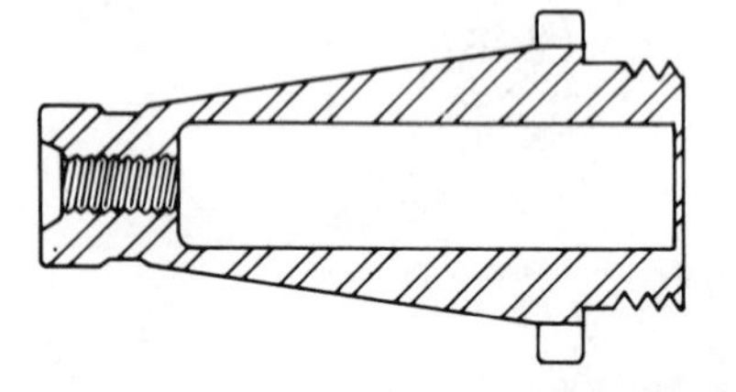

Courtesy Brown & Sharpe Manufacturing Company

Fig. 18-15. Chuck adapter used as a means for attaching chucks on milling machines having a standard spindle end.

SUMMARY

Three basic devices used on all milling machines are the arbor, collet, and adapter. The arbor is a tapered, cylindrical shaft designed to hold the milling cutters. The holding part of an arbor is cylindrical, and clamps the cutter by the use of a nut. The precision and trueness of the arbor influences the accuracy of the milling operations.

The collet is a sleeve bushing reducing the size of the hole in the milling machine spindle so that an arbor with a smaller shank can be used. A collet holder may be used to take the place of an arbor for holding the cutter.

The adapter is a form of collet having a standardized spindle end for use on the milling machines. Many adapters are available and are styled to hold various arbors and cutters. A standard-taper shank adapter may be used for single-end mills.

REVIEW QUESTIONS

1. What is an adapter?
2. What is a collet?
3. How are cutters attached to the arbor?
4. What is an arbor?
5. What is the purpose for bushings on an arbor?

CHAPTER 19

Broaches and Broaching

A broach is a straight tool with series of cutting teeth that gradually increase in size. The broach is used to cut metal and is especially adapted to finishing square, rectangular, or irregularly shaped holes, and for cutting keyways in pulleys, hubs, etc.

BROACHING PRINCIPLE

Considerable power is required to operate a broach. The broaching operation is used to machine cored or drilled holes to the required shape; one or more broaches are pulled or pushed through these holes. The broach, a special tapered cutter, is forced through an opening to enlarge a hole, or alongside a piece of work to shape an exterior. Broaching differs from other machining processes in that a long tool with a series of teeth is used.

A small portion of the metal along the entire cut is removed by each tooth on the broach. The first teeth are smaller, for entering or beginning the cut. The intermediate teeth remove most of the metal, and the last few teeth on the broach finish the hole to proper size (Fig. 19-1).

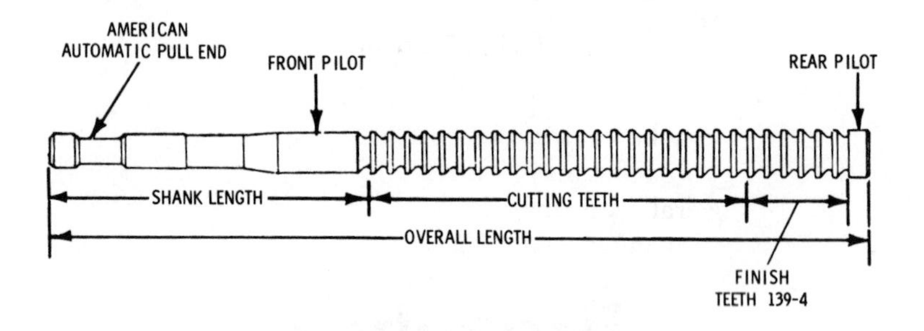

Fig. 19-1. Diagram of a typical broach.

A side view of the teeth of a typical broach is shown in Fig. 19-2. The *face angle* enables the tooth to cut properly and varies with the material to be cut; it is usually 0° to 20°. The span between the teeth is the *pitch*. The *land* should be of sufficient strength to withstand the cutting strain (may be approximately 25 percent of the pitch). A *straight land* is generally used only at the finishing end of the broach to retain the broach size. The *radius* at the bottom of the teeth curls the chip and strengthens the broach. The *clearance angle* reduces friction; it should be held to a minimum (about 2°), so as to prevent excessive wear. The *tooth depth* is proportional to the pitch and should be sufficient to accommodate the chip.

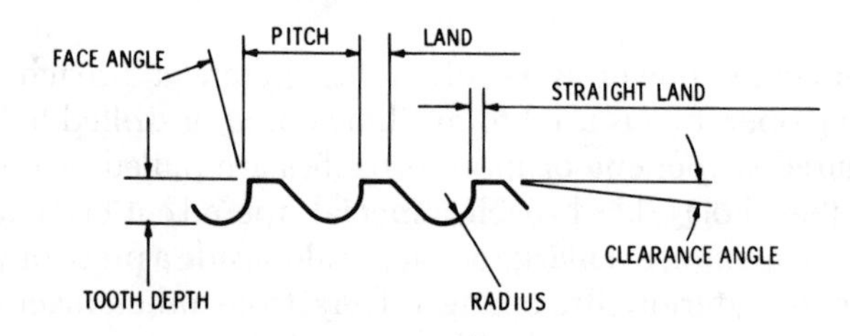

Fig. 19-2. Side view of the teeth of a typical broach.

TYPES OF BROACHES

A great variety of broaches are available for all types of broaching requirements. Broaching has reduced machining costs and has replaced milling in many situations.

Broaches are commonly made of high-speed steel; carbide-tipped broaches are also used. Molybdenum steel is generally used in broaches.

Broaches may be made either in the solid form or in sections. The sectional broaches may be made in a variety of ways. In some instances, several teeth may be made on a single section and several sections can be used to form the broach.

SHAPES OF BROACHES

Broaches are made in numerous shapes to meet the requirements of all kinds of broaching operations (Fig. 19-3). Proportioning of the teeth is important because the cutting operation is progressive. The pitch and other details of tooth design depend on the kind of work for which the broach is intended. In general, two teeth of the broach should be in contact with the work at the same time. Although most broaches have cutting teeth similar to those on milling cutters, a broach may have rounded or smooth teeth. The most common shapes of broaches are as follows:

1. Round.
2. Square.
3. Keyway.
   a. Single.
   b. Double.
4. Spline.
5. Helical.

INTERNAL AND EXTERNAL BROACHES

Internal broaching of round holes is a common commercial broaching operation. Cutting of splines in the hubs of gears, or propellers, is commonly done by broaching.

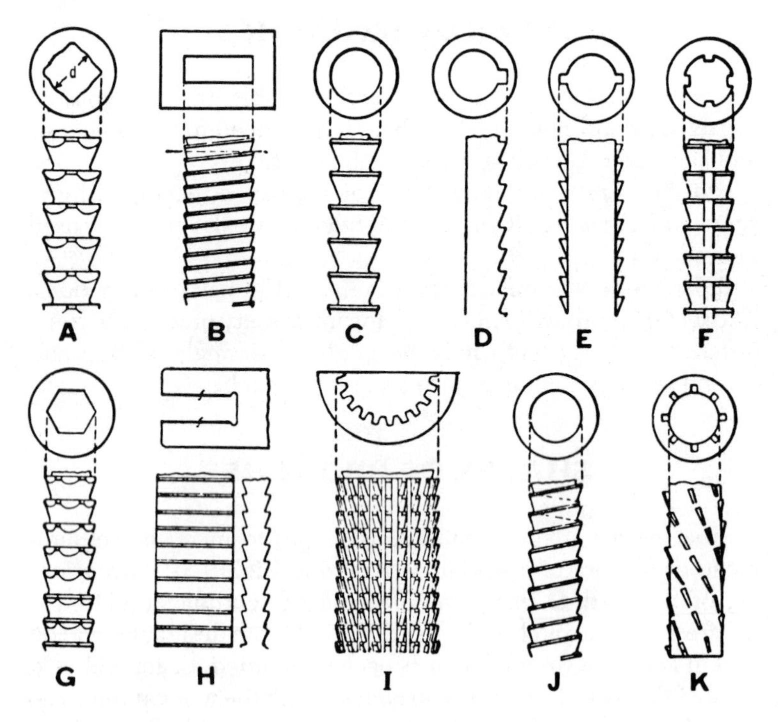

Fig. 19-3. The shapes of various types of broaches: (A) square, (B) rectangular, (C) round, (D) single keyway, (E) double keyway, (F) four spline, (G) hexagon, (H) double cut, (I) internal gear, (J) round with helical teeth, (K) helical groove.

In external (surface) broaching, the broached surface is on the outside of the piece. Broaching takes the place of milling in this type of work. Parts that have irregular external surfaces are well adapted to surface broaching. This type of surface frequently can be finished faster and at less cost by broaching than by milling with a formed milling cutter.

Extreme accuracy can be obtained with a surface broach. Accuracy depends on whether the part is heavy enough to withstand heavy cutting pressures, and on the length of the broach. Accuracy is also greater with a larger number of cutting teeth. The quality of finish is determined by the number of finishing teeth on the broach.

PULL OR PUSH BROACHES

Much commercial broaching work is done with the pull broach or the stationary broach. Pull broaching is used on parts that require a great amount of metal to be removed and on parts that have long finished surfaces. A regular broaching machine is used to pull a long, slender broach through the work. The job is usually completed with one passage of the broach; however, in some instances, two or more broaches of graduated size may be used, especially if considerable stock is to be removed.

Push broaching may be done on an ordinary press by pushing a short, stout broach through the work. Push broaches are shorter and less expensive. This method is used on parts where a small amount of metal is to be removed from the hole and a good finish with close limits is desired.

CARE AND SHARPENING OF BROACHES

To prevent possible injury to broaches, the teeth should not be allowed to strike a hard object in handling. Minute fractures of the cutting edges of the teeth may result from improper handling. The contour of the cutting teeth should be maintained; that is, original face angle, depth, radius, and straight land should be retained for the tool to cut correctly.

A broach requires sharpening when it no longer provides the finish of which it is capable. Broaches having a small cross section should be sharpened more often than larger broaches to prevent breakage. When sharpening is necessary, only the front face of the teeth is usually ground. Enough grinding to remove only the slightly rounded edge is necessary. The same amount of stock should be removed from the face of each tooth; thus the step from tooth to tooth is retained.

All sharpening operations should be performed on a rigid machine. Chatter or any tendency of the grinding wheel to vibrate results in a poor job of sharpening. If a tooth should become weakened after numerous sharpenings—or if a portion of a tooth should break for some other reason—several teeth should be retapered to take the added burden, rather than permit the following tooth to assume the additional work.

When internal broaches are used in abrasive metals, considerable land is sometimes formed on the tops of the teeth. Thus, grinding the face of the teeth does not reduce the land sufficiently; then the backoff (clearance) angle of the teeth must be ground. This operation is quite delicate, because the broach must be made to run true between centers. It is recommended that only the roughing teeth should receive this treatment because any attempt at grinding the finishing teeth in this manner may result in a reduction in size.

Almost all internal broaches are provided with a series of straight teeth to maintain nominal size. Only the first two teeth of this series should be sharpened. The balance are not sharpened until the first two teeth have worn; then the following two teeth are sharpened, etc.

When a broach is new, the straight teeth do not have chip breakers. As the broach wears, and the original straight teeth become smaller and are required to remove metal, chip breakers should be added to prevent a wide chip. Wide chips are difficult to remove, and if they are allowed to collect, may cause either tooth breakage or torn holes.

Surface broaches may be resharpened readily by grinding the hook angle or tooth face; occasionally, they should be ground on the land when they show wear. After resharpening the land, the teeth should again be backed off near the cutting edge of the tooth.

BROACHING MACHINES

A great reduction in velocity must occur between the driving pulley and the broach, in order to convert the power applied into tremendous force necessary to pull or to push a broach in the metal being cut. This great reduction in velocity is obtained by a drive screw and nut (Fig. 19-4), and compounded by other gearing.

The drive screw has a broach holder head at one end. A nut is free to turn on the other end of the screw. The screw can move lengthwise only; turning is prevented by suitable guides. The nut is free to turn on the screw, but endwise movement is prevented by

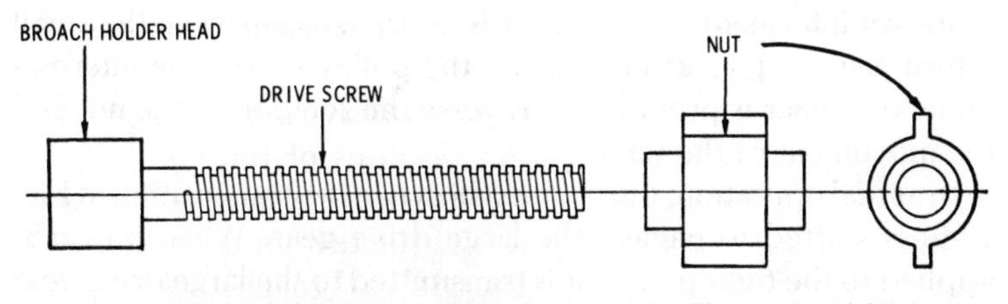

Fig. 19-4. Detail of elementary broaching machine illustrating drive screw, broach holder head, and nut.

stops or thrust bearings placed at the end. Thus, when the nut turns, the drive screw will move one way or the other, depending upon the direction of rotation of the nut; the action is similar to that of the old-fashioned letter press. This combination gives tremendous leverage, similar to that of a differential hoist. Power is transmitted to turn the nut through a clutch, as shown in detail in Fig. 19-5.

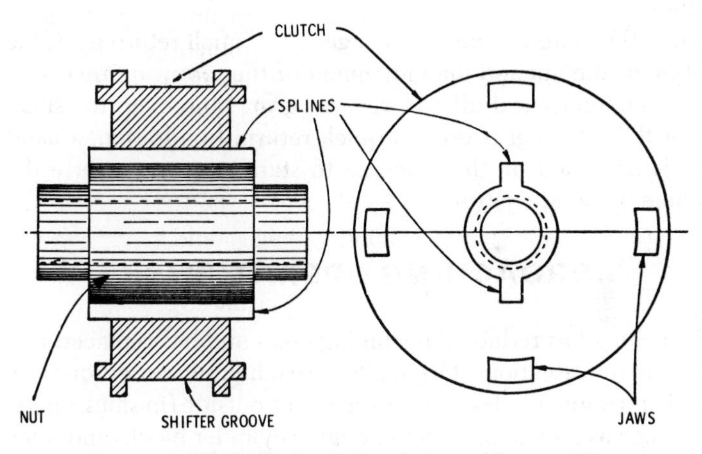

Fig. 19-5. A nut and clutch assembly of a broaching machine.

The clutch has jaws on each end, and slides on spines, so that it may engage either a large or a small gear for both slow drive and quick return. This arrangement is elementary and is not intended to include actual construction details, but only to illustrate principal parts of the elementary machine.

A large gear is on one side of the clutch, and a small gear is on the other side. The large gear meshes with a pinion on the belt-pulley shaft. An intermediate idler gear is interposed between the small return gear and a second gear on the pulley shaft. The intermediate idler gear is provided to reverse the rotation of the nut and the movement of the screw.

In actual operation, the clutch (shown in neutral position in Fig. 19-6) is shifted to engage the large drive gear. When power is applied to the tight pulley, it is transmitted to the large drive gear by the pinion and then to the clutch and nut. With this hookup providing first- and second-stage reduction, the pinion makes many revolutions for each revolution of the drive screw. Thus, the high rotative speed and the comparatively weak torque of the pinion is converted into the slow rotative speed and great torque of the nut; the tremendous force necessary to pull or push the broach through the piece being machined is transmitted to the screw. Adjustable tappets and clutch control gear, the means by which the length of stroke may be regulated, are not shown in the diagram.

When the clutch is shifted to engage the small return gear, the rotation of the nut and the movement of the screw are reversed, due to the interposed idler or reverse gear. Moreover, the small size of the return gear gives a quick return. In addition, a hand lever is provided on the machine to start, stop, or reverse the machine by hand control.

BROACHING OPERATIONS

Broaching has reduced machining costs and has replaced milling in many situations. Originally, broaching was used only for round or irregular holes. It is now used for outside finishing operations, such as outside surfaces of engine cylinder blocks and other pieces used in automobile and aviation engines and parts.

Pull Broaching

In this operation, a part having irregularly shaped holes may be drilled slightly undersized, and a combination broach may be used to finish the inner surfaces. Any desired shape may be machined from this point. Pull broaching is also used to finish parts that have thin walls with a limited amount of resistance to pressure, or on

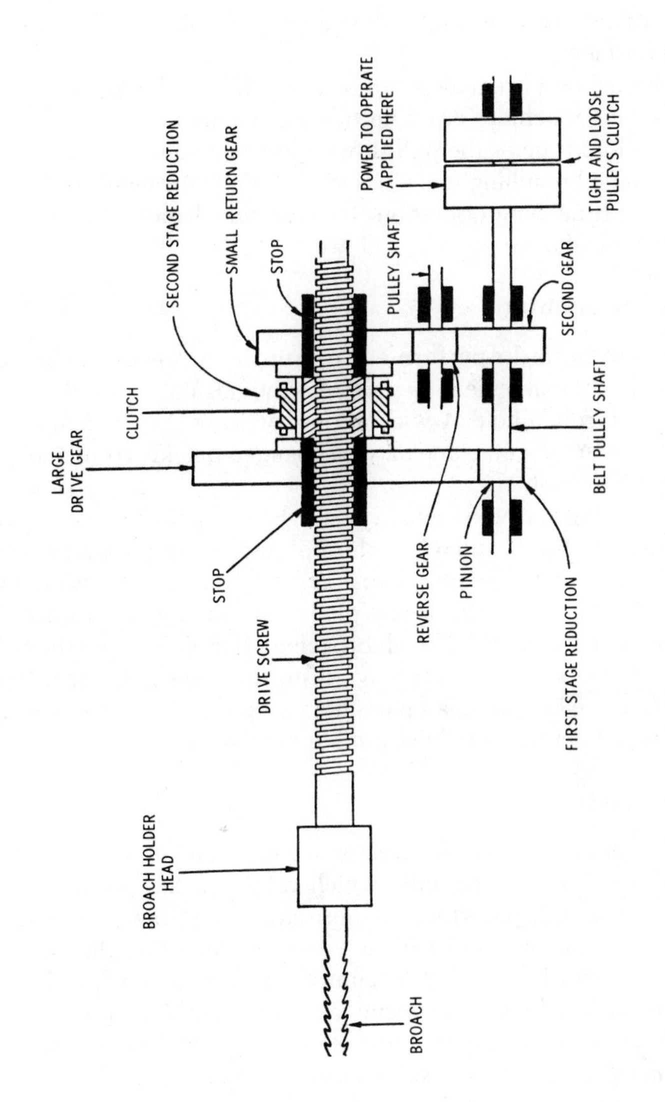

Fig. 19-6. Diagram of the essential parts of an elementary broaching machine.

parts that have irregular wall thicknesses. This method is also used extensively for finishing both straight and spiral splines in gears and bushings.

Internal pull broaching can be performed either on horizontal broaching machines or on vertical broaching machines of either the pull-up type or the pull-down type. Semiautomatic and fully automatic broaching machines are desirable for high-production internal broaching operations because they handle the broaches automatically.

Push Broaching

Because it can be performed on an ordinary press, push broaching is a very convenient method of removing only a small amount of metal from a hole. A wide variety of work may be handled in this manner, as a machine may be changed quickly from one type of work to another.

Broaching of round holes in parts where reaming does not give a satisfactory finish, and where the tool cost is too high, is adapted to push broaching. Irregular shapes in die castings and parts made of steel, bronze, Babbitt, or any of the machinable metals may be made with this method. Push broaching is used in sizing the holes in heat-treated gears to remove distortion caused by heat treatment, as well as for small parts having splines or square, hexagonal, and other irregularly shaped openings.

Surface Broaching

This operation permits greater accuracy and closer limits than other production methods, which results in the production of interchangeable parts that are more easily assembled. Because the surface-broaching tool has an extremely long life, the cost per piece is greatly reduced in comparison with other production methods. The longer life is due to the substantial support back of the surface-broaching tool, the elimination of vibration, the slow cutting speed, and its positive cutting action.

A shear angle is used to facilitate the cutting action wherever possible. Chip breakers are essential on the roughing teeth for breaking the heavy cut and for making narrow chips that will readily fall out of the broach teeth on completion of the cut.

Several types of surface broaches are available for unusual types of surface-broaching jobs. Surface broaches are made in plain slab or special types for many irregularly shapes and formed cuts that could not be machined by any other method. These broaches are usually made in sections and then placed in a substantial holder. The holder is attached to a subholder or to the machine slide itself.

The surface broach has the cutting quality of both a roughing tool and a finishing tool built into a single cutting unit. The roughing section of the broach removes the major portion of the metal. The teeth are evenly graduated for size (or amount of metal to be removed) from the roughing section to the finishing teeth which are reserved for light-duty cutting, for accurate sizing of the work, and for producing a fine finish at the same time. The work has an opportunity to cool slightly as it passes from the roughing teeth to the finishing teeth; therefore, any error due to temperature change or springing of the work is eliminated.

Round Broaching

The chief advantages of round broaching are speed, uniformity of size of holes, and long tool life. In broaching round holes and other shapes, each tooth of the broach passes the work only once. The chip rarely exceeds .003 inch. This explains the long tool life of the broach.

If an extremely fine finish is desired, a burnishing section can be added to the broach. Pilot sections placed between the broach cutting teeth make it possible to finish the broached hole concentric with the starting hole. These pilots permit the removal of a small amount of stock, removing it from all sides, as in sizing operations. Of course, the smaller the amount of stock there is to be removed from the hole, the finer should be the finish before broaching. Likewise, the broach may be shorter and cheaper for the smaller amounts of stock to be removed.

SUMMARY

A broach is a straight tool with a series of teeth that gradually increase in size. The broach is used to cut metal, especially for square, rectangular, and irregularly shaped holes. Considerable power is needed to operate a broach.

Broaches are made in various shapes to meet the requirements of all kinds of operations. The most common shapes of broaches are round, square, keyway, spline and helical. Broaches may be made either in the solid form or in sections. In some cases, several teeth may be made on a single section and several sections can be used to form the broach.

Great reduction in velocity must occur between the driving pulley and the broach in order to convert the power applied into the tremendous force necessary to pull or push the broach through the metal being cut. This force is obtained by a drive screw and nut, and is compounded by other gearing.

REVIEW QUESTIONS

1. What is broaching?
2. Name the most common shapes of broaches.
3. How are broaches made?
4. Explain why broaching has cut operation cost in machine shops.

Appendix

METRIC MEASURES

The metric unit of length is the meter = 39.37 inches.
The metric unit of weight is the gram = 15.432 grains.
The following prefixes are used for subdivisions and multiples: milli = $\frac{1}{1000}$, centi = $\frac{1}{100}$, deci = $\frac{1}{10}$, deca = 10, hecto = 100, kilo = 1000, myria = 10,000.

METRIC AND ENGLISH EQUIVALENT MEASURES

Measures of Length

Metric	English
1 meter	= 39.37 inches, or 3.28083 feet, or 1.09361 yards
0.3048 meter	= 1 foot
1 centimeter	= 0.3937 inch
2.54 centimeters	= 1 inch
1 millimeter	= 0.03937 inch, or nearly $\frac{1}{25}$ inch
25.4 millimeters	= 1 inch
1 kilometer	= 1093.61 yards, or 0.62137 mile14

The ratio 25.4mm = 1 inch is used to convert millimeters to inches.

Measures of Weight

Metric	English
1 gram	= 15.432 grains
0.0648 gram	= 1 grain
28.35 grams	= 1 ounce avoirdupois
1 kilogram	= 2.2046 pounds
0.4536 kilogram	= 1 pound
1 metric ton 1000 kilograms	= 0.9842 ton of 2240 pounds (long tons) 19.68 cwt. 2204.6 pounds
1.016 metric tons 1016 kilograms	= 1 ton of 2240 pounds long ton

Measures of Capacity

Metric		English
1 liter (= 1 cubic decimeter)	=	61.023 cubic inches 0.03531 cubic foot 0.2642 gal. (American) 2.202 lbs. of water at 62°F.
28.317 liters	=	1 cubic foot
3.785 liters	=	1 gallon (American)
4.543 liters	=	1 gallon (Imperial)

ENGLISH CONVERSION TABLE

Length

Inches	×	0.0833	= feet
Inches	×	0.02778	= yard
Inches	×	0.00001578	= miles
Feet	×	0.3333	= yards
Feet	×	0.0001894	= miles
Yards	×	36.00	= inches
Yards	×	3.00	= feet
Yards	×	0.0005681	= miles
Miles	×	63360.00	= inches
Miles	×	5280.00	= feet
Miles	×	1760.00	= yards
Circumference of circle	×	0.3188	= diameter
Diameter of circle	×	3.1416	= circumference

Area

Square inches	×	0.00694	= square feet
Square inches	×	0.0007716	= square yards
Square feet	×	144.00	= square inches
Square feet	×	0.11111	= square yards
Square yards	×	1296.00	= square inches
Square yards	×	9.00	= square feet
Dia. of circle squared	×	0.7854	= area
Dia. of sphere squared	×	3.1416	= surface

Volume

Cubic inches	×	0.0005787	= cubic feet
Cubic inches	×	0.00002143	= cubic yards
Cubic inches	×	0.004329	= U.S. gallons
Cubic feet	×	1728.00	= cubic inches
Cubic feet	×	0.03704	= cubic yards
Cubic feet	×	7.4805	= U.S. gallons
Cubic yards	×	27.00	= cubic feet
Dia. of sphere cubed	×	0.5236	= volume

ENGLISH CONVERSION TABLE, CONT.

Weight

Grains (avoirdupois)	×	0.002286	= ounces
Ounces (avoirdupois)	×	0.0625	= pounds
Ounces (avoirdupois)	×	0.00003125	= tons
Pounds (avoirdupois)	×	16.00	= ounces
Pounds (avoirdupois)	×	0.01	= hundredweight
Pounds (avoirdupois)	×	0.0005	= tons
Tons (avoirdupois)	×	32000.00	= ounces
Tons (avoirdupois)	×	2000.00	= pounds

Energy

Horsepower	×	33000.00	= ft.-lbs. per min.
Btu	×	778.26	= ft.-lb.
Ton of refrigeration	×	200.00	= Btu per min.

Pressure

Lbs. per sq. in.	×	2.31	= ft. of water (60°F.)
Ft. of water (60°F.)	×	0.433	= lbs. per sq. in.
Ins. of water (60°F.)	×	0.0361	= lbs. per sq. in.
Lbs. per sq. in.	×	27.70	= ins. of water (60°F.)
Lbs. per sq. in.	×	2.041	= ins. of Hg. (60°F.)
Ins. of Hg. (60°F.)	×	0.490	= lbs. per sq.

Power

Horsepower	×	746.	= watts
Watts	×	0.001341	= horsepower
Horsepower	×	42.4	= Btu per min.

Water Factors

(at point of greatest density—39.2°F.)

Miners inch (of water)	×	8.976	= U.S. gal. per min.
Cubic inches (of water)	×	0.57798	= ounces
Cubic inches (of water)	×	0.036124	= pounds
Cubic inches (of water)	×	0.004329	= U.S. gallons
Cubic inches (of water)	×	0.003607	= English gallons
Cubic feet (of water)	×	62.425	= pounds
Cubic feet (of water)	×	0.03121	= tons
Cubic feet (of water)	×	7.4805	= U.S. gallons
Cubic feet (of water)	×	6.232	= English gallons
Cubic feet of ice	×	57.2	= pounds
Ounces (of water)	×	1.73	= cubic inches
Pounds (of water)	×	26.68	= cubic inches
Pounds (of water)	×	0.01602	= cubic feet
Pounds (of water)	×	0.1198	= U.S. gallons
Pounds (of water)	×	0.0998	= English gallons
Tons (of water)	×	32.04	= cubic feet

Tons (of water)	×	239.6	= U.S. gallons
Tons (of water)	×	199.6	= English gallons
U.S. gallons	×	231.00	= cubic inches
U.S. gallons	×	0.13368	= cubic feet
U.S. gallons	×	8.345	= pounds
U.S. gallons	×	0.8327	= English gallons
U.S. gallons	×	3.785	= liters
English gallons (Imperial)	×	277.41	= cubic inches
English gallons (Imperial)	×	0.1605	= cubic feet
English gallons (Imperial)	×	10.02	= pounds
English gallons (Imperial)	×	1.201	= U.S. gallons
English gallons (Imperial)	×	4.546	= liters

METRIC CONVERSION TABLE

Length

Millimeters	×	0.03937	= inches
Millimeters	÷	25.4	= inches
Centimeters	×	0.3937	= inches
Centimeters	÷	2.54	= inches
Meters	×	39.37	= inches
Meters	×	3.281	= feet
Meters	×	1.0936	= yards
Kilometers	×	0.6214	= miles
Kilometers	÷	1.6093	= miles
Kilometers	×	3280.8	= feet

Area

Sq. Millimeters	×	0.00155	= sq. in.
Sq. Millimeters	÷	645.2	= sq. in.
Sq. Centimeters	×	0.155	= sq. in.
Sq. Centimeters	÷	6.452	= sq. in.
Sq. Meters	×	10.764	= sq. ft.
Sq. Kilometers	×	247.1	= acres
Hectares	×	2.471	= acres

Volume

Cu. Centimeters	÷	16.387	= cu. in.
Cu. Centimeters	÷	3.69	= fl. drs. (U.S.P.)
Cu. Centimeters	÷	29.57	= fl. oz. (U.S.P.)
Cu. Meters	×	35.314	= cu. ft.
Cu. Meters	×	1.308	= cu. yards
Cu. Meters	×	264.2	= gals. (231 cu. in.)
Liters	×	61.023	= cu. in.

Liters	×	33.82	= fl. oz. (U.S.J.)
Liters	×	0.2642	= gal. (231 cu. in.)
Liters	÷	3.785	= gal. (231 cu. in.)
Liters	÷	28.317	= cu. ft.
Hectoliters	×	3.531	= cu. ft.
Hectoliters	×	2.838	= bu. (2150.42 cu. in.)
Hectoliters	×	0.1308	= cu. yds.
Hectoliters	×	26.42	= gal. (231 cu. in.)

Weight

Grams	×	15.432	= grains
Grams	÷	981.	= dynes
Grams (water)	÷	29.57	= fl. oz.
Grams	÷	28.35	= oz. avoirdupois
Kilograms	×	2.2046	= lb.
Kilograms	×	35.27	= oz. avoirdupois
Kilograms	×	0.0011023	= tons (2000 lbs.)
Tonne (Metric ton)	×	1.1023	= tons (2000 lbs.)
Tonne (Metric ton)	×	2204.6	= lbs.

Unit Weight

Grams per cu. cent.	÷	27.68	= lb. per cu. in.
Kilogram per meter	×	0.672	= lb. per ft.
Kilogram per cu. meter	×	0.06243	= lb. per cu. ft.
Kilogram per cheval	×	2.235	= lb. per hp.
Grams per liter	×	0.06243	= lb. per cu. ft.

Pressure

Kilograms per sq. cm.	×	14.223	= lbs. per sq. in.
Kilograms per sq. cm.	×	32.843	= ft. of water (60°F.)
Atmospheres (international)	×	14.696	= lbs. per sq. in.

Energy

Joule	×	0.7376	= ft.-lbs.
Kilogram-meters	×	7.233	= ft.-lbs.

Power

Cheval vapeur	×	0.9863	= hp
Kilowatts	×	1.341	= hp
Watts	÷	746.	= hp
Watts	×	0.7373	= ft.-lb.per sec.

Miscellaneous

Kilogram calorie	×	3.968	= Btu
Standard gravity (sea level 45° lat.)	÷	980.665	= cm. per sec. per sec.
Frigories/hr. (French)	÷	3023.9	= tons refrigeration

WEIGHTS OF STEEL AND BRASS BARS

WEIGHT OF BAR ONE FOOT LONG

STEEL—Weights cover hot worked steel about 0.50 percent carbon. One cubic inch weights 0.2833 lbs. High speed steel 10 percent heavier.

BRASS—One cubic inch weighs 0.3074 lbs.

Actual weight of stock may be expected to vary somewhat from these figures because of variations in manufacturing processes.

	Steel			Brass		
SIZE Inches	lbs.	lbs.	lbs.	lbs.	lbs.	lbs.
1/16	0.0104	0.013	0.0115	0.0113	0.0144	0.0125
1/8	0.042	0.05	0.046	0.045	0.058	0.050
3/16	0.09	0.12	0.10	0.102	0.130	0.112
1/4	0.17	0.21	0.19	0.18	0.23	0.20
5/16	0.26	0.33	0.29	0.28	0.36	0.31
3/8	0.38	0.48	0.42	0.41	0.52	0.45
7/16	0.51	0.65	0.56	0.55	0.71	0.61
1/2	0.67	0.85	0.74	0.72	0.92	0.80
9/16	0.85	1.08	0.94	0.92	1.17	1.01
5/8	1.04	1.33	1.15	1.13	1.44	1.25
11/16	1.27	1.61	1.40	1.37	1.74	1.51
3/4	1.50	1.92	1.66	1.63	2.07	1.80
13/16	1.76	2.24	1.94	1.91	2.43	2.11
7/8	2.04	2.60	2.25	2.22	2.82	2.45
15/16	2.35	2.99	2.59	2.55	3.24	2.81
1	2.67	3.40	2.94	2.90	3.69	3.19
1 1/16	3.01	3.84	3.32	3.27	4.16	3.61
1 1/8	3.38	4.30	3.73	3.67	4.67	4.04
1 3/16	3.77	4.80	4.16	4.08	5.20	4.51
1 1/4	4.17	5.31	4.60	4.53	5.76	4.99
1 5/16	4.60	5.86	5.07	4.99	6.35	5.50
1 3/8	5.04	6.43	5.56	5.48	6.97	6.04
1 7/16	5.52	7.03	6.08	5.99	7.62	6.60
1 1/2	6.01	7.65	6.63	6.52	8.30	7.19

USEFUL INFORMATION

To find the circumference of a circle, multiply the diameter by 3.1416.

To find the diameter of a circle, multiply the circumference by 0.31831.

To find the area of a circle, multiply the square of the diameter by 0.7854.

The radius of a circle × 6.283185 = the circumference.

The square of the circumference of a circle × 0.07958 = the area.

Half the circumference of a circle × half its diameter = the area.

The circumference of a circle × 0.159155 = the radius.

The square root of the area of a circle × 0.56419 = the radius.

The square root of the area of a circle × 1.12838 = the diameter.

To find the diameter of a circle equal in area to a given square, multiply a side of the square by 1.12838.

To find the side of a square equal in area to a given circle, multiply the diameter by 0.8862.

To find the side of a square inscribed in a circle, multiply the diameter by 0.7071.

To find the side of a hexagon inscribed in a circle, multiply the diameter of the circle by 0.500.

To find the diameter of a circle inscribed in a hexagon, multiply a side of the hexagon by 1.7321.

To find the side of an equilateral triangle inscribed in a circle, multiply the diameter of the circle by 0.866.

To find the diameter of a circle inscribed in an equilateral triangle, multiply a side of the triangle by 0.57735.

To find the area of the surface of a ball (sphere), multiply the square of the diameter by 3.1416.

To find the volume of a ball (sphere), multiply the cube of the diameter by 0.5236.

Index

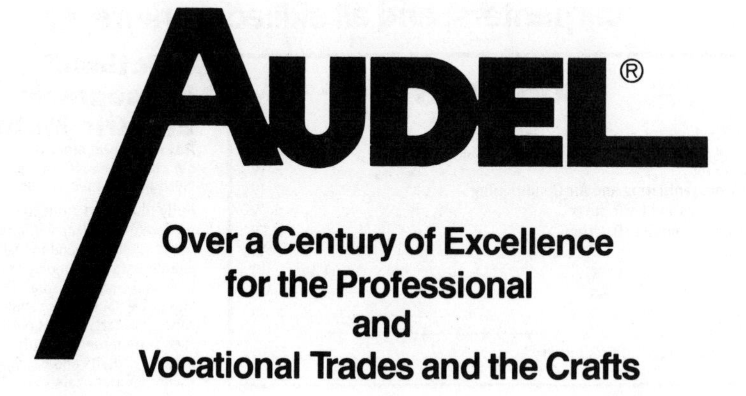
AUDEL®
Over a Century of Excellence
for the Professional
and
Vocational Trades and the Crafts

AUDEL

These fully illustrated, up-to-date guides and manuals mean a better job done for mechanics, engineers, electricians, plumbers, carpenters, and all skilled workers.

Contents

Electrical

House Wiring sixth edition
Roland E. Palmquist

5½ x 8¼ Hardcover 256 pp. 150 illus.
ISBN: 0-672-23404-1 $14.95

Rules and regulations of the current National Electrical Code® for residential wiring, fully explained and illustrated: • basis for load calculations • calculations for dwellings • services • nonmetallic-sheathed cable • underground feeder and branch-circuit cable • metal-clad cable • circuits required for dwellings • boxes and fittings • receptacle spacing • mobile homes • wiring for electric house heating.

Practical Electricity fourth edition
Robert G. Middleton; revised by L. Donald Meyers

5½ x 8¼ Hardcover 504 pp. 335 illus.
ISBN: 0-672-23375-4 $14.95

Complete, concise handbook on the principles of electricity and their practical application: • magnetism and electricity • conductors and insulators • circuits • electromagnetic induction • alternating current • electric lighting and lighting calculations • basic house wiring • electric heating • generating stations and substations.

Guide to the 1984 Electrical Code®
Roland E. Palmquist

5½ × 8¼ Hardcover 664 pp. 225 illus.
ISBN: 0-672-23398-3 $13.95

Authoritative guide to the National Electrical Code® for all electricians, contractors, inspectors, and homeowners: • terms and regulations for wiring design and protection • wiring methods and materials • equipment for general use • special occupancies • special equipment and conditions • and communication systems. Guide to the 1987 NEC® will be available in mid-1987.

Mathematics for Electricians and Electronics Technicians
Rex Miller

5½ x 8¼ Hardcover 312 pp. 115 illus.
ISBN: 0-8161-1700-4 $14.95

Mathematical concepts, formulas, and problem solving in electricity and electronics: • resistors and resistance • circuits • meters • alternating current and inductance • alternating current and capacitance • impedance and phase angles • resonance in circuits • special-purpose circuits. Includes mathematical problems and solutions.

Fractional Horsepower Electric Motors
Rex Miller and Mark Richard Miller

5½ x 8¼ Hardcover 436 pp. 285 illus.
ISBN: 0-672-23410-6 $15.95

Fully illustrated guide to small-to-moderate-size electric motors in home appliances and industrial equipment: • terminology • repair tools and supplies • small DC and universal motors • split-phase, capacitor-start, shaded pole, and special motors • commutators and brushes • shafts and bearings • switches and relays • armatures • stators • modification and replacement of motors.

Electric Motors
Edwin P. Anderson; revised by Rex Miller

5½ x 8¼ Hardcover 656 pp. 405 illus.
ISBN: 0-672-23376-2 $14.95

Complete guide to installation, maintenance, and repair of all types of electric motors: • AC generators • synchronous motors • squirrel-cage motors • wound rotor motors • DC motors • fractional-horsepower motors • magnetic contractors • motor testing and maintenance • motor calculations • meters • wiring diagrams • armature windings • DC armature rewinding procedure • and stator and coil winding.

Home Appliance Servicing fourth edition
Edwin P. Anderson; revised by Rex Miller

5½ x 8¼ Hardcover 640 pp. 345 illus.
ISBN: 0-672-23379-7 $15.95

Step-by-step illustrated instruction on all types of household appliances: • irons • toasters • roasters and broilers • electric coffee makers • space heaters • water heaters • electric ranges and microwave ovens • mixers and blenders • fans and blowers • vacuum cleaners and floor polishers • washers and dryers • dishwashers and garbage disposals • refrigerators • air conditioners and dehumidifiers.

Television Service Manual

fifth edition

Robert G. Middleton; revised by Joseph G. Barrile

$5^1/_2$ x $8^1/_4$ Hardcover 512 pp. 395 illus.

ISBN: 0-672-23395-9 $15.95

Practical up-to-date guide to all aspects of television transmission and reception, for both black and white and color receivers: • step-by-step maintenance and repair • broadcasting • transmission • receivers • antennas and transmission lines • interference • RF tuners • the video channel • circuits • power supplies • alignment • test equipment.

Electrical Course for Apprentices and Journeymen

second edition

Roland E. Palmquist

$5^1/_2$ x $8^1/_4$ Hardcover 478 pp. 290 illus.

ISBN:0-672-23393-2 $14.95

Practical course on operational theory and applications for training and re-training in school or on the job: • electricity and matter • units and definitions • electrical symbols • magnets and magnetic fields • capacitors • resistance • electromagnetism • instruments and measurements • alternating currents • DC generators • circuits • transformers • motors • grounding and ground testing.

Questions and Answers for Electricians Examinations eighth edition

Roland E. Palmquist

$5^1/_2$ x $8^1/_4$ Hardcover 320 pp. 110 illus.

ISBN: 0-672-23399-1 $12.95

Based on the current National Electrical Code®, a review of exams for apprentice, journeyman, and master, with explanations of principles underlying each test subject: • Ohm's Law and other formulas • power and power factors • lighting • branch circuits and feeders • transformer principles and connections • wiring • batteries and rectification • voltage generation • motors • ground and ground testing.

Machine Shop and Mechanical Trades

Machinists Library

fourth edition 3 vols

Rex Miller

$5^1/_2$ x $8^1/_4$ Hardcover 1,352 pp. 1,120 illus.

ISBN: 0-672-23380-0 $44.85

Indispensable three-volume reference for machinists, tool and die makers, machine operators, metal workers, and those with home workshops.

Volume I, Basic Machine Shop

$5^1/_2$ x $8^1/_4$ Hardcover 392 pp. 375 illus.

ISBN: 0-672-23381-9 $14.95

• Blueprint reading • benchwork • layout and measurement • sheet-metal hand tools and machines • cutting tools • drills • reamers • taps • threading dies • milling machine cutters, arbors, collets, and adapters.

Volume II, Machine Shop

$5^1/_2$ x $8^1/_4$ Hardcover 528 pp. 445 illus

ISBN: 0-672-23382-7 $14.95

• Power saws • machine tool operations • drilling machines • boring • lathes • automatic screw machine • milling • metal spinning.

Volume III, Toolmakers Handy Book

$5^1/_2$ x $8^1/_4$ Hardcover 432 pp. 300 illus.

ISBN: 0-672-23383-5 $14.95

• Layout work • jigs and fixtures • gears and gear cutting • dies and diemaking • toolmaking operations • heat-treating furnaces • induction heating • furnace brazing • cold-treating process.

Mathematics for Mechanical Technicians and Technologists

John D. Bies

$5^1/_2$ x $8^1/_4$ Hardcover 392 pp. 190 illus.

ISBN: 0-02-510620-1 $17.95

Practical sourcebook of concepts, formulas, and problem solving in industrial and mechanical technology: • basic and complex mechanics • strength of materials • fluidics • cams and gears • machine elements • machining operations • management controls • economics in machining • facility and human resources management.

Millwrights and Mechanics Guide

third edition

Carl A. Nelson

$5^1/_2$ x $8^1/_4$ Hardcover 1,040 pp. 880 illus.

ISBN: 0-672-23373-8 $24.95

Most comprehensive and authoritative guide available for millwrights and mechanics at all levels of work or supervision: • drawing and sketching • machinery and equipment installation • principles of mechanical power transmission • V-belt drives • flat belts • gears • chain drives • couplings • bearings • structural steel • screw threads • mechanical fasteners • pipe fittings and valves • carpentry • sheet-metal work • blacksmithing • rigging • electricity • welding • pumps • portable power tools • mensuration and mechanical calculations.

Welders Guide third edition

James E. Brumbaugh

$5^1/_2$ x $8^1/_4$ Hardcover 960 pp. 615 illus.

ISBN: 0-672-23374-6 $23.95

Practical, concise manual on theory, operation, and maintenance of all welding machines: • gas welding equipment, supplies, and process • arc welding equipment, supplies, and process • TIG and MIG welding • submerged-arc and other shielded-arc welding processes • resistance, thermit, and stud welding • solders and soldering • brazing and braze welding • welding plastics • safety and health measures • symbols and definitions • testing and inspecting welds. Terminology and definitions as standardized by American Welding Society.

Welder/Fitters Guide

John P. Stewart

$8^1/_2$ x 11 Paperback 160 pp. 195 illus.

ISBN: 0-672-23325-8 $7.95

Step-by-step instruction for welder/fitters during training or on the job: • basic assembly tools and aids • improving blueprint reading skills • marking and alignment techniques • using basic tools • simple work practices • guide to fabricating weldments • avoiding mistakes • exercises in blueprint reading • clamping devices • introduction to using hydraulic jacks • safety in weld fabrication plants • common welding shop terms.

Sheet Metal Work

John D. Bies

$5^1/_2$ x $8^1/_4$ Hardcover 456 pp. 215 illus.

ISBN: 0-8161-1706-3 $19.95

On-the-job sheet metal guide for manufacturing, construction, and home workshops: • mathematics for sheet metal work • principles of drafting • concepts of sheet metal drawing • sheet metal standards, specifications, and materials • safety practices • layout • shear cutting • holes • bending and folding • forming operations • notching and clipping • metal spinning • mechanical fastening • soldering and brazing • welding • surface preparation and finishes • production processes.

Power Plant Engineers Guide

third edition

Frank D. Graham; revised by Charlie Buffington

5½ x 8¼ Hardcover 960 pp. 530 illus.
ISBN: 0-672-23329-0 $27.50

All-inclusive question-and-answer guide to steam and diesel-power engines: • fuels • heat • combustion • types of boilers • shell or fire-tube boiler construction • strength of boiler materials • boiler calculations • boiler fixtures, fittings, and attachments • boiler feed pumps • condensers • cooling ponds and cooling towers • boiler installation, startup, operation, maintenance and repair • oil, gas, and waste-fuel burners • steam turbines • air compressors • plant safety.

Mechanical Trades Pocket Manual

second edition

Carl A. Nelson

4 × 6 Paperback 364 pp. 255 illus.
ISBN: 0-672-23378-9 $10.95

Comprehensive handbook of essentials, pocket-sized to fit in the tool box: • mechanical and isometric drawing • machinery installation and assembly • belts • drives • gears • couplings • screw threads • mechanical fasteners • packing and seals • bearings • portable power tools • welding • rigging • piping • automatic sprinkler systems • carpentry • stair layout • electricity • shop geometry and trigonometry.

Plumbing

Plumbers and Pipe Fitters Library

third edition 3 vols

Charles N. McConnell; revised by Tom Philbin

5½x8¼ Hardcover 952 pp. 560 illus.
ISBN: 0-672-23384-3 $34.95

Comprehensive three-volume set with up-to-date information for master plumbers, journeymen, apprentices, engineers, and those in building trades.

Volume 1, Materials, Tools, Roughing-In

5½ x 8¼ Hardcover 304 pp. 240 illus.
ISBN: 0-672-23385-1 $12.95

• Materials • tools • pipe fitting • pipe joints • blueprints • fixtures • valves and faucets.

Volume 2, Welding, Heating, Air Conditioning

5½ x 8¼ Hardcover 384 pp. 220 illus.
ISBN: 0-672-23386-x $13.95

• Brazing and welding • planning a heating system • steam heating systems • hot water heating systems • boiler fittings • fuel-oil tank installation • gas piping • air conditioning.

Volume 3, Water Supply, Drainage, Calculations

5½ x 8¼ Hardcover 264 pp. 100 illus.
ISBN: 0-672-23387-8 $12.95

• Drainage and venting • sewage disposal • soldering • lead work • mathematics and physics for plumbers and pipe fitters.

Home Plumbing Handbook

third edition

Charles N. McConnell

8½ x 11 Paperback 200 pp. 100 illus.
ISBN: 0-672-23413-0 $13.95

Clear, concise, up-to-date fully illustrated guide to home plumbing installation and repair: • repairing and replacing faucets • repairing toilet tanks • repairing a trip-lever bath drain • dealing with stopped-up drains • working with copper tubing • measuring and cutting pipe • PVC and CPVC pipe and fittings • installing a garbage disposals • replacing dishwashers • repairing and replacing water heaters • installing or resetting toilets • caulking around plumbing fixtures and tile • water conditioning • working with cast-iron soil pipe • septic tanks and disposal fields • private water systems.

The Plumbers Handbook

seventh edition

Joseph P. Almond, Sr.

4 × 6 Paperback 352 pp. 170 illus.
ISBN: 0-672-23419-x $10.95

Comprehensive, handy guide for plumbers, pipe fitters, and apprentices that fits in the tool box or pocket: • plumbing tools • how to read blueprints • heating systems • water supply • fixtures, valves, and fittings • working drawings • roughing and repair • outside sewage lift station • pipes and pipelines • vents, drain lines, and septic systems • lead work • silver brazing and soft soldering • plumbing systems • abbreviations, definitions, symbols, and formulas.

Questions and Answers for Plumbers Examinations

second edition

Jules Oravetz

5½ x 8¼ Paperback 256 pp. 145 illus.
ISBN: 0-8161-1703-9 $9.95

Practical, fully illustrated study guide to licensing exams for apprentice, journeyman, or master plumber: • definitions, specifications, and regulations set by National Bureau of Standards and by various state codes • basic plumbing installation • drawings and typical plumbing system layout • mathematics • materials and fittings • joints and connections • traps, cleanouts, and backwater valves • fixtures • drainage, vents, and vent piping • water supply and distribution • plastic pipe and fittings • steam and hot water heating.

HVAC

Air Conditioning: Home and Commercial

second edition

Edwin P. Anderson; revised by Rex Miller

5½ x 8¼ Hardcover 528 pp. 180 illus.
ISBN: 0-672-23397-5 $15.95

Complete guide to construction, installation, operation, maintenance, and repair of home, commercial, and industrial air conditioning systems, with troubleshooting charts: • heat leakage • ventilation requirements • room air conditioners • refrigerants • compressors • condensing equipment • evaporators • water-cooling systems • central air conditioning • automobile air conditioning • motors and motor control.

Heating, Ventilating and Air Conditioning Library

second edition 3 vols

James E. Brumbaugh

5½ x 8¼ Hardcover 1,840 pp. 1,275 illus.
ISBN: 0-672-23388-6 $47.95

Authoritative three-volume reference for those who install, operate, maintain, and repair HVAC equipment commercially, industrially, or at home. Each volume fully illustrated with photographs, drawings, tables and charts.

Volume I, Heating Fundamentals, Furnaces, Boilers, Boiler Conversions

5½ x 8¼ Hardcover 656 pp. 405 illus.
ISBN: 0-672-23389-4 $16.95

• Insulation principles • heating calculations • fuels • warm-air, hot water, steam, and electrical heating systems • gas-fired, oil-fired, coal-fired, and electric-fired furnaces • boilers and boiler fittings • boiler and furnace conversion.

Volume II, Oil, Gas and Coal Burners, Controls, Ducts, Piping, Valves

5½ x 8¼ Hardcover 592 pp. 455 illus.
ISBN: 0-672-23390-8 $15.95

• Coal firing methods • thermostats and humidistats • gas and oil controls and other automatic controls •

ducts and duct systems • pipes, pipe fittings, and piping details • valves and valve installation • steam and hot-water line controls.

Volume III, Radiant Heating, Water Heaters, Ventilation, Air Conditioning, Heat Pumps, Air Cleaners
5 1/2 x 8 1/4 Hardcover 592 pp. 415 illus.
ISBN: 0-672-23391-6 $17.95

• Radiators, convectors, and unit heaters • fireplaces, stoves, and chimneys • ventilation principles • fan selection and operation • air conditioning equipment • humidifiers and dehumidifiers • air cleaners and filters.

Oil Burners fourth edition
Edwin M. Field
5 1/2 x 8 1/4 Hardcover 360 pp. 170 illus.
ISBN: 0-672-23394-0 $15.95

Up-to-date sourcebook on the construction, installation, operation, testing, servicing, and repair of all types of oil burners, both industrial and domestic: • general electrical hookup and wiring diagrams of automatic control systems • ignition system • high-voltage transportation • operational sequence of limit controls, thermostats, and various relays • combustion chambers • drafts • chimneys • drive couplings • fans or blowers • burner nozzles • fuel pumps.

Refrigeration: Home and Commercial second edition
Edwin P. Anderson; revised by Rex Miller
5 1/2 x 8 1/4 Hardcover 768 pp. 285 illus.
ISBN: 0-672-23396-7 $17.95

Practical, comprehensive reference for technicians, plant engineers, and homeowners on the installation, operation, servicing, and repair of everything from single refrigeration units to commercial and industrial systems: • refrigerants • compressors • thermoelectric cooling • service equipment and tools • cabinet maintenance and repairs • compressor lubrication systems • brine systems • supermarket and grocery refrigeration • locker plants • fans and blowers • piping • heat leakage • refrigeration-load calculations.

Pneumatics and Hydraulics

Hydraulics for Off-the-Road Equipment second edition
Harry L. Stewart; revised by Tom Philbin
5 1/2 x 8 1/4 Hardcover 256 pp. 175 illus.
ISBN: 0-8161-1701-2 $13.95

Complete reference manual for those who own and operate heavy equipment and for engineers, designers, installation and maintenance technicians, and shop mechanics: • hydraulic pumps, accumulators, and motors • force components • hydraulic control components • filters and filtration, lines and fittings, and fluids • hydrostatic transmissions • maintenance • troubleshooting.

Pneumatics and Hydraulics fourth edition
Harry L. Stewart; revised by Tom Philbin
5 1/2 x 8 1/4 Hardcover 512 pp. 315 illus.
ISBN: 0-672-23412-2 $19.95

Practical guide to the principles and applications of fluid power for engineers, designers, process planners, tool men, shop foremen, and mechanics: • pressure, work and power • general features of machines • hydraulic and pneumatic symbols • pressure boosters • air compressors and accessories • hydraulic power devices • hydraulic fluids • piping • air filters, pressure regulators, and lubricators • flow and pressure controls • pneumatic motors and tools • rotary hydraulic motors and hydraulic transmissions • pneumatic circuits • hydraulic circuits • servo systems.

Pumps fourth edition
Harry L. Stewart; revised by Tom Philbin
5 1/2 x 8 1/4 Hardcover 508 pp. 360 illus.
ISBN: 0-672-23400-9 $15.95

Comprehensive guide for operators, engineers, maintenance workers, inspectors, superintendents, and mechanics on principles and day-to-day operations of pumps: • centrifugal, rotary, reciprocating, and special service pumps • hydraulic accumulators • power transmission • hydraulic power tools • hydraulic cylinders • control valves • hydraulic fluids • fluid lines and fittings.

Carpentry and Construction

Carpenters and Builders Library
fifth edition 4 vols
John E. Ball; revised by Tom Philbin
5 1/2 x 8 1/4 Hardcover 1,224 pp. 1,010 illus.
ISBN: 0-672-23369-x $43.95
Also available in a new boxed set at no extra cost:
ISBN: 0-02-506450-9 $43.95

These profusely illustrated volumes, available in a handsome boxed edition, have set the professional standard for carpenters, joiners, and woodworkers.

Volume 1, Tools, Steel Square, Joinery
5 1/2 x 8 1/4 Hardcover 384 pp. 345 illus.
ISBN: 0-672-23365-7 $10.95

• Woods • nails • screws • bolts • the workbench • tools • using the steel square • joints and joinery • cabinetmaking joints • wood patternmaking • and kitchen cabinet construction.

Volume 2, Builders Math, Plans, Specifications
5 1/2 x 8 1/4 Hardcover 304 pp. 205 illus.
ISBN: 0-672-23366-5 $10.95

• Surveying • strength of timbers • practical drawing • architectural drawing • barn construction • small house construction • and home workshop layout.

Volume 3, Layouts, Foundations, Framing
5 1/2 x 8 1/4 Hardcover 272 pp. 215 illus.
ISBN: 0-672-23367-3 $10.95

• Foundations • concrete forms • concrete block construction • framing, girders and sills • skylights • porches and patios • chimneys, fireplaces, and stoves • insulation • solar energy and paneling.

Volume 4, Millwork, Power Tools, Painting
5 1/2 x 8 1/4 Hardcover 344 pp. 245 illus.
ISBN: 0-672-23368-1 $10.95

• Roofing, miter work • doors • windows, sheathing and siding • stairs • flooring • table saws, band saws, and jigsaws • wood lathes • sanders and combination tools • portable power tools • painting.

Complete Building Construction
second edition
John Phelps; revised by Tom Philbin
5 1/2 x 8 1/4 Hardcover 744 pp. 645 illus.
ISBN: 0-672-23377-0 $19.95

Comprehensive guide to constructing a frame or brick building from the footings to the ridge: • laying out building and excavation lines • making concrete forms and pouring fittings and foundation • making concrete slabs, walks, and driveways • laying concrete block, brick, and tile • building chimneys and fireplaces • framing, siding, and roofing • insulating • finishing the inside • building stairs • installing windows • hanging doors.

Complete Roofing Handbook

James E. Brumbaugh

$5^1/_2 \times 8^1/_4$ Hardcover 536 pp. 510 illus.

ISBN: 0-02-517850-4 $29.95

Authoritative text and highly detailed drawings and photographs, on all aspects of roofing: • types of roofs • roofing and reroofing • roof and attic insulation and ventilation • skylights and roof openings • dormer construction • roof flashing details • shingles • roll roofing • built-up roofing • roofing with wood shingles and shakes • slate and tile roofing • installing gutters and downspouts • listings of professional and trade associations and roofing manufacturers.

Complete Siding Handbook

James E. Brumbaugh

$5^1/_2 \times 8^1/_4$ Hardcover 512 pp. 450 illus.

ISBN: 0-02-517880-6 $23.95

Companion to *Complete Roofing Handbook*, with step-by-step instructions and drawings on every aspect of siding: • sidewalls and siding • wall preparation • wood board siding • plywood panel and lap siding • hardboard panel and lap siding • wood shingle and shake siding • aluminum and steel siding • vinyl siding • exterior paints and stains • refinishing of siding, gutter and downspout systems • listings of professional and trade associations and siding manufacturers.

Masons and Builders Library

second edition 2 vols

Louis M. Dezettel; revised by Tom Philbin

$5^1/_2 \times 8^1/_4$ Hardcover 688 pp. 500 illus.

ISBN: 0-672-23401-7 $27.95

Two-volume set on practical instruction in all aspects of materials and methods of bricklaying and masonry: • brick • mortar • tools • bonding • corners, openings, and arches • chimneys and fireplaces • structural clay tile and glass block • brick walks, floors, and terraces • repair and maintenance • plasterboard and plaster • stone and rock masonry • reading blueprints.

Volume 1, Concrete, Block, Tile, Terrazzo

$5^1/_2 \times 8^1/_4$ Hardcover 304 pp. 190 illus.

ISBN: 0-672-23402-5 $13.95

Volume 2, Bricklaying, Plastering, Rock Masonry, Clay Tile

$5^1/_2 \times 8^1/_4$ Hardcover 384 pp. 310 illus.

ISBN: 0-672-23403-3 $12.95

Woodworking

Woodworking and Cabinetmaking

F. Richard Boller

$5^1/_2 \times 8^1/_4$ Hardcover 360 pp. 455 illus.

ISBN: 0-02-512800-0 $18.95

Compact one-volume guide to the essentials of all aspects of woodworking: • properties of softwoods, hardwoods, plywood, and composition wood • design, function, appearance, and structure • project planning • hand tools • machines • portable electric tools • construction • the home workshop • and the projects themselves – stereo cabinet, speaker cabinets, bookcase, desk, platform bed, kitchen cabinets, bathroom vanity.

Wood Furniture: Finishing, Refinishing, Repairing second edition

James E. Brumbaugh

$5^1/_2 \times 8^1/_4$ Hardcover 352 pp. 185 illus.

ISBN: 0-672-23409-2 $12.95

Complete, fully illustrated guide to repairing furniture and to finishing and refinishing wood surfaces for professional woodworkers and do-it-yourselfers: • tools and supplies • types of wood • veneering • inlaying • repairing, restoring, and stripping • wood preparation • staining • shellac, varnish, lacquer, paint and enamel, and oil and wax finishes • antiquing • gilding and bronzing • decorating furniture.

Maintenance and Repair

Building Maintenance second edition

Jules Oravetz

$5^1/_2 \times 8^1/_4$ Hardcover 384 pp. 210 illus.

ISBN: 0-672-23278-2 $9.95

Complete information on professional maintenance procedures used in office, educational, and commercial buildings: • painting and decorating • plumbing and pipe fitting • concrete and masonry • carpentry • roofing • glazing and caulking • sheet metal • electricity • air conditioning and refrigeration • insect and rodent control • heating • maintenance management • custodial practices.

Gardening, Landscaping and Grounds Maintenance

third edition

Jules Oravetz

$5^1/_2 \times 8^1/_4$ Hardcover 424 pp. 340 illus.

ISBN: 0-672-23417-3 $15.95

Practical information for those who maintain lawns, gardens, and industrial, municipal, and estate grounds: • flowers, vegetables, berries, and house plants • greenhouses • lawns • hedges and vines • flowering shrubs and trees • shade, fruit and nut trees • evergreens • bird sanctuaries • fences • insect and rodent control • weed and brush control • roads, walks, and pavements • drainage • maintenance equipment • golf course planning and maintenance.

Home Maintenance and Repair: Walls, Ceilings and Floors

Gary D. Branson

$8^1/_2 \times 11$ Paperback 80 pp. 80 illus.

ISBN: 0-672-23281-2 $6.95

Do-it-yourselfer's step-by-step guide to interior remodeling with professional results: • general maintenance • wallboard installation and repair • wallboard taping • plaster repair • texture paints • wallpaper techniques • paneling • sound control • ceiling tile • bath tile • energy conservation.

Painting and Decorating

Rex Miller and Glenn E. Baker

$5^1/_2 \times 8^1/_4$ Hardcover 464 pp. 325 illus.

ISBN: 0-672-23405-x $18.95

Practical guide for painters, decorators, and homeowners to the most up-to-date materials and techniques: • job planning • tools and equipment needed • finishing materials • surface preparation • applying paint and stains • decorating with coverings • repairs and maintenance • color and decorating principles.

Tree Care second edition

John M. Haller

8½ x 11 Paperback 224 pp. 305 illus.
ISBN: 0-02-062870-6 $16.95

New edition of a standard in the field, for growers, nursery owners, foresters, landscapers, and homeowners: • planting • pruning • fertilizing • bracing and cabling • wound repair • grafting • spraying • disease and insect management • coping with environmental damage • removal • structure and physiology • recreational use.

Upholstering

updated

James E. Brumbaugh

5½ x 8¼ Hardcover 400 pp. 380 illus.
ISBN: 0-672-23372-x $14.95

Essentials of upholstering for professional, apprentice, and hobbyist: • furniture styles • tools and equipment • stripping • frame construction and repairs • finishing and refinishing wood surfaces • webbing • springs • burlap, stuffing, and muslin • pattern layout • cushions • foam padding • covers • channels and tufts • padded seats and slip seats • fabrics • plastics • furniture care.

Automotive and Engines

Diesel Engine Manual fourth edition

Perry O. Black; revised by William E. Scahill

5½ x 8¼ Hardcover 512 pp. 255 illus.
ISBN: 0-672-23371-1 $15.95

Detailed guide for mechanics, students, and others to all aspects of typical two- and four-cycle engines: • operating principles • fuel oil • diesel injection pumps • basic Mercedes diesels • diesel engine cylinders • lubrication • cooling systems • horsepower • engine-room procedures • diesel engine installation • automotive diesel engine • marine diesel engine • diesel electrical power plant • diesel engine service.

Gas Engine Manual third edition

Edwin P. Anderson; revised by Charles G. Facklam

5½ x 8¼ Hardcover 424 pp. 225 illus.
ISBN: 0-8161-1707-1 $12.95

Indispensable sourcebook for those who operate, maintain, and repair gas engines of all types and sizes: • fundamentals and classifications of engines · engine parts • pistons • crankshafts • valves • lubrication, cooling, fuel, ignition, emission control and electrical systems • engine tune-up • servicing of pistons and piston rings, cylinder blocks, connecting rods and crankshafts, valves and valve gears, carburetors, and electrical systems.

Small Gasoline Engines

Rex Miller and Mark Richard Miller

5 ½ x 8¼ Hardcover 640 pp. 525 illus.
ISBN: 0-672-23414-9 $16.95

Practical information for those who repair, maintain, and overhaul two- and four-cycle engines – with emphasis on one-cylinder motors – including lawn mowers, edgers, grass sweepers, snowblowers, emergency electrical generators, outboard motors, and other equipment up to ten horsepower: • carburetors, emission controls, and ignition systems • starting systems • hand tools • safety • power generation • engine operations • lubrication systems • power drivers • preventive maintenance • step-by-step overhauling procedures • troubleshooting • testing and inspection • cylinder block servicing.

Truck Guide Library 3 vols

James E. Brumbaugh

5½ x 8¼ Hardcover 2,144 pp. 1,715 illus.
ISBN: 0-672-23392-4 $45.95

Three-volume comprehensive and profusely illustrated reference on truck operation and maintenance.

Volume 1, Engines

5½ x 8¼ Hardcover 416 pp. 290 illus.
ISBN: 0-672-23356-8 $16.95

• Basic components · engine operating principles • troubleshooting • cylinder blocks • connecting rods, pistons, and rings • crankshafts, main bearings, and flywheels • camshafts and valve trains • engine valves.

Volume 2, Engine Auxiliary Systems

5½ x 8¼ Hardcover 704 pp. 520 illus.
ISBN: 0-672-23357-6 $16.95

• Battery and electrical systems • spark plugs • ignition systems, charging and starting systems • lubricating, cooling, and fuel systems • carburetors and governors • diesel systems • exhaust and emission-control systems.

Volume 3, Transmissions, Steering, and Brakes

5½ x 8¼ Hardcover 1,024 pp. 905 illus.
ISBN: 0-672-23406-8 $16.95

• Clutches • manual, auxiliary, and automatic transmissions • frame and suspension systems • differentials and axles, manual and power steering • front-end alignment • hydraulic, power, and air brakes • wheels and tires • trailers.

Drafting

Answers on Blueprint Reading

fourth edition

Roland E. Palmquist; revised by Thomas J. Morrisey

5½ x 8¼ Hardcover 320 pp. 275 illus.
ISBN: 0-8161-1704-7 $12.95

Complete question-and-answer instruction manual on blueprints of machines and tools, electrical systems, and architecture: • drafting scale • drafting instruments • conventional lines and representations • pictorial drawings • geometry of drafting • orthographic and working drawings • surfaces • detail drawing • sketching • map and topographical drawings • graphic symbols • architectural drawings • electrical blueprints • computer-aided design and drafting. Also included is an appendix of measurements • metric conversions • screw threads and tap drill sizes • number and letter sizes of drills with decimal equivalents • double depth of threads • tapers and angles.

Hobbies

Complete Course in Stained Glass

Pepe Mendez

8½ x 11 Paperback 80 pp. 50 illus.
ISBN: 0-672-23287-1 $8.95

Guide to the tools, materials, and techniques of the art of stained glass, with ten fully illustrated lessons: • how to cut glass • cartoon and pattern drawing • assembling and cementing • making lamps using various techniques • electrical components for completing lamps • sources of materials • glossary of terminology and techniques of stained glasswork.